Bayesian Approaches to Clinical Trials and Health-Care Evaluation

STATISTICS IN PRACTICE

Advisory Editor

Stephen Senn
University of Glasgow, UK

Founding Editor

Vic Barnett
Nottingham Trent University, UK

Statistics in Practice is an important international series of texts which provide detailed coverage of statistical concepts, methods and worked case studies in specific fields of investigation and study.

With sound motivation and many worked practical examples, the books show in down-to-earth terms how to select and use an appropriate range of statistical techniques in a particular practical field within each title's special topic area.

The books provide statistical support for professionals and research workers across a range of employment fields and research environments. Subject areas covered include: medicine and pharmaceutics; industry, finance and commerce; public services; the earth and environmental sciences, and so on.

The books also provide support to students studying statistical courses applied to the above areas. The demand for graduates to be equipped for the work environment has led to such courses becoming increasingly prevalent at universities and colleges.

It is our aim to present judiciously chosen and well-written workbooks to meet everyday practical needs. Feedback of views from readers will be most valuable to monitor the success of this aim.

A complete list of titles in this series appears at the end of the volume.

Bayesian Approaches to Clinical Trials and Health-Care Evaluation

David J. Spiegelhalter
MRC Biostatistics Unit, Cambridge, UK

Keith R. Abrams
University of Leicester, UK

Jonathan P. Myles
Cancer Research UK, London, UK

John Wiley & Sons, Ltd

The work is based on an original NHS Health Technology Assessment funded project (93/50/05).
Adapted with kind permission of the National Coordinating Centre for Health Technology Assessment.

Other Wiley Editorial Offices

John Wiley & Sons Inc., 111 River Street, Hoboken, NJ 07030, USA

Jossey-Bass, 989 Market Street, San Francisco, CA 94103-1741, USA

Wiley-VCH Verlag GmbH, Boschstr. 12, D-69469 Weinheim, Germany

John Wiley & Sons Australia Ltd, 33 Park Road, Milton, Queensland 4064, Australia

John Wiley & Sons (Asia) Pte Ltd, 2 Clementi Loop #02-01, Jin Xing Distripark, Singapore 129809

John Wiley & Sons Canada Ltd, 22 Worcester Road, Etobicoke, Ontario, Canada M9W 1L1

Wiley also publishes its books in a variety of electronic formats. Some content that appears in print
may not be available in electronic books.

British Library Cataloguing in Publication Data

A catalogue record for this book is available from the British Library

ISBN 0-471-49975-7

Typeset in 10/12 pt Photina from LATEX files supplied by the author, processed
by Kolam Information Services Pvt. Ltd, Pondicherry, India.

This book is printed on acid-free paper responsibly manufactured from sustainable forestry in which
at least two trees are planted for each one used for paper production.

Contents

7 Observational Studies   251

8 Evidence Synthesis   267

9 Cost-effectiveness, Policy-Making and Regulation   305

10 Conclusions and Implications for Future Research   349

A Websites and Software   353

Preface

This book began life as a review of Bayesian methods in health technology assessment commissioned by the UK National Health Service Research and Development Programme, which appeared as Spiegelhalter *et al.* (2000). It was then thought to be a good idea to change the review into a basic introduction to Bayesian methods which also tried to cover the field of clinical trials and health-care evaluation. We did not realise the amount of work this would involve.

We are very grateful to all those who have read all or part of the manuscript and given such generous comments, particularly David Jones, Laurence Freedman, Mahesh Parmar, Tony Ades, Julian Higgins, Nicola Cooper, Cosetta Minelli, Alex Sutton and Denise Kendrick. Unfortunately, by tradition, we must take full responsibility for all errors and idiosyncrasies. Our particular thanks go to Daniel Farewell for writing the BANDY program, and Nick Freemantle for providing data. The University of Leicester provided the second author with study leave, during which part of this work was carried out. Finally, we must thank Rob Calver and Siân Jones at Wiley for being so patient with the repeated excuses for delay: in the words of Douglas Adams (1952–2001). "I love deadlines. I especially like the whooshing sound they make as they go flying by". We hope it has been worth the wait.

List of Examples

1

Introduction

1.1 WHAT ARE BAYESIAN METHODS?

Bayesian statistics began with a posthumous publication in 1763 by Thomas Bayes, a Nonconformist minister from the small English town of Tunbridge Wells. His work was formalised as *Bayes theorem* which, when expressed mathematically, is a simple and uncontroversial result in probability theory. However, specific uses of the theorem have been the subject of continued controversy for over a century, giving rise to a steady stream of polemical arguments in a number of disciplines. In recent years a more balanced and pragmatic perspective has developed and this more ecumenical attitude is reflected in the approach taken in this book: we emphasise the benefits of Bayesian analysis and spend little time criticising more traditional statistical methods.

The basic idea of Bayesian analysis is reasonably straightforward. Suppose an unknown quantity of interest is the median years of survival gained by using an innovative rather than a standard therapy on a defined group of patients: we shall call this the 'treatment effect'. A clinical trial is carried out, following which conventional statistical analysis of the results would typically produce a *P*-value for the null hypothesis that the treatment effect is zero, as well as a point estimate and a confidence interval as summaries of what this particular trial tells us about the treatment effect. A Bayesian analysis supplements this by focusing on how the trial should change our opinion about the treatment effect. This perspective forces the analyst to explicitly state

- a reasonable opinion concerning the plausibility of different values of the treatment effect *excluding* the evidence from the trial (known as the prior distribution),
- the support for different values of the treatment effect based *solely* on data from the trial (known as the likelihood),

and to combine these two sources to produce

- a final opinion about the treatment effect (known as the posterior distribution).

Bayesian Approaches to Clinical Trials and Health-Care Evaluation   D. J. Spiegelhalter, K. R. Abrams and J. P. Myles
© 2004 John Wiley & Sons, Ltd   ISBN: 0-471-49975-7

The final combination is done using Bayes theorem, which essentially weights the likelihood from the trial with the relative plausibilities defined by the prior distribution. This basic idea forms the entire foundation of Bayesian analysis, and will be developed in stages throughout the book.

One can view the Bayesian approach as a formalisation of the process of learning from experience, which is a fundamental characteristic of all scientific investigation. Advances in health-care typically happen through incremental gains in knowledge rather than paradigm-shifting breakthroughs, and so this domain appears particularly amenable to a Bayesian perspective.

1.2 WHAT DO WE MEAN BY 'HEALTH-CARE EVALUATION'?

Our concern is with the evaluation of 'health-care interventions', which is a deliberately generic term chosen to encompass all methods used to improve health, whether drugs, medical devices, health education programmes, alternative systems for delivering care, and so on. The appropriate evaluation of such interventions is clearly of deep concern to individual consumers, health-care professionals, organisations delivering care, policy-makers and regulators: such evaluations are commonly called 'health-technology assessments', but we feel this term carries connotations of 'high' technology that we wish to avoid.

A wide variety of research designs have been used in evaluation, and it is not the purpose of this book to argue the benefits of one design over another. Rather, we are concerned with appropriate methods for analysing and interpreting evidence from one or multiple studies of possibly varying designs. Many of the standard methods of analysis revolve around the classical randomised controlled trial (RCT): these include power calculations at the design stage, methods for controlling Type I error within sequential monitoring, calculation of P-values and confidence intervals at the final analysis, and meta-analytic techniques for pooling the results of multiple studies. Such methods have served the medical research community well.

The increasing sophistication of evaluations is, however, highlighting the limitations of these traditional methods. For example, when carrying out a clinical trial, the many sources of evidence and judgement available beforehand may be inadequately summarised by a single 'alternative hypothesis', monitoring may be complicated by simultaneous publication of related studies, and multiple subgroups may need to be analysed and reported. Randomised trials may not be feasible or may take a long time to reach conclusions. A single clinical trial will also rarely be sufficient to inform a policy decision, such as embarking or continuing on a research programme, regulatory approval of a drug or device, or recommendation of a treatment at an individual or population level. Standard statistical methods are designed for summarising the evidence from single studies or pooling evidence from similar studies, and have difficulties dealing with the pervading complexity of multiple sources of evidence. Many have argued that a fresh, Bayesian, approach is worth investigating.

1.3　A BAYESIAN APPROACH TO EVALUATION

We may define a Bayesian approach as 'the explicit quantitative use of external evidence in the design, monitoring, analysis, interpretation and reporting of a health-care evaluation'. The argument of this book is that such a perspective can be more *flexible* than traditional methods in that it can adapt to each unique situation, more *efficient* in using all available evidence, more *useful* in providing predictions and inputs for making decisions for specific patients, for planning research or for public policy, and more *ethical* in both clarifying the basis for randomisation and fully exploiting the experience provided by past patients.

For example, a Bayesian approach allows evidence from diverse sources to be pooled through assuming that their underlying probability models (their likelihoods) share parameters of interest: thus the 'true' underlying effect of an intervention may feature in models for both randomised trials and observational data, even though there may be additional adjustments for potential biases, different populations, crossovers between treatments, and so on.

Attitudes have changed since Feinstein (1977) claimed that 'a statistical consultant who proposes a Bayesian analysis should therefore be expected to obtain a suitably informed consent from the clinical client whose data are to be subjected to the experiment'. Increasing attention to the Bayesian approach is shown by the medical and statistical literature, the popular scientific press, pharmaceutical companies and regulatory agencies. However, many important outstanding questions remain: in particular, to what extent will the scientific community, or the regulatory authorities, allow the explicit introduction of evidence that is not totally derived from observed data, or the formal pooling of data from studies of differing designs? Indeed, Berry (2001) warns that 'There is as much Bayesian junk as there is frequentist junk. Actually, there's probably more of the former because, to the uninitiated, the Bayesian approach seems like it provides a free lunch'. External evidence must therefore be introduced with caution, and used in a clear, explicit and transparent manner that can be challenged by those who need to critique any analysis: this balanced approach should help resolve these complex questions.

1.4　THE AIM OF THIS BOOK AND THE INTENDED AUDIENCE

This book is intended to provide:

- a review of the essential ideas of Bayesian analysis as applied to the evaluation of health-care interventions, without obscuring the essential message with undue technicalities;
- a suggested 'template' for reporting a Bayesian analysis;

- a critical commentary on similarities and differences between Bayesian and conventional approaches;
- a structured review of published work in the areas covered;
- a wide range of stand-alone examples of Bayesian methods applied to real data, mainly in a common format, with accompanying software which will allow the reader to reproduce all analyses;
- a guide to potential areas where Bayesian methods might be particularly valuable, and where further research may be necessary;
- an indication of appropriate methods that may be applied in different contexts (although this is not intended as a 'cookbook');
- a range of exercises suitable for use in a course based on the material in this book.

Our intended audience comprises anyone with a good grasp of quantitative methods in health-care evaluation, and whose mathematical and statistical training includes basic calculus and probability theory, use of normal tables, clinical trial design, and familiarity with hypothesis testing, estimation, confidence intervals, and interpretation of odds and hazard ratios, up to the level necessary to use standard statistical packages. Bayesian statistics has a (largely deserved) reputation for being mathematically challenging and difficult to put into practice, although we recommend O'Hagan and Luce (2003) as a good non-technical preliminary introduction to the basic ideas. In this book we deliberately try to use the simplest possible analytic methods, largely based on normal distributions, without distorting the conclusions: more technical aspects are placed in starred sections that can be omitted without loss of continuity. There is a steady progression throughout the book in terms of analytic complexity, so that by the final chapters we are dealing with methods that are at the research frontier. We hope that readers will find their own level of comfort and make some effort to transcend it.

1.5   STRUCTURE OF THE BOOK

We have struggled to decide on an appropriate structure for the material in this book. It could be ordered by *stage of evaluation* and so separate, for example, initial observational studies, RCTs possibly for licensing purposes, cost-effectiveness analysis and monitoring interventions in routine use. Alternatively, we might structure by *study design*, with discussion of randomised trials, databases, case–control studies, and so on. Finally, we could identify the *modelling issue*, for example prior distributions, alternative forms for likelihoods, and loss functions. We have, after much deliberation, made a compromise and used aspects of all three proposals, using extensive examples to weave together analytic techniques with evaluation problems.

Chapter 2 is a brief *revision* of important aspects of traditional statistical analysis, covering issues such as probability distributions, normal tables, parameterisation of outcomes, summarising results by estimates and confidence intervals, hypothesis testing and sample-size assessment. There is a particular emphasis on normal likelihoods, since they are an important prerequisite for much of the subsequent Bayesian analysis, but we also provide a fairly detailed catalogue of other distributions and their use.

Chapter 3 forms the core of the book, being an *overview* of the main features of the Bayesian approach. Topics include the subjective interpretation of probability, use of prior to posterior analysis in a clinical trial, assessing the evidence in reported clinical trial results, comparing hypotheses, predictions, decision-making, exchangeability and hierarchical models, and computation: these topics are then applied to substantive problems in later chapters. Differing perspectives on prior distributions and loss functions are shown to lead to different schools of Bayesianism. A proposed checklist for reporting Bayesian health-care evaluations forms the basis for all further examples in the book.

Chapter 4 briefly critiques the 'classical' statistical approach to health-care evaluation and makes a *comparison* with the Bayesian approach. Hypothesis tests, *P*-values, Bayes factors, stopping rules and the 'likelihood principle' are discussed with examples. This chapter can be skipped without loss of continuity.

Chapter 5 deals in detail with sources of *prior distributions*, such as expert opinion, summaries of evidence, 'off-the-shelf' default priors and hierarchical priors based on exchangeability assumptions. The criticism of prior opinions in the light of data is featured, and a detailed taxonomy provided of ways of using historical data as a basis for prior opinion.

Chapter 6 attempts to structure the substantial work on Bayesian approaches to all aspects of RCTs, including design, monitoring, reporting, and interpretation. The many worked examples emphasise the need for analysis of sensitivity to alternative prior assumptions.

Chapter 7 covers *observational studies*, such as case–control and other non-randomised designs. Particular aspects emphasised include the explicit modelling of potential biases with such designs, and non-randomised comparisons of institutions including ranking into 'league tables'.

Chapter 8 considers the *synthesis of evidence* from multiple studies, starting from 'standard' meta-analysis and then considering various extensions such as potential dependence of treatment effects on baseline risk. We particularly focus on examples of 'generalised evidence synthesis', which might feature studies of different designs, or 'indirect' comparison of treatments that have never been directly compared in a trial.

Chapter 9 examines how Bayesian analyses may be used to inform *policy*, including cost-effectiveness analysis, research planning and regulatory affairs. The view of alternative stakeholders is emphasised, as is the integration of evidence synthesis and cost-effectiveness in a single unified analytic model.

Chapter 10 includes a final summary, general discussion and some suggestions for future research. Appendix A briefly describes available software and Internet sites of interest.

Most of the chapters finish with a list of key points and questions/exercises, and some have a further guide to the literature.

This structure will inevitably mean some overlap in methodological questions, such as the appropriate form of the prior distribution, and whether it is reasonable to adopt an explicit loss function. For example, a particular issue that arises in many contexts is the appropriate means of including historical data. This will be introduced as a general issue and a list of different approaches provided (Section 3.16), and then these approaches will be illustrated in four different contexts in which one might wish to use historical data: first, obtaining a prior distribution from historical studies (Section 5.4); second, historical controls in randomised trials (Section 6.9); third, modelling the potential biases in observational studies (Section 7.3), and fourth, pooling data from many sources in an evidence synthesis (Section 8.2). This overlap means that a considerable amount of cross-referencing is inevitable and ideally there would be hypertext links, but a traditional book format forces us into a linear structure.

Different audiences may want to focus on different parts of the book. The material up to Chapter 5 comprises a basic short course in Bayesian analysis, suitable for both students and researchers. After that, Chapter 6 may be of more interest to statisticians working with clinical trials in the pharmaceutical industry or the public sector, while Chapters 7–9 may be more appropriate for those exploring policy decisions. However, there are no clear boundaries and we hope that most of the material is relevant for much of the potential readership.

In order to avoid disappointment, we should make clear what this book does *not* contain:

- There is almost no guidance on data analysis, model checking and many other essential ingredients of professional statistical practice. Our discussion of study design is limited to sample-size calculations, and there is little contribution to the debate concerning the relative importance of observational and randomised studies.

- There is no rigorous mathematical or philosophical development of the Bayesian approach, and the technical development is limited entirely to the level required for the examples.

- The examples are almost all taken from published work by ourselves and others, and although they deal with real problems and use real data, there is necessarily a degree of simplification in the presentation. In addition, while the Bayesian approach emphasises the formal use of substantive knowledge and subjective opinion, it is inevitable that judgements are introduced in a somewhat stylised manner into such 'second-hand' examples. We should also point out that numbers given in the text have been rounded, and the accompanying programs should be used for a more accurate analysis.

- There is limited development of the decision-theoretic approach to evaluation, and many will feel this is a serious omission. This bias arises from two related reasons. First, our personal experience has been almost entirely concerned with problems of inference, and so that is what we feel qualified to write about. Second, it will become clear that we have some misgivings concerning the application of decision theory in this context, and so prefer to emphasise the more immediately relevant material.

- There is very limited exploration of more general Bayesian approaches to modelling data that arise in health-care evaluations, such as applications to survival analysis, longitudinal models, non-compliance in trials, drop-outs and other missing data, and so on.

The accompanying website will be found at `http://www.mrc-bsu.cam.ac.uk/bayeseval/`, which provides code for most of the examples in the book, either using the BANDY spreadsheet program for simple analysis of odds and hazard ratios, or WinBUGS code for more complex examples. The website will also contain a list of any errors detected.

Finally, we should emphasise that this book is not intended as a polemic in favour of Bayesianism – there have been enough of those – and we shall try to avoid making exaggerated claims as to the benefits of this new 'treatment' for statistical problems. Our hope is that we can contribute to the responsible use of Bayesian methods and hence help in a small way towards the development of cost-effective health-care.

2

Basic Concepts from Traditional Statistical Analysis

The Bayesian approach, to a considerable extent, supplements rather than replaces the kind of analyses traditionally carried out in assessing health-care interventions, and in this chapter we shall briefly review some of the basic ideas that will subsequently be found useful. In particular, probability theory is fundamental to Bayesian analysis, and we therefore revise the basic concepts with a natural emphasis on Bayes theorem. We also consider random variables and probability distributions with particular emphasis on the normal distribution, which plays a vital role in summarising what the observed data can tell us about unknown quantities of interest. A particularly important practical aspect is the transformation of output from standard statistical packages into a form amenable to Bayesian interpretation.

Bayesian analysis makes a much wider use of probability distributions than traditional statistical methods, in that not only are sampling distributions required for summaries of data, but also a wide range of distributions are used to represent prior opinion about proportions, event rates, and other unknown quantities. The *shapes* of distributions therefore become particularly important, as they are intended to represent the plausibility of different values, and so we shall provide (in starred sections) extensive graphical displays as well the usual formulae.

Most of the issues addressed in this chapter are covered in a concise and readable manner in standard textbooks such as Altman (2001) and Berry *et al.* (2001b). In addition, Clayton and Hills (1993) consider a likelihood-based approach to many of the models that are frequently encountered in epidemiology and health-care evaluation.

Bayesian Approaches to Clinical Trials and Health-Care Evaluation   D. J. Spiegelhalter, K. R. Abrams and J. P. Myles
© 2004 John Wiley & Sons, Ltd   ISBN: 0-471-49975-7

2.1 PROBABILITY

2.1.1   What is probability?

Suppose a is some event which may or may not take place, such as the next toss of a coin coming up heads. Although we may casually speak of the 'probability' of a occurring, and give it a mathematical notation $p(a)$, it is perhaps remarkable that there is no universally agreed definition of what this term means. Perhaps the currently most accepted interpretation is the following: $p(a)$ is the proportion of times a will occur in an infinitely long series of repeated identical situations. This is known as the 'frequentist' perspective, as it rests on the frequency with which specific events occur. However, a number of other interpretations of probability have been made throughout history, and we shall consider a different, 'subjective', definition in Section 3.1.

There is little dispute, however, about the mathematical properties of probability. Let a and b be events, and H represent the context in which a and b might arise, and let $p(a|H)$ denote the probability of a given the context H: the vertical line represents 'conditioning'. Then $p(a|H)$ is a number that satisfies the following three basic rules:

1. *Bounds.*

$$0 \leq p(a|H) \leq 1,$$

   where $p(a|H) = 0$ if a is impossible and $p(a|H) = 1$ if a is certain in the context H.

2. *Addition rule.* If a and b are mutually exclusive (*i.e.* one at most can occur),

$$p(a \text{ or } b|H) = p(a|H) + p(b|H).$$

   (We note that, for technical reasons, it is helpful if Rule 2 is taken as holding for an infinite set of mutually exclusive events.)

3. *Multiplication rule.* For any events a and b,

$$p(a \text{ and } b|H) = p(a|b,H)p(b|H).$$

We say that a and b are independent if $p(a \text{ and } b|H) = p(a|H)p(b|H)$ or equivalently $p(a|b,H) = p(a|H)$: thus the fact that b has occurred does not alter the probability of a. The multiplication rule can equivalently be expressed as the definition of conditional probability,

$$p(a|b, H) = \frac{p(a \text{ and } b|H)}{p(b|H)},$$

provided $p(b|H) \neq 0$.

The explicit introduction of the context H is unusual in standard texts and we shall subsequently drop it to avoid accusations of pedantry: however, it is always useful to keep in mind that *all probabilities are conditional* and so, if the situation changes, then probabilities may change. We shall see in Section 3.1 that this notion forms the basis of *subjective probability*, in which H, the context, represents the information on which an individual bases their *own* subjective assessment of the *degree of belief, i.e.* probability, of an event occurring.

Example 2.1 illustrates that these rules can be given an immediate intuitive justification by comparison with a standard experiment.

Example 2.1   *Dice: Illustration of rules of probability*

Suppose H denotes the roll of two perfectly balanced six-sided dice, and let '≡' denote 'is equivalent to'.

Rule 1. For a single die: if $a \equiv$ 'throw 7', then $p(a) = 0$; if $a \equiv$ 'throw ≤ 6', then $p(a) = 1$. If c is the sum of the two dice: then if $c \equiv$ '13', then $p(c) = 0$; if $c \equiv$ '≤ 12', then $p(c) = 1$.

Rule 2. For a single die: if $a \equiv$ 'throw 3', $b \equiv$ 'throw 4', then

$$p(a \text{ or } b) = p(a) + p(b) \text{ since } a \text{ and } b \text{ are mutually exclusive}$$
$$= 1/6 + 1/6 = 1/3.$$

Rule 3. If we throw two dice: if $a \equiv$ 'first die throw 2', $b \equiv$ 'second die throw 5', then

$$p(a \text{ and } b) = p(a)p(b) \text{ since } a \text{ and } b \text{ are independent}$$
$$= 1/6 \times 1/6 = 1/36.$$

If $a \equiv$ 'total score of the two throws is greater than or equal to 6', $b \equiv$ 'first die throw 1', then

$$p(a \text{ and } b) = p(a|b)p(b)$$
$$= 1/3 \times 1/6 = 1/18.$$

Suppose we also consider the events 'a and b' and 'a and $\bar{b}$', where $\bar{b}$ represents the event 'not b'. Then 'a and b' and 'a and $\bar{b}$' are mutually exclusive and together form the event a, and hence, using Rule 2, we have the identity

$$p(a) = p(a \text{ and } b) + p(a \text{ and } \bar{b}) \tag{2.1}$$

which is known as 'marginalisation'. Further, by using Rule 3, we obtain

$$p(a) = p(a|b)p(b) + p(a|\bar{b})p(\bar{b}), \tag{2.2}$$

which is known by the curious title of 'extending the conversation' (or 'extending the argument'). Example 2.2 shows these expressions follow naturally from considering the full 'joint' distribution over all possible combinations of events.

Example 2.2 *Prognosis: Marginalisation and extending the conversation*

Suppose we wish to determine the probability of survival (up to a specified point in time) following a particular cancer diagnosis, given that it depends on the stage of disease at diagnosis amongst other factors. Whilst directly specifying the probability of surviving, denoted b, may be difficult, by extending the conversation to include whether the cancer was at an early stage, denoted a, or not, denoted $\bar{a}$, we obtain from (2.1),

$$p(b) = p(b|a)p(a) + p(b|\bar{a})p(\bar{a}).$$

For example, suppose patients with early stage disease have a good prognosis, say $p(b|a) = 0.80$, but for late stage it is poor, say $p(b|\bar{a}) = 0.20$, and that of new diagnoses the majority, 90%, are early stage, *i.e.* $p(a) = 0.90$ and $p(\bar{a}) = 0.10$. Then the marginal probability of surviving is $p(b) = 0.80 \times 0.90 + 0.20 \times 0.10 = 0.74$.

Table 2.1 shows all possible combinations of events and their probabilities, as well as the marginal probabilities that, appropriately, appear in the margin of the table. The joint probabilities of events have been obtained by Rule 2 so that, for example, $p(b$ and $a) = p(b|a)p(a) = 0.80 \times 0.90 = 0.72$.

Table 2.1 Probabilities of all combinations of survival and stage, including marginal probabilities.

	Early stage a	Late stage $\bar{a}$	
Survive b	0.72	0.02	0.74
Not survive $\bar{b}$	0.18	0.08	0.26
	0.90	0.10	1.00

2.1.2   Odds and log-odds

Any probability p can also be expressed in terms of 'odds' O, where

$$O = \frac{p}{1-p}$$

and

$$p = \frac{O}{1+O},$$

so that, for example, a probability of 0.20 (20% chance) corresponds to odds of $O = 0.20/0.80 = 0.25$ or, in betting parlance, '4 to 1 against'. Conversely, betting odds of '7 to 4 against' correspond to $O = 4/7$, or a probability of $p = 4/11 = 0.36$.

The natural logarithm (denoted log) of the odds is termed the 'logit', so that

$$\text{logit}(p) = \log\left[\frac{p}{1 - p}\right].$$

2.1.3 Bayes theorem for simple events

A number of properties can immediately be derived from Rules 1 to 3 of Section 2.1.1. Since $p(b \text{ and } a) = p(a \text{ and } b)$, Rule 3 implies that $p(b|a)p(a) = p(a|b)p(b)$, or equivalently

$$p(b|a) = \frac{p(a|b)}{p(a)} \times p(b). \tag{2.3}$$

We have proved Bayes theorem! In words, this vital result tells us how an initial probability $p(b)$ is changed into a conditional probability $p(b|a)$ when taking into account the event a occurring: it should be clear by this description that we are interpreting Bayes theorem as providing a formal mechanism for learning from experience.

Equation (2.3) also holds for $\bar{b}$, so that

$$p(\bar{b}|a) = \frac{p(a|\bar{b})}{p(a)} \times p(\bar{b}), \tag{2.4}$$

and dividing (2.3) by (2.4) we obtain the *odds form* for Bayes theorem:

$$\frac{p(b|a)}{p(\bar{b}|a)} = \frac{p(a|b)}{p(a|\bar{b})} \times \frac{p(b)}{p(\bar{b})}. \tag{2.5}$$

Thus $p(b)/p(\bar{b}) = p(b)/(1 - p(b))$, the odds on b before taking into account the event a, which is changed into the new odds $p(b|a)/p(\bar{b}|a)$ after conditioning on a. Equation (2.5) shows how Bayes theorem accomplishes this transformation without even explicitly calculating $p(a)$, and this insight is exploited in Section 3.2.

Example 2.3 *Prognosis (continued): Bayes theorem for single events*

Suppose we were given Table 2.1, and wanted to use Bayes theorem to tell us how knowing the stage of the disease at diagnosis revises our probability for survival a. Initially, before we know the stage, $p(b) = 0.74$ from the marginal probability in Table 2.1. Suppose we find out that the disease is at an early stage, *i.e. a*, where we know from Table 2.1 that $p(a|b) = 0.72/0.74 = 0.97$ and $p(a) = 0.9$. Hence from (2.3) we obtain a revised probability of survival

$$p(b|a) = \frac{0.97}{0.9} \times 0.74 = 0.80,$$

matching what, in fact, we knew already.

To use the odds form of Bayes theorem (2.5) we first require the initial odds for survival, *i.e.* $p(b)/p(\overline{b}) = 0.74/0.26 = 2.85$, and the ratio $p(a|b)/p(a|\overline{b}) = 0.97/0.69 = 1.405$. Then from (2.5) we obtain the final odds on survival as $2.85 \times 1.41 = 4.01$, corresponding to a probability $p(b|a) = 0.80$ (up to rounding error).

The two forms of Bayes theorem both give the required results and can be thought of as a means of moving from a marginal probability in a table to a conditional probability having taken into account some evidence. As we shall see in Section 3.2, it is this use of Bayes theorem that is used in many diagnostic testing situations without any controversy.

2.2   RANDOM VARIABLES, PARAMETERS AND LIKELIHOOD

2.2.1   Random variables and their distributions

Random variables have a somewhat complex formal definition, but it is sufficient to think of them as unknown quantities that may take on one of a set of values: traditionally a random variable is denoted by a capital Latin letter, say Y, before being observed and by a lower-case letter y as a specific observed value. This convention tends to be broken in Bayesian analysis, in which all unknown quantities are considered as random variables, but we shall try to keep to it where it clarifies the exposition.

Loosely speaking, $p(y)$ denotes the probability of a random variable Y taking on each of its possible values y. $p(y)$ is formally known as the *probability density function*, and the probability that Y does not exceed y, $P(Y \leqslant y)$, is termed the *probability distribution function*. We shall tend to use 'probability distribution' as a generic term, hopefully without causing confusion.

Probability distributions may be:

Binary. When Y can take on one of two values, we shall generally use the notation $Y = 1$ for when an event of interest occurs, and $Y = 0$ when it does not: this is known as a *Bernoulli trial*, after Jakob Bernoulli (1654–1705). The corresponding probability distribution obeys the rules $p(Y = 1) = 1 - p(Y = 0)$, and is said to have a Bernoulli distribution (Section 2.6.1); see Example 2.4.

Discrete. $p(y)$ forms a discrete distribution when Y can take on one of a list of values, say 0, 1, 2, 3, ... The binomial (Section 2.6.1) and Poisson (Section 2.6.2) distributions are used in this book.

Continuous. Suppose Y can, in theory, take on values measured to an arbitrary degree of precision (of course, in practice, rounding of measurements prevents this). This means that calculus is needed, and the probability of Y lying in any specified interval I is obtained by the integral $\int_I p(y) \, dy$. The continuous distributions met most often in this book are the normal (Section 2.3) and the uniform (Section 2.6.4), although a wide range of others are discussed in Section 2.6: many of these are useful as prior distributions for unknown quantities.

Following Rule 1 in Section 2.1.1, all probability distributions should assign total probability 1 to the set of all possible events – these are known as 'proper' probability distributions. For continuous distributions this would mean that they integrated to 1, *i.e.* $\int p(y) \, dy = 1$. In some theoretical exercises it can be useful to imagine 'improper' distributions that do not obey this rule, for example uniform distributions over the entire range $-\infty$ to ∞. In practice, however, all distributions used in our examples will be proper (this can in any case always be achieved by truncating such a distribution at very low and high values).

The expressions derived in Section 2.1 for simple events have their counterparts for continuous random variables x, y. To express how the probability of y is changed when taking into account an observation x, we write Bayes theorem as

$$p(y|x) = \frac{p(x|y)}{p(x)} \times p(y). \tag{2.6}$$

To obtain the (marginal) distribution $p(x)$ from the joint distribution $p(x,y)$, we require the continuous counterpart to (2.1),

$$p(x) = \int p(x,y) \, dy; \tag{2.7}$$

shows how this is particularly important in Bayesian analysis as there may be many unknown quantities but we may only be interested in one at a time. Finally, the notion of extending the conversation (see (2.2)), given by

$$p(x) = \int p(x|y) \, p(y) \, dy, \tag{2.8}$$

expresses how a conditional distribution $p(x|y)$ is 'averaged over' by a distribution $p(y)$ in order to produce a distribution on x.

Bayesian methods make repeated use of such integrations, and indeed the technical problems of carrying them out has, in the past, hampered the development of the approach. Fortunately, in subsequent chapters their use will be implicit and intuitive, with the necessary integrations made reasonably straightforward either by simplifying assumptions of normal distributions, or by using modern simulation methodology.

2.2.2   Expectation, variance, covariance and correlation

If we have a distribution, $p(y)$, for an unknown quantity, Y, and we require the expectation (mean) of Y then this is given by

$$E(Y) = \sum_{i=1}^{k} y_i \, p(y_i) \qquad (2.9)$$

if the distribution is discrete, and by

$$E(Y) = \int y \, p(y) \, dy \qquad (2.10)$$

if the distribution is continuous.

The variance of Y is defined as

$$V(Y) = E(Y - E(Y))^2$$
$$= E(Y^2) - E(Y)^2,$$

which may be calculated, for example, using $E(Y^2) = \int y^2 p(y) \, dy$. The standard deviation is then defined as $SD(Y) = \sqrt{V(Y)}$.

The 'covariance' of X and Y is defined as

$$Cov(X,Y) = E(XY) - E(X)E(Y) \qquad (2.11)$$

and measures the association between X and Y. However the covariance is not generally easy to interpret, and a better summary measure is the correlation, which is the covariance scaled by the standard deviations of the variables:

$$Corr(X,Y) = \frac{Cov(X,Y)}{SD(X)SD(Y)}. \qquad (2.12)$$

$Corr(X,Y)$ is a number between -1 and 1 which, loosely speaking, expresses how close X and Y are to lying on a straight line: $Corr(X,Y)$ is near 1 for a positive relationship, near 0 when X and Y are unrelated, and near -1 for a negative relationship.

Conditional expectation and variance*

We return to the relationship between joint and marginal distributions introduced in (2.7). X has both a *conditional* mean and variance defined for each value y, *i.e.* $E(X|y)$ and $V(X|y)$, and a *marginal* mean and variance defined for the marginal distribution of X alone, *i.e.* $E(X)$ and $V(X)$. Their relationship can be shown to be as follows:

$$E(X) = E_Y[E_X(X|Y)], \tag{2.13}$$

$$V(X) = V_Y[E_X(X|Y)] + E_Y[V_X(X|Y)], \tag{2.14}$$

where the subscripts indicate the relevant variable for the expectation or variance. Some interpretation of these expressions might be obtained by assuming that Y will be the interim results of a study, and X will be the final results. Then (2.13) shows that our overall expectation of the final results can be calculated by first conditioning on the interim data as if they were known, and then taking our expectations (with respect to the interim data) of those conditional expectations. Equation (2.14) is more complex and says that our overall uncertainty about the final outcomes can be broken down into two components: our uncertainty about its conditional expectation given the interim data, and our expectation of its conditional variance.

We shall use these expressions in the context of prediction: first for normal variables in Section 3.13, and then in Section 9.8.3 within the context of microsimulation in complex cost-effectiveness models.

2.2.3   Parametric distributions and conditional independence

A central aspect of statistical inference is learning about the assumed underlying distribution of quantities we observe, and this is generally carried out by assuming that the probability distributions follow a particular *parametric* form $p(y|\theta)$, *i.e.* the distribution of Y depends on some currently unknown parameter θ. Parameters are usually given Greek letters: in Bayesian inference they are considered as random variables but the usual convention of capital and lower-case letters is ignored, to no apparent detriment.

For example, for a Bernoulli variable Y such that $p(Y = 0) = 1 - \theta$, $p(Y = 1) = \theta$, we may write this likelihood in the form

$$p(y|\theta) = \theta^y (1 - \theta)^{1-y}; \qquad y = 0, \ 1. \tag{2.15}$$

A standard assumption in traditional statistics is that a set of random variables $Y_1, \ldots, Y_n$ are independent and identically distributed (i.i.d.). If we are willing to adopt a parametric distribution, this corresponds to assuming that each is drawn independently from a probability distribution $p(y|\theta)$ where θ is some unknown parameter or parameters, and hence by Rule 3 of Section 2.1.1 their joint distribution is

$$p(y_1, \ldots, y_n|\theta) = \prod_{i=1}^{n} p(y_i|\theta). \tag{2.16}$$

This is an example of what is known as *conditional independence*, since each Y_i is independent of the others, *conditional* on θ. We shall discuss in Section 3.4 how this expression can be derived rather than directly assumed.

2.2.4   Likelihoods

Much of traditional statistical inference is based on noting that, once data y have been observed, $p(y|\theta)$ can be considered as being a function of θ, and can tell us the extent to which different values of θ are *supported* by the data. When $p(y|\theta)$ is considered in this way it is known as the *likelihood*, and plays a very important role in Bayesian analysis, as it summarises all the information that the data y can provide about the parameter θ. It is important to note that any function of θ that is proportional to $p(y|\theta)$ can be considered as the likelihood, since multiplying $p(y|\theta)$ by any value that does not depend on θ does not affect the range of values of θ being supported.

The *likelihood function* expresses the relative plausibility of different values of θ, with the value of θ for which the likelihood is a maximum is referred to as the *maximum likelihood estimate*. We can use a range of values which are *best* supported by the data as an interval estimate for θ, and it can be argued (Clayton and Hills, 1993) that a reasonable range is defined by values of the likelihood above $\exp(-1.96^2/2) = 14.7\%$ of the maximum value – the reason for this choice will become apparent in Section 2.4.1. In practice, constructing intervals in such a manner is laborious, and in general we try to approximate likelihood functions by the normal distribution, as discussed in Section 2.4. Consider, for example, n individuals in a study; we measure whether the ith individual responds to treatment, $Y_i = 1$, or not, $Y_i = 0$. If we assume a set of independent Bernoulli trials such that the probability of response is θ, then, using (2.15) and (2.16), we can obtain the joint distribution for all n individuals as

$$
\begin{aligned}
p(y_1, \ldots, y_n|\theta) &= \prod_{i=1}^{n} p(y_i|\theta) \\
&= \prod_{i=1}^{n} \theta^{y_i}(1-\theta)^{1-y_i} \tag{2.17} \\
&= \theta^{y_1+\ldots+y_n}(1-\theta)^{(1-y_1)+\ldots+(1-y_n)} \\
&= \theta^{y_1+\ldots+y_n}(1-\theta)^{n-(y_1+\ldots+y_n)} \\
&= \theta^r(1-\theta)^{n-r}, \tag{2.18}
\end{aligned}
$$

where $r = \sum_i y_i$ is the number of responders. This likelihood is maximised at $\hat{\theta} = r/n$; hence the maximum likelihood estimate is the proportion of responders. The independence of the individual responses means that the probability (2.18) is the same regardless of the actual sequence, and hence if we were told that there were 3 successes out of 10 trials, our likelihood would be precisely the same.

Example 2.4   *Response: Combining Bernoulli likelihoods*

Suppose we observed the responses of 10 individuals to a drug, and the particular sequence observed is 0,1,0,0,0,1,0,1,0,0. Let θ be the probability of a random patient responding to the drug. There are 3 successes and 7 failures, and the probability of the data, *i.e.* the likelihood, is given by

$$p(y_1, \ldots, y_{10}|\theta) = \theta^3(1 - \theta)^{10-3} = \theta^3(1 - \theta)^7. \qquad (2.19)$$

Figure 2.1 shows this likelihood plotted for different values of θ and scaled to have maximum value 1. We return to this example in Section 2.4.1.

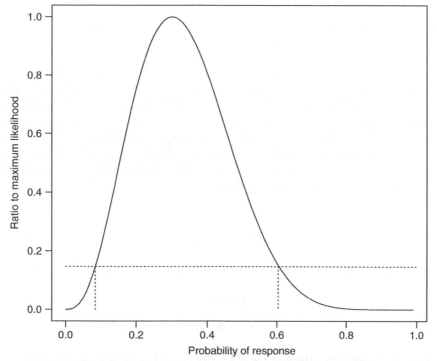

Figure 2.1   Likelihood function for the probability θ of response, after observing 10 individuals of whom 3 responded. The likelihood is scaled relative to its maximum value obtained at the maximum likelihood estimate $\hat{\theta} = 0.3$, and the interval (0.09, 0.61) is based on values with relative likelihood above $\exp(-1.96^2/2) = 0.147$.

2.3   THE NORMAL DISTRIBUTION

The normal (Gaussian) probability distribution is fundamental to much of statistical analysis and features in the majority of the examples covered in this book. We shall make frequent reference to properties of the normal distribution, and therefore it is worth some revision.

We shall use the expression

$$Y \sim N[\theta, \sigma^2]$$

to represent the assumption that the random quantity Y comes from a normal distribution with mean θ and variance σ^2 (standard deviation σ), which means that

$$p(y) = \frac{1}{\sqrt{2\pi}\sigma} \exp\left(-\frac{1}{2}\frac{(y-\theta)^2}{\sigma^2}\right); \qquad -\infty < y < \infty. \qquad (2.20)$$

We also occasionally make use of the notation $p(y) = N[y|\theta, \sigma^2]$. We note that the inverse of the variance, $1/\sigma^2$, is known as the *precision* of the distribution.

We shall often want to make use of areas under a normal distribution, for example the probability that Y is greater than 0 (a 'tail area'), or the range that comprises, say, 95% of the distribution (a '95% interval'). Let $Z \sim N[0, 1]$ denote a standard normal variable with mean $\theta = 0$ and standard deviation $\sigma = 1$: the shape of its probability distribution is given in Figure 2.2. Tables or computer programs generally provide the standard normal 'distribution function' $\Phi(z) = P(Z \leqslant z)$, the probability that Z is less than or equal to z, and Table 2.2 displays some useful values for $\Phi(z)$.

We note the useful property

$$\Phi(z) = 1 - \Phi(-z). \qquad (2.21)$$

For any tail area ϵ, we denote the corresponding normal deviate by z_ϵ, so that

$$P(Z \leqslant z_\epsilon) = \epsilon \qquad (2.22)$$

$$z_\epsilon = \Phi^{-1}(\epsilon), \qquad (2.23)$$

where Φ^{-1} represents the inverse of Φ. Hence (2.21) leads to the identity

$$z_\epsilon = -z_{1-\epsilon}.$$

Perhaps the most familiar value is $\Phi^{-1}(0.025) = z_{0.025} = -1.96 = -z_{0.975}$.

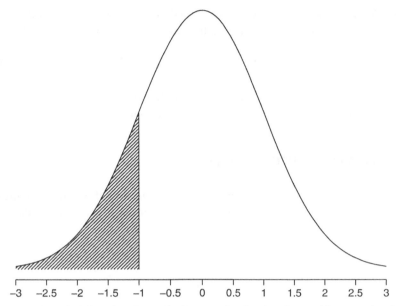

Figure 2.2 Probability distribution of a standard normal variable $Z \sim N[0,1]$. The shaded area represents $\Phi(-1) = P(Z \leqslant -1) = 0.159$.

For a general normal quantity we can easily derive tail areas and intervals from $\Phi(z)$, using the fact that if $Y \sim N[\theta, \sigma^2]$, then $(Y - \theta)/\sigma$ is a standard normal variable $Z \sim N[0, 1]$. Hence

$$P(Y \leqslant y) = P\left(\frac{Y - \theta}{\sigma} \leqslant \frac{y - \theta}{\sigma}\right) = P\left(Z \leqslant \frac{y - \theta}{\sigma}\right) = \Phi\left(\frac{y - \theta}{\sigma}\right). \qquad (2.24)$$

Thus, if we want to know $P(Y \leqslant y)$ we calculate the standardised statistic $z = (y - \theta)/\sigma$ and consult a table such as Table 2.2 to obtain $\Phi(z)$.

Alternatively, if we want, say, a 99% interval for Y, we use a table to find that the 99% interval for Z is $(-2.576, 2.576)$, and then transform this to an interval for Y of $(\theta - 2.576\sigma, \theta + 2.576\sigma)$.

An important property of normally distributed quantities is that they retain normality under addition or subtraction. For example, if Y_1 and Y_2 are independent quantities such that $Y_1 \sim N[\mu_1, \sigma_1^2]$, and $Y_2 \sim N[\mu_2, \sigma_2^2]$, then their sum has distribution

$$Y_1 + Y_2 \sim N[\mu_1 + \mu_2, \sigma_1^2 + \sigma_2^2], \qquad (2.25)$$

i.e. their sum is normally distributed with mean equal to the sum of the means, and variance equal to the sum of the variances. We shall find this property very helpful when making predictions (Section 3.13). In many health-care applications we also frequently consider the difference between two independent quantities; when they are both normally distributed we have

$$Y_1 - Y_2 \sim N[\mu_1 - \mu_2, \sigma_1^2 + \sigma_2^2], \qquad (2.26)$$

i.e. their difference is normally distributed with mean equal to the difference of the means, and variance equal to the *sum* of the variances.

2.4   NORMAL LIKELIHOODS

In many contexts it will be reasonable to assume that the data relevant to a parameter θ will be, after m 'observations', summarised by a statistic Y_m with a normal distribution

$$Y_m \sim N\left[\theta, \frac{\sigma^2}{m}\right], \qquad (2.27)$$

where θ is the parameter of interest, generally a treatment effect defined on a suitable scale, and σ^2 is assumed known: note that 'observations' is in quotes as we will find it convenient to use this form even when m is an 'effective' number of observations. After having observed a particular y_m, in traditional statistical terms y_m can be considered as an estimate of the true treatment effect θ, with standard error $\sigma/\sqrt{m}$.

Table 2.2   Some normal tail areas, expressed as percentages, where $100\epsilon = 100\Phi(z_\epsilon) = 100P(Z \leqslant z_\epsilon)$. From this table we can read, for example, that a symmetric 90% interval for Z would be $(-1.645, 1.645)$, while a one-sided 90% interval could be $(-\infty, 1.282)$ or $(-1.282, \infty)$.

z_ϵ	$100 \times \Phi(z_\epsilon)$	z_ϵ	$100 \times \Phi(z_\epsilon)$
		0.00	50.0
−0.50	30.8	0.50	69.2
−0.842	20.0	0.842	80.0
−1.00	15.9	1.00	84.1
−1.282	10.0	1.282	90.0
−1.50	6.7	1.50	93.3
−1.645	5.0	1.645	95.0
−1.960	2.5	1.960	97.5
−2.00	2.3	2.00	97.7
−2.326	1.0	2.326	99.0
−2.50	0.6	2.50	99.4
−2.576	0.5	2.576	99.5
−3.00	0.1	3.00	99.9
−3.090	0.1	3.090	99.9

Much of our approximate analysis is based on assuming a normal likelihood (2.27) in quite general contexts. These can be characterised as situations in which it is considered reasonable to quote the results of fitting a statistical model in terms of estimates and standard errors, for example after using standard statistical packages. This can, unfortunately, involve some effort transforming forwards and backwards between the quantities of interest and the somewhat unintuitive scales on which a normal likelihood is more appropriate. However, the examples in this book should demonstrate the value of becoming familiar with this process. It is worth emphasising that, since the likelihood is a function of θ and not a distribution for θ, it is not appropriate to speak, for example, of the mean, variance or tail-area of a likelihood.

We now consider a range of types of data on which the results of different interventions may be compared, detailing the parameters for which it may be appropriate to assume a normal likelihood, and describing how the results of standard regression analyses can be exploited. Obviously there are many areas, particularly with small samples, which cannot be adequately modelled assuming normality. This generally indicates a computational shift away from closed-form analysis and into simulation methodology, which will be discussed in Section 3.19.2.

2.4.1 Normal approximations for binary data

Suppose our data comprise a series of observations in which an event has occurred or not, and we wish to compare the probability of such events under two different interventions. For two events with probabilities p_1 and p_2, the odds ratio (OR) is

$$\text{OR} = \frac{p_1}{1 - p_1} \bigg/ \frac{p_2}{1 - p_2}, \tag{2.28}$$

which is a standard way of reporting changes in the chances of events due to an intervention, on a scale between 0 and ∞. In many circumstances the event is 'negative' (*e.g.* death or disease recurrence) and the 'new' intervention is in the numerator of (2.28), making odds ratios less than 1 favour the new. However, this will not always be the case and care must be taken. We note that for rare events, $(1 - p_1)$ and $(1 - p_2)$ are near 1, and hence the odds ratio is approximately the relative risk or risk ratio (RR) $= p_1/p_2$, and an odds ratio of, say, 0.7 can also be referred to as a 30% risk reduction. However, we shall try to avoid the term 'relative risk' due to potential confusion.

In order to make the assumption of a normal likelihood more plausible, it is convenient to work with the natural logarithm of the odds ratio so that it takes values on the whole range between $-\infty$ and $+\infty$. Thus

$$\log(\text{OR}) = \theta = \log\left[\frac{p_1}{1 - p_1}\right] - \log\left[\frac{p_2}{1 - p_2}\right], \tag{2.29}$$

and so the interventions are compared through their difference on the logit scale (Section 2.1.2). This is the standard scale underlying logistic regression analysis. In our analyses we will tend to perform calculations on the log(OR) scale, but report results as odds ratios, which are more intuitive. To assist slightly in the interpretation of log(odds ratios), we note that for small values of $\theta = \log(\text{OR})$, we have the approximation

$$\theta \approx \log(1 + \theta)$$

so that, for example, $\log(\text{OR}) = -0.1$ corresponds roughly to OR = 0.9, or a 10% risk reduction (the exact figure is OR = 0.905). So for small treatment effects, $100 \times \log(\text{OR})$ is approximately the percentage change in risk.

Use of the logit scale has the effect of improving the normal approximation of the likelihood. For example, Figure 2.3 shows the likelihood from Example 2.4 plotted on both the original probability scale and on the log(odds) scale, and the improvement is clear. We now argue why it might be appropriate for likelihood-based intervals to comprise all parameter values with support greater than 14.7% of the maximum, as already quoted in Section 2.2.4 – the following paragraph may be skipped without loss of continuity.

First, note that if the likelihood really *were* $N[\theta, \sigma^2/m]$, then from (2.20) it has a maximum of $\sqrt{m}/(\sqrt{2\pi}\sigma)$. Hence, relative to its maximum, the likelihood has ordinate $\exp[-(y-\theta)^2/2\sigma^2]$. Second, a 95% interval would comprise values $\theta \pm 1.96\sigma/\sqrt{m}$. Plugging these values into the formula for the normal distribution (2.20) therefore reveals that the boundaries for the 95% interval would have ordinate relative to the maximum of $e^{-1.96^2/2} = 0.147$. Transforming the x-scale of the likelihood does not change the relative ordinates in any way, and hence exactly the same interval is obtained by using this value of 14.7% on the original likelihood on the untransformed scale. Therefore, as long as there is some transformation that can give a reasonable normal approximation, the value of 14.7% of the maximum is justified.

Suppose N observations have been cross-classified by two binary factors, say intervention and response, leading to the following 2×2 table:

		Intervention		
		New	Control	
Event	Death	a	b	$a+b$
	No death	c	d	$c+d$
		$a+c$	$b+d$	N

The maximum likelihood estimate of the odds of death under the new intervention is a/c (the number of deaths divided by the number of survivors), under the control is b/d, and of the odds ratio OR is $(a/c)/(b/d)$. $\theta = \log(\text{OR})$ could be estimated by $\log[(a/c)/(b/d)]$, but in fact the estimator of choice is

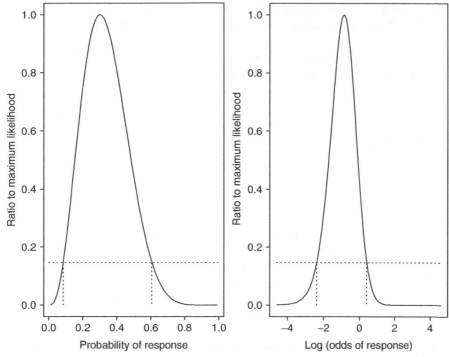

Figure 2.3 Likelihood function for the probability of disease, after treating 10 individuals of whom 3 were successes, plotted on both probability and log(odds) scale. The improvement to the normal approximation is clear.

$$\hat{\theta} = \log\left[\frac{(a + \frac{1}{2})(d + \frac{1}{2})}{(b + \frac{1}{2})(c + \frac{1}{2})}\right], \tag{2.30}$$

where $\hat{\theta}$ represents an estimate of θ. Lower mortality with the new intervention is represented by OR < 1, or negative values of θ. The estimator has approximate variance

$$V(\hat{\theta}) = \frac{1}{a + \frac{1}{2}} + \frac{1}{b + \frac{1}{2}} + \frac{1}{c + \frac{1}{2}} + \frac{1}{d + \frac{1}{2}}. \tag{2.31}$$

The $\frac{1}{2}$s have the effect of lessening the bias of the estimator and preventing problems with small numbers of events, and will generally have a negligible effect with reasonable sample sizes. Adjustment for confounding factors, using either a Mantel–Haenszel analysis or logistic regression, will also provide an estimate $\hat{\theta}$ with estimated standard error s, and provided N is not too small it will be reasonable to assume a normal likelihood with $V(\hat{\theta}) = s^2$.

In the notation of (2.27), we need to set $y_m = \hat{\theta}$ and $\sigma^2/m = V(\hat{\theta})$. Strictly speaking, it is unnecessary to select appropriate values of σ^2 and m since we could just use $V(\hat{\theta})$ in any analysis, but we shall find that this formulation is useful both for calculation and interpretation. There are two options:

1. We might fix m as the sample size N and so obtain $\sigma^2 = N\,V(\hat{\theta})$.
2. We might fix σ at some specific value, and choose m such that $m = \sigma^2/V(\hat{\theta})$. It turns out that in many contexts $\sigma = 2$ is a suitable choice. For example, consider a balanced randomised trial with a rare event occurring approximately equally often in each arm, so that $a \approx b$ and c and d are very large compared to a and b. Then, from (2.31),

$$V(\hat{\theta}) \approx \frac{2}{a} \approx \frac{4}{m},$$

where $m = a + b$ is the number of events. Thus if we take $\sigma = 2$ and $m = \sigma^2/V(\hat{\theta})$, we should find that m has an approximate interpretation as the number of events underlying the estimate of θ. This is likely to be easier to interpret than a variance on a log(OR) scale, which is fairly incomprehensible. We shall find in Section 2.4.2 that $\sigma = 2$ is also an appropriate choice in survival analysis, in that it also leads to m representing the effective number of events underlying the estimate.

If we are parameterising in terms of differences in proportions rather than the log(odds ratio), it may still be possible to assume a normal likelihood with large sample sizes, where y_m is the difference in sample response rates. Strictly speaking, σ^2 then depends upon the unknown response rates, but an estimate of σ^2 may be used.

Example 2.5   *GREAT: Normal likelihood from a 2 × 2 table*

The GREAT trial of early treatment for myocardial infarction, to be described in greater detail in Example 3.6, gave rise to the following data:

		Treatment New	Control	
Event	Death	13	23	36
	No death	150	125	275
		163	148	311

Using (2.30) gives an estimated log(OR) of $y_m = -0.736$, with estimated variance (2.31) of $0.131 = 0.362^2$. Taking $\sigma = 2$, we obtain $m = 4/0.131 = 30.5$, which is reasonably near the observed number of events (36) and gives an intuitive idea of the amount of evidence underlying the estimate.

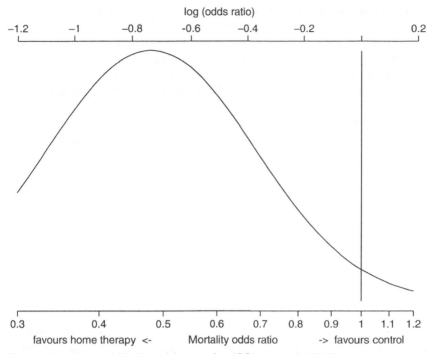

Figure 2.4   Normal likelihood for $\theta = \log(\text{OR})$ in the GREAT trial, with the upper axis labelled on the log(OR) scale. The lower scale is marked in terms of $\text{OR} = e^{\theta}$ for ease of interpretation.

Assuming a normal sampling distribution $y_m \sim N[\theta, \sigma^2/m]$ leads to the likelihood shown in Figure 2.4, which is plotted on the log(OR) scale but with axes labelled on both OR and log(OR) scales.

2.4.2   Normal likelihoods for survival data

Suppose we have a set of measurements of time to some event, say death or disease recurrence, often referred to as *survival data*. This event is assumed to occur with hazard rate $h(t)$, which is the chance of an event in a short interval of time following t. Survival under two different interventions with hazard rates $h_1(t)$ and $h_2(t)$ may be compared by their hazard ratio, $\text{HR} = h_1(t)/h_2(t)$: the common 'proportional hazards' assumption assumes HR is constant with time. The hazard ratio varies between 0 and ∞, and once again it is convenient to work with its natural logarithm,

$$\log(\text{HR}) = \theta = \log\left[\frac{h_1(t)}{h_2(t)}\right]. \tag{2.32}$$

In our analyses we will tend to perform calculations on the log(HR) scale, but report results as hazard ratios: generally events will be 'negative', such as death or disease recurrence, and so HR < 1 or $\theta < 0$ will favour the treatment in the numerator, which is usually the new intervention.

We note an important connection between hazard ratios and survival probabilities (although this derivation can be skipped). Let T be a random survival time with probability density $p(t)$, and let $S(t) = P(T > t)$ be the chance of surviving beyond t. The hazard rate $h(t)$ is the instantaneous chance of dying, given survival until t, and hence $h(t) = p(t)/S(t)$. Thus the cumulative hazard $H(t)$ obeys

$$H(t) = \int h(t)dt = \int p(t)/S(t) \; dt = -\log S(t).$$

Thus if we assume a proportional hazard model with HR $= h_1(t)/h_2(t)$, then we have

$$\mathrm{HR} = \frac{h_1(t)}{h_2(t)} = \frac{H_1(t)}{H_2(t)} = \frac{\log S_1(t)}{\log S_2(t)}.$$

From this it follows that if p_1 and p_2 are the chances of surviving until some fixed time under the two interventions being compared, then under the proportional hazards assumption

$$\mathrm{HR} = \frac{\log p_1}{\log p_2}, \tag{2.33}$$

$$\log(\mathrm{HR}) = \theta = \log\left[\frac{\log p_1}{\log p_2}\right]. \tag{2.34}$$

This means that if we know the two survival proportions and are willing to assume proportional hazards, then we can transform onto a log(HR) scale. This relationship is shown in Figure 2.5, from which can be read approximate values of log(HR) corresponding to changes in survival probabilities. For example, if a new treatment is thought to change 5-year survival from $p_2 = 20\%$ to $p_1 = 40\%$, then Figure 2.5 suggests this corresponds to a log(hazard ratio) of around -0.5, or HR $= 0.61$. The precise value is given by $\theta = \log[\log(p_1)/\log(p_2)] = -0.56$, corresponding to HR $= 0.57$.

Suppose that the first intervention corresponds to an active treatment T, and the second to a control C. Often the results of a survival analysis may be given in terms of an observed log-rank test statistic L_m, which is defined as the excess of events under T, compared to that expected were there no treatment effect, where m is the total number of events observed. L_m is often denoted as $O - E$ (observed minus expected). Assuming proportional hazards, we have the

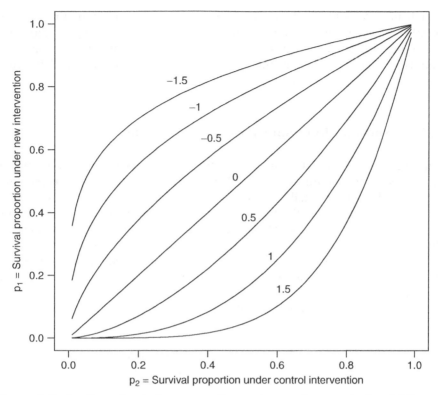

Figure 2.5   Log(hazard ratios) corresponding to changes from survival probability p_2 under a control treatment, to p_1 under a new treatment, where $\log(\mathrm{HR}) = \theta = \log[\log(p_1)/\log(p_2)]$.

following approximation in the particular case of equal allocation and follow-up. If there have been O_T events on treatment, and O_C events on control, then the expected number of events in the treatment group under the null hypothesis is approximately $m/2$, and hence the log-rank statistic is $L_m = O_T - m/2 = (O_T - O_C)/2$. It can be shown (Tsiatis, 1981) that, for large trials, $y_m = 4L_m/m = 2(O_T - O_C)/m$ is an approximate estimate of the log(hazard ratio) θ, and

$$y_m \sim \mathrm{N}[\theta, 4/m].$$

Hence we can set $\sigma = 2$ and adopt a normal likelihood.

If the estimated variance of the log-rank statistic, denoted $V[O - E]$, is provided in the report of the study, this will take into account different censoring, follow-up and so on. Now

$$V[O - E] = V[L_m] = V[my_m/4] = m^2 V[y_m]/16 \approx m/4,$$

and hence $V[O - E]$ can be equated to $m/4$ in order to obtain the effective number of events m. In more general circumstances we might adjust for covariates using a Cox regression analysis, and hence obtain an estimate $\hat{\theta}$ and its standard error s: if we then set $\sigma = 2$ we may obtain an 'implicit' event count $m = \sigma^2/s^2$, in the same manner as in Section 2.4.1.

2.4.3   Normal likelihoods for count responses

Suppose events occur at a rate λ per unit of population or time. Then our responses will be a count y of the number of events in, say, T units of population or time, which will usually be assumed to have a Poisson distribution with mean λT (Section 2.6.2). For two series of events with rates λ_1 and λ_2, the rate ratio (RaR) λ_1/λ_2 is a standard way of reporting changes in the rates of events due to an intervention. The rate ratio varies between 0 and ∞.

It is again convenient to work with the natural logarithm of a rate ratio, $\theta = \log(\lambda_1/\lambda_2)$, which may be estimated either directly from observed rates or from a Poisson regression.

Suppose we have observed the following data:

	Treatment New	Control
Events	r_1	r_2
Patient-years of follow-up	n_1	n_2

Here n_1 and n_2 are assumed to be large. The maximum likelihood estimate of the rate ratio is $(r_1/n_1)/(r_2/n_2)$, and $\theta = \log(\text{RaR})$ can be estimated by

$$\hat{\theta} = \log\frac{(r_1 + \frac{1}{2})/n_1}{(r_2 + \frac{1}{2})/n_2}. \tag{2.35}$$

RaR < 1, or negative values of θ, indicate a lower event rate with the new treatment. The estimator has approximate variance

$$V(\hat{\theta}) = \frac{1}{r_1 + \frac{1}{2}} + \frac{1}{r_2 + \frac{1}{2}}. \tag{2.36}$$

As with binary and survival data, a normal likelihood can be assumed provided the number of events is not too small, and once again we shall generally set $\sigma = 2$.

2.4.4   Normal likelihoods for continuous responses

Suppose that difference in mean response is the outcome measure of interest, m individuals are allocated to each treatment in a trial, and their individual responses are assumed normal with variance $\sigma^2/2$. Let θ be the true difference in mean response, and y_m be the difference in group sample means. Then $y_m \sim N[\theta, \sigma^2/m]$. (If σ^2 is unknown, then a full Bayesian analysis with a prior on σ^2 is possible: with a specific choice of prior one obtains the standard Student's t distribution for y_m (Section 5.5.1).)

2.5   CLASSICAL INFERENCE

In this section we give the briefest of summaries of standard statistical analysis when normal likelihoods can be assumed: for a comparative discussion of the basis for these and Bayesian techniques, we refer to Chapter 4.

The normal likelihood

$$ y_m \sim N\left[\theta, \frac{\sigma^2}{m}\right] $$

leads to θ being estimated by $\hat{\theta} = y_m$ with an accompanying two-sided 95% confidence interval of $y_m \pm 1.96 \times \sigma/\sqrt{m}$; this may be given the standard sampling-theory interpretation that 95% of the intervals produced using this procedure will contain the true parameter. If we wish to test a null hypothesis, say H_0: $\theta = 0$, we may examine whether the two-sided 95% interval excludes H_0, or equivalently use $z_m = y_m\sqrt{m}/\sigma$ as a standardised test statistic to refer to normal tables and, for example, declare the result 'statistically significant at the two-sided 5% level' if $|z_m| > 1.96$. We may also calculate the 'P-value' P_m associated with z_m, which is the probability of observing data as extreme as z_m under the null hypothesis. This can be taken as

$$ P_m = \min\left(P(Z \geqslant z_m),\ P(Z \leqslant z_m)\right) = \min\left(\Phi(-z_m),\ \Phi(z_m)\right), $$

although generally the 'two-sided' P-value is considered a more appropriate summary of 'extremeness' for H_0: $\theta = 0$, being

$$ 2P_m = P(Z > |z_m|) = \Phi(-|z_m|). $$

Suppose we are designing a clinical trial with proposed size n to detect an alternative hypothesis H_1: $\theta = \theta_A > 0$, and we decide that the result will be declared statistically significant and in favour of H_1 if a two-sided $100(1 - 2\epsilon)\%$ interval based on a future estimate Y_n lies wholly above 0, corresponding to the future standardised statistic $Z_n > -z_\epsilon$: typically $\epsilon = 0.025$ and so $-z_\epsilon = -z_{0.025} = 1.96$.

In this context this event is equivalent to $P_n \leqslant 2\epsilon$, and 2ϵ is therefore the probability of obtaining a statistically significant conclusion in either direction if the null hypothesis is in fact true. 2ϵ may be termed the 'significance level', the 'size', or the Type I error of the study, and is often denoted α. The null hypothesis will be rejected in favour of H_1 provided $Y_n > -z_\epsilon \sigma / \sqrt{n}$, which from (2.21) and (2.24) will occur with probability

$$1 - \Phi\left(\frac{-z_\epsilon \sigma / \sqrt{n} - \theta}{\sigma / \sqrt{n}}\right) = 1 - \Phi\left(-z_\epsilon - \frac{\theta \sqrt{n}}{\sigma}\right) = \Phi\left(\frac{\theta \sqrt{n}}{\sigma} + z_\epsilon\right).$$

The probability that a trial of n observations will lead to a statistically significant conclusion at the 2ϵ level, given that the alternative hypothesis is true, is known as the *power* of the study, conventionally denoted $1 - \beta$, and hence

$$1 - \beta = \Phi\left(\frac{\theta_A \sqrt{n}}{\sigma} + z_\epsilon\right). \tag{2.37}$$

From (2.37) we can easily see that the sample size necessary to obtain a specified power, say $100(1 - \beta)\%$, will obey

$$\frac{\theta_A \sqrt{n}}{\sigma} + z_\epsilon = \Phi^{-1}(1 - \beta) = z_{1-\beta},$$

and therefore

$$n = (z_{1-\beta} - z_\epsilon)^2 \frac{\sigma^2}{\theta_A^2}. \tag{2.38}$$

Typical values might be $\epsilon = 0.025$, $1 - \beta = 0.80$ and so, from Table 2.2, $(z_{1-\beta} - z_\epsilon)^2 = (0.842 + 1.96)^2 = 7.85$.

Note that some care is required in specifying σ and n. Our formulation is based on assuming that the estimate of the treatment effect has distribution $y_n \sim N[\theta, \sigma^2/n]$. Suppose, however, that we are performing a two-arm study with n patients per group, in which $y_n = \overline{y_2} - \overline{y_1}$, the difference in group means. Then σ^2 must be the variance of the *difference* between the responses from a random pair of patients, one from each arm. This will be the sum of the sampling variances in the two arms.

Example 2.6   *Power: Choosing the sample size for a trial*

Suppose we are designing a trial for a new cancer treatment which it is hoped will raise 5-year survival from 20% to 40%. From the analysis in Section 2.4.2, this is equivalent to a hazard ratio of $\log(0.40)/\log(0.20) = 0.57$ when assuming proportional hazards, or a log(hazard ratio) of $\theta_A = -0.56$. We note the above discussion of power has assumed an alternative hypothesis $\theta_A > 0$, whereas our θ_A is negative. However, we may simply reverse the role of null and alternative hypotheses and take $\theta_A = 0.56$: this is equivalent to redefining the hazard ratio as control hazard divided by new intervention hazard instead of its inverse. Taking $\sigma = 2$, the power of a study in which n events occur is given by (2.37): assuming $\epsilon = 0.025$ generates the power curve shown in Figure 2.6. From (2.38), 80% power is achieved at $n = 7.85 \times 2^2/(0.56)^2 = 100$: power rises slowly above this size of trial. Under the alternative hypothesis we expect about a 30% overall 5-year mortality in the trial, and so to observe 100 deaths we might recruit about 330 patients, 165 in each arm, and follow them for approximately 5 years.

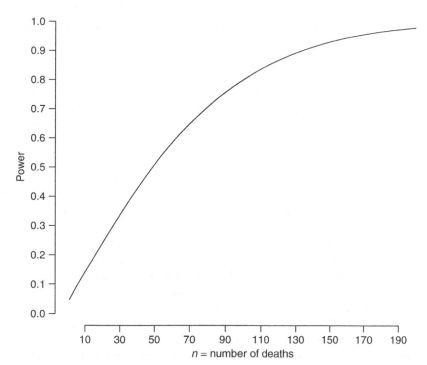

Figure 2.6   Power of a clinical trial in which n events are to be observed, and the alternative hypothesis is a rise from 20% survival to 40% survival, equivalent to a hazard ratio (control/new) of 1/0.57 (log(hazard ratio) $= \theta_A = 0.56$): Power $= \Phi(\theta_A\sqrt{n}/\sigma + z_\epsilon)$. 80% power is achieved at $n = 100$.

In Example 2.6 we took the alternative hypothesis as $\theta > 0$, leading to a power curve that rises for increasing values of θ. However, we shall be using many examples where low values of θ correspond to benefit of the new intervention, and hence care must be taken in using the equations. This rather technical point is considered in detail in Section 6.5, where we also show how to take into account uncertainty about parameters when conducting power calculations.

2.6   A CATALOGUE OF USEFUL DISTRIBUTIONS*

Bayesian analysis makes use of a wide range of standard, and not so standard, parametric probability distributions in two contexts:

- *Sampling distributions for individual data points or summary statistics* form the basis for likelihoods, just as in classical statistical inference. We shall make use of standard distributional families such as the normal, binomial, and Poisson, but also more unusual choices such as the log-normal for cost data.

- *Prior distributions for parameters* form the very core of Bayesian inference, and the shape of the chosen distribution becomes vital as it represents the relative plausibility for different parameter values. It is therefore important to have a supply of flexible parametric families that can express properties such as skewness and having heavy tails, and so although many of the prior opinions used in this book can be approximated by a normal distribution, we shall also require less standard forms such as the beta, root-inverse-gamma, and half-normal.

These two contexts come together in the use of 'conjugate' distributions, which are families of prior distributions that 'fit together' with particular sampling distributions. These are discussed in Section 3.6.2 and are useful for illustrating Bayesian analysis in simple examples, but modern computational techniques have reduced their importance.

A familiarity with the uses, shapes and properties of different families of distributions can be very valuable, and Bayesian texts contain extensive catalogues of distributions and their mathematical properties: see, for example, Lee (1997), Bernardo and Smith (1994), Gelman *et al.* (1995) and Carlin and Louis (2000). Here we focus on the distributions that will be used in the examples in this book. We shall first discuss their derivation and give formal expressions for their distributional form, expectation and variance, but our primary focus will be on displaying their shapes and discussing their possible use in practical circumstances. We omit explicit restrictions on ranges of parameters when they are clear from the context.

This section might best be used as a reference throughout the book.

2.6.1   Binomial and Bernoulli

A discrete binomial variable Y arises as the sampling distribution of the total number of 'successes' in n independent Bernoulli trials, each with probability θ of success. The likelihood $\theta^y(1 - \theta)^{n-y}$ gives the probability for a specific sequence of

$n - y$ 'failures' and y 'successes' (Section 2.2.3), and there are $\binom{n}{y}$ such sequences. Thus $Y \sim \text{Bin}[n, \ \theta]$ represents a binomial distribution with properties:

$$p(y|n, \ \theta) = \binom{n}{y} \theta^y (1 - \theta)^{n-y}; \qquad y = 0, 1, \ldots, n, \qquad (2.39)$$

$$E(Y|n, \ \theta) = n\theta, \qquad (2.40)$$

$$V(Y|n, \ \theta) = n\theta(1 - \theta). \qquad (2.41)$$

The binomial with $n = 1$ is simply a Bernoulli distribution, denoted $Y \sim \text{Bern}[\theta]$.

Shape.   The examples in Figure 2.7 illustrate the decreasing relative variability and the tendency to a normal distribution that occurs when sample size increases.

Use.   The binomial is used as a sampling distribution for empirical counts that occur as proportions. Uses in this book include preference studies (Section 4.4.4), meta-analysis (Section 8.2.2, Example 8.2), and evidence synthesis (Example 8.6).

2.6.2   Poisson

Suppose there are a large number of opportunities for an event to occur, but the chance of any particular event occurring is very low. Then the total number of events occurring may often be represented by a discrete variable Y, where $Y \sim \text{Poisson}[\theta]$ represents a Poisson distribution with properties:

$$p(y|\theta) = \frac{\theta^y e^{-\theta}}{y!}; \quad y = 0, 1, 2, 3, \ldots, \qquad (2.42)$$

$$E(Y|\theta) = \theta, \qquad (2.43)$$

$$V(Y|\theta) = \theta. \qquad (2.44)$$

In many applications it will arise as a total number of events occurring in a period of time T, where the events occur at an unknown rate λ per unit of time, in which case the expected value of Y is $\theta = \lambda T$.

Shape.   The examples in Figure 2.8 show that if events happen with a constant rate, observing for longer periods of time leads to smaller relative variability and a tendency towards a normal shape. Comparison of Figure 2.8 with Figure 2.7 shows that, when sample size increases, a binomial might be approximated by a Poisson with the same mean.

Use.   The Poisson distribution is used for count data, as in Example 8.3.

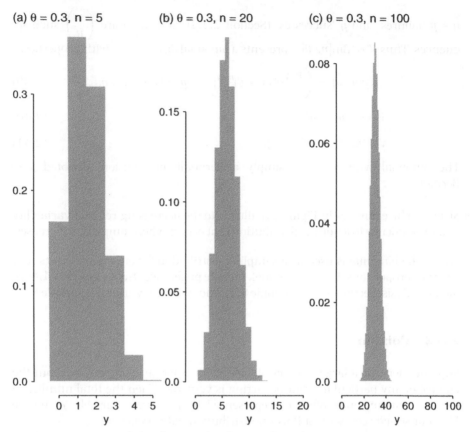

Figure 2.7 Binomial distributions for the number of successes in $n = 5$, 20, 100 Bernoulli trials, each with probability $\theta = 0.3$ of success.

2.6.3 Beta

Beta distributions form a flexible and mathematically convenient class for quantities constrained to lie between 0 and 1, and so can be used as a prior distribution for unknown proportions. $Y \sim \text{Beta}[a, b]$ represents a distribution with properties:

$$p(y|a,b) = \frac{\Gamma(a+b)}{\Gamma(a)\Gamma(b)} y^{a-1}(1-y)^{a-1}; \quad y \in (0, 1), \tag{2.45}$$

$$E(Y|a,b) = \frac{a}{a+b}, \tag{2.46}$$

$$V(Y|a,b) = \frac{ab}{(a+b)^2(a+b+1)}. \tag{2.47}$$

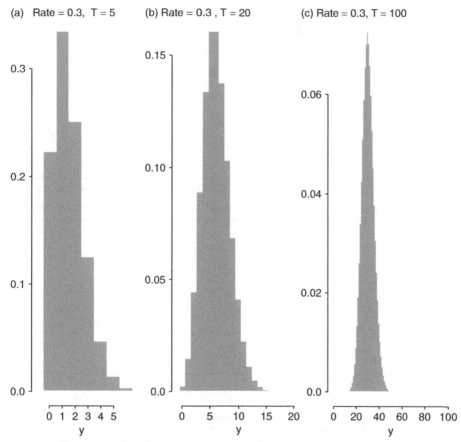

(a) Rate = 0.3, T = 5   (b) Rate = 0.3 , T = 20   (c) Rate = 0.3, T = 100

Figure 2.8   Poisson distributions representing the number of events occurring in time $T = 5$, 20, 100, when the rate at which an event occurs in a unit of time is $r = 0.3$: the Poisson distributions therefore correspond to $\theta = 1.5$, 6 and 30.

$\Gamma(a)$ represents the gamma function, a generalisation of the factorial for non-integers, in that $\Gamma(a) = (a - 1)!$ if a is an integer. A Beta[1,1] distribution is uniform between 0 and 1 (see Figure 2.9(b) and Section 2.6.4).

Shape.   The examples in Figure 2.9 show the flexibility of the family, with a tendency to normal as both parameters become larger.

Use.   The sole use of beta distributions is for uncertain proportions where they are 'conjugate' to the binomial family of sampling distributions (Section 3.6) and hence make the necessary computations straightforward. However, we saw in Section 2.4.1 that in most applications with binary data it is much more flexible and convenient to transform the quantity of interest from a proportion (defined on a (0,1) scale) to log(odds) (defined on the full range of $-\infty$ to ∞).

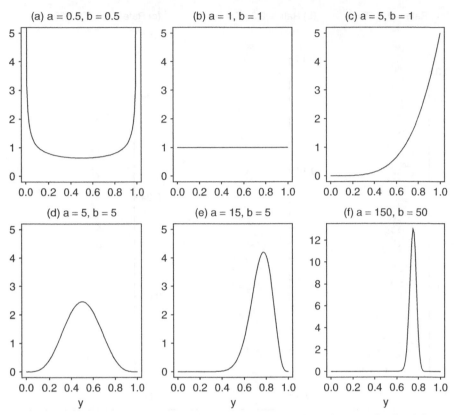

Figure 2.9   Beta distributions for different parameter values showing the flexibility of the family: note change in y-axis for (f).

Therefore, we shall find limited use for the beta except in tutorial examples (see Examples 3.3 and 8.6).

2.6.4   Uniform

Like the beta distribution, a uniform distribution on a range (a, b) is generally adopted for an unknown parameter. $Y \sim \text{Unif}[a, b]$ means that:

$$p(y|a,b) = \frac{1}{b-a}; \quad y \in (a, b), \tag{2.48}$$

$$E(Y|a,b) = \frac{a+b}{2}, \tag{2.49}$$

$$V(Y|a,b) = \frac{(b-a)^2}{12}. \tag{2.50}$$

Shape. The shape of this distribution hardly needs plotting, but an example is given in Figure 2.9(b). Uniform distributions can also be given over a discrete set of values (see Example 3.2).

Use. The only use in this book is as a means of expressing indifference concerning the prior plausibility of a range of values – a so-called 'non-informative' or reference prior (Section 5.5.1). We shall frequently use it in this manner and merely refer to a 'uniform prior', which means uniform over a range that is large enough to encompass all plausible values of θ.

2.6.5 Gamma

Gamma distributions form a flexible and mathematically convenient class for quantities constrained to be positive. $Y \sim \text{Gamma}[a, b]$ represents a gamma distribution with properties:

$$p(y|a, b) = \frac{b^a}{\Gamma(a)} y^{a-1} e^{-by}; \qquad y \in (0, \infty), \tag{2.51}$$

$$E(Y|a,b) = \frac{a}{b}, \tag{2.52}$$

$$V(Y|a,b) = \frac{a}{b^2}. \tag{2.53}$$

Particular cases include the Gamma$[1, b]$ distribution, which is exponential with mean $1/b$, and the Gamma$[\frac{1}{2}v, \frac{1}{2}]$, which is the same as the chi-squared distribution χ_v^2 on v degrees of freedom. A useful piece of distribution theory is that if $Y_1, \ldots, Y_n$ are a set of i.i.d. $N[\theta, \sigma^2]$ variables with mean $\overline{Y}$ and sample variance $S^2 = \sum_i (Y_i - \overline{Y})^2 / n$, then $\sum_i (Y_i - \theta)^2 / \sigma^2 \sim \chi_n^2$, and $nS^2 / \sigma^2 \sim \chi_{n-1}^2$. We shall use this in Example 8.4.

Shape. The examples in Figure 2.10 show the family to be reasonably flexible.

Use. One justification is that the gamma distribution 'conjugate' to the Poisson family (Section 3.6.2). However, as with binary data, we shall see in Section 2.4.3 that in most applications it is much more flexible and convenient to transform the quantity of interest from a rate (defined on a $(0, \infty)$ scale) to a log-rate (defined on the full range of $-\infty$ to ∞), and then use normal approximations.

An alternative popular use has been as a prior distribution for the precision parameter (1/variance) of a normal distribution, for which it is also conjugate (Section 3.6.2). This is equivalent to using a root-inverse-gamma distribution for the standard deviation (see Section 2.6.6).

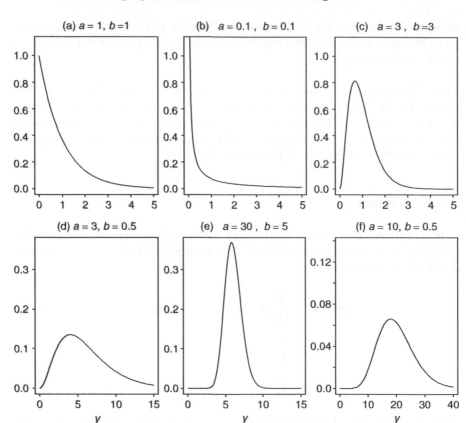

Figure 2.10 Gamma distributions. (a) is exponential with mean 1, (a), (b) and (c) all have the same mean but different shapes, (d) is a χ_6^2 distribution with mean 6, while (e) has the same mean as (a) but a different shape and becomes increasingly close to normal as the parameters both increase. (f) is a χ_{20}^2 distribution.

2.6.6   Root-inverse-gamma

If $X \sim$ Gamma$[a,b]$, then $1/\sqrt{X} \sim$ RIG$[a,b]$. $Y \sim$ RIG$[a,b]$ represents a root-inverse-gamma distribution with properties (Bernardo and Smith, 1994, p. 431):

$$p(y|a, b) = \frac{2b^a}{\Gamma(a)} \frac{1}{y^{2a+1}} e^{-b/y^2}; \qquad y \in (0, \infty), \tag{2.54}$$

$$E(Y|a,b) = \frac{\sqrt{b}\ \Gamma(a - \frac{1}{2})}{\Gamma(a)}, \tag{2.55}$$

$$V(Y|a,b) = \frac{b}{a-1} - E^2(Y|a, b). \tag{2.56}$$

We note that the variance is only defined for $a > 1$.

Shape. The examples in Figure 2.11 show that the family can have the somewhat curious property of forcing the quantity away from 0.

Use. The RIG is the implied prior distribution for a standard deviation when a gamma distribution is used for a precision, and so is frequently implicitly adopted in Bayesian analysis. However, it is almost never plotted, and the shape is perhaps not what was intended in many applications, given its property of rejecting low values. We shall therefore adopt it with some caution in Section 5.7.3 and in Example 8.1.

2.6.7   Half-normal

The half-normal arises by folding a normal distribution around 0: formally, if $X \sim N[0, \sigma^2]$, then $|X| \sim HN[\sigma^2]$. Thus $Y \sim HN[\sigma^2]$ represents a half-normal distribution with properties:

$$p(y|\sigma^2) = \sqrt{\frac{2}{\pi\sigma^2}}\, e^{\frac{-y^2}{2\sigma^2}}; \quad y \in (0, \infty), \tag{2.57}$$

$$E(Y|\sigma^2) = \sqrt{\frac{2}{\pi}}\, \sigma, \tag{2.58}$$

$$V(Y|\sigma^2) = \sigma^2\left(1 - \frac{2}{\pi}\right), \tag{2.59}$$

and a median of $\Phi^{-1}(0.75)\,\sigma = z_{0.75}\,\sigma = 0.773\,\sigma$, using the notation of Section 2.3.

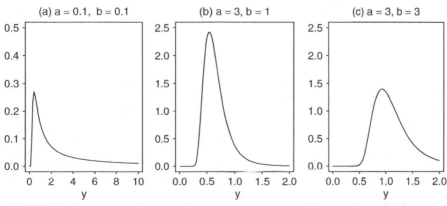

Figure 2.11 Root-inverse-gamma distributions. Note the different scale for (a), which has a very long right-hand tail. Comparing (c) with (b) shows that increasing *b* retains the shape but multiplies the mean and standard deviation by *b*.

Shape. The examples in Figure 2.12 show the family to express maximum support for 0, with the rate of decline governed by σ.

Use. The half-normal is useful to express support for values near 0, with σ controlling the upper range of support. This is applied to standard deviations in Section 5.7.3, and illustrated in Examples 8.1 and 8.5.

2.6.8 Log-normal

The log-normal is a distribution on positive values, like the gamma, root-inverse-gamma, and half-normal. It is defined as the *exponential* of a normal variable (this can cause confusion). Thus if $Y \sim \text{LN}[\mu, \sigma^2]$, then $\log(Y) \sim \text{N}[\mu, \sigma^2]$. $Y \sim \text{LN}[\mu, \sigma^2]$ represents a log-normal distribution with properties:

$$p(y|\mu,\sigma^2) = \frac{1}{\sqrt{2\pi}\sigma y}e^{-(\log y - \mu)^2/2\sigma^2}; \qquad y \in (0,\infty), \qquad (2.60)$$

$$E(Y|\mu,\sigma^2) = e^{\mu+\sigma^2/2}, \qquad (2.61)$$

$$V(Y|\mu,\sigma^2) = e^{2\mu+\sigma^2}(e^{\sigma^2} - 1). \qquad (2.62)$$

Shape. The examples in Figure 2.13 show that a range of skewed distributions can be represented, although the right-hand tail is remarkably long. For example, Figure 2.13(b) has a broadly similar shape to the Gamma[0.1, 0.1] shown in Figure 2.11(a): however, while the latter has mean 1 and standard deviation $\sqrt{10} = 3.2$, the LN[0, 3] has mean $e^{4.5} = 90$, and standard deviation

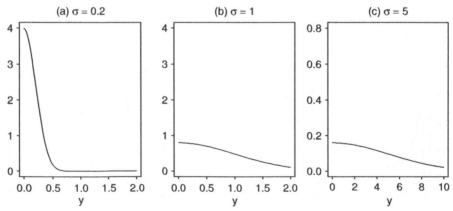

Figure 2.12   Half-normal distributions, with maximum at 0 and declining support for increasing y.

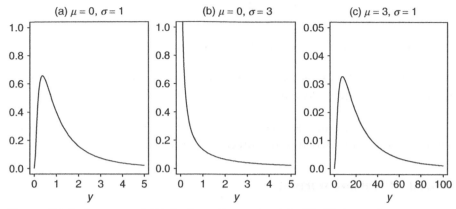

Figure 2.13   Log-normal distributions. Comparing (c) with (b) shows that μ acts as a scale parameter and does not change the shape of the distribution.

$\sqrt{e^9(e^9 - 1)} = 8100$. Thus although the gamma and log-normal are sometimes considered as alternative options for skewed distributions, the much heavier tail of the log-normal should be kept in mind.

Use.   The log-normal can be used as a sampling distribution for positive observations such as costs (Example 9.2), or as a prior distribution for positive parameters such as variances (Examples 6.10 and 9.2). We have seen in Section 2.4 that in many situations we carry out inferences on logarithms of quantities, and then transform results back to a more interpretable scale. Thus in our examples that use normal theory, our posterior distributions of odds ratios, hazard ratios and rate ratios are in fact log-normal distributions.

2.6.9   Student's *t*

A standardised Student's *t* distribution arises as the ratio of a standard normal variable to the square root of an independent χ^2 variable divided by its degrees of freedom, and has a prominent role in classical statistics as the sampling distribution of a sample mean divided by its estimated standard error. It also occurs as a posterior distribution for the mean of a normal distribution given a specific choice of prior for the unknown variance (DeGroot, 1970). $Y \sim t[\mu, \sigma^2, v]$ represents a Student's *t* distribution with v degrees of freedom, which has properties:

$$p(y|\mu,\sigma^2,v) = \frac{\Gamma(\frac{v+1}{2})}{\Gamma(\frac{v}{2})\sqrt{\pi v}\sigma} \frac{1}{\left(1 + \frac{(y-\mu)^2}{v\sigma^2}\right)^{\frac{v+1}{2}}}; \quad y \in (\infty, \infty), \tag{2.63}$$

$$E(Y|\mu,\sigma^2,v) = \mu, \tag{2.64}$$

$$V(Y|\mu,\sigma^2,v) = \sigma^2 \frac{v}{v-2}; \tag{2.65}$$

the mean only exists if $v > 1$, and the variance only exists if $v > 2$.

Shape. Figure 2.14 shows the heavy-tailed nature of the t distribution, with high degrees of freedom looking increasingly normal.

Use. Apart from arising as a posterior distribution, it can also be used as a sampling distribution when some outliers are expected.

2.6.10   Bivariate normal

X and Y are said to have a bivariate normal distribution, denoted $X,Y \sim \text{BN}[\theta_X,\theta_Y,\sigma_X,\sigma_Y,\rho]$, if

$$p(x,y|\theta_X,\theta_Y,\sigma_X,\sigma_Y,\rho) = \frac{1}{2\pi\sigma_X\sigma_Y\sqrt{1-\rho^2}} \exp\left(-\frac{Q}{2(1-\rho^2)}\right); \quad x, y \in (\infty,\infty), \tag{2.66}$$

where Q is the quadratic expression

$$Q = \frac{(x-\theta_X)^2}{\sigma_X^2} - \frac{2\rho(x-\theta_X)(y-\theta_Y)}{\sigma_X\sigma_Y} + \frac{(y-\theta_Y)^2}{\sigma_Y^2}.$$

The distribution has properties

$$E(X) = \theta_X, \quad E(Y) = \theta_Y, \quad V(X) = \sigma_X^2, \quad V(Y) = \sigma_Y^2,$$

and covariance and correlation

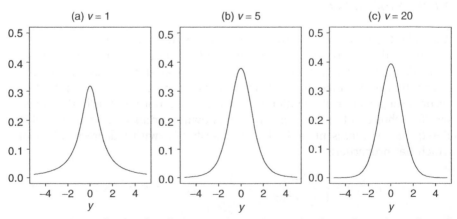

Figure 2.14   Student's t distributions with $\mu = 0$, $\sigma = 1$: other values of μ and σ will change the location and scale but not the shape.

$$\text{Cov}(X,Y) = \rho \sigma_X \sigma_Y, \quad \text{Corr}(X,Y) = \rho.$$

In addition, the conditional distribution of $Y|x$ is normal with mean and variance

$$E(Y|x) = \theta_Y + \frac{\rho \sigma_Y}{\sigma_X}(x - \theta_X),$$

$$V(Y|x) = \sigma_Y^2(1 - \rho^2). \tag{2.67}$$

The conditional variance $\sigma_Y^2(1 - \rho^2)$ is never more than the unconditional variance σ_Y^2, showing that knowing the value of X never increases our uncertainty about Y. In addition, the conditional mean is a linear function of x – this is known as the 'regression' of Y on X. The bivariate normal generalises naturally to higher dimensions but we shall not require this extension for this book.

Shape. Figure 2.15 shows a 'contour plot' of a bivariate normal distribution, where contours are ellipses obtained as solutions of $Q = $ constant.

Use. The bivariate normal can be used as a sampling distribution of two correlated quantities, such as in Example 9.1 where it is used to describe the joint

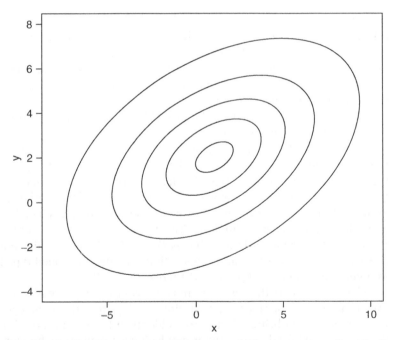

Figure 2.15 A bivariate normal distribution with parameters $\theta_X = 1$, $\theta_Y = 2$, $\sigma_X = 3$, $\sigma_Y = 2$, $\rho = 0.5$, with expanding ellipses enclosing 5%, 25%, 50%, 75% and 95% of the probability distribution.

distribution of costs and benefits. It also arises naturally as a prior distribution for two possibly correlated unknown parameters, such as the baseline rate and treatment effect in a clinical trial or epidemiological study (Section 8.2.3): see Example 8.3 for an example in a meta-analysis of observational studies.

2.7 KEY POINTS

1. Bayesian analysis rests wholly on probability theory, and all inferences can be derived from three basic rules.
2. The sampling distributions for data are used to derive likelihoods for unknown parameters, and so familiarity with classical methods helps in Bayesian analysis.
3. Normal approximations for likelihoods play a very important role.
4. Bayesian analysis makes use of a wide range of parametric probability distributions, both as a basis for likelihoods and as prior distributions.

EXERCISES

2.1. A coin is tossed and lands 'heads'.
   (a) What is *your* assessment of the probability that a second toss of the coin will also yield a 'head'?
   Before the coin was tossed for the first time it was randomly selected from two possible coins, one a 'fair' coin, *i.e.* with with both 'head' and 'tail', and the other a 'double-headed' coin.
   (b) What is your assessment of the probability that the second toss of the coin will now yield a 'head'?
2.2. Consider a case of disputed paternity, and the blood groups of the mother, the child and the alleged father. The mother has blood type O and the alleged father has blood type AB: let F denote the event that he is the true father. If the child has blood group O then the alleged father can be excluded from the paternity case. After testing, the child has blood type B, and Mendelian genetics implies $P(B|F) = 0.5$. The blood bank gives $P(B|\bar{F}) = 0.09$ for Caucasians. What is $P(F|B)$, *i.e.* the probability that the alleged father really is the father given that the child has blood type B, (a) as a general function of $P(F)$, and (b) when $P(F) = 0.5$?
2.3. Lee (1997) considers the case of twins and whether they are monozygotic (M) or dizygotic (D). Monozygotic twins develop from the same egg, look very similar (often being referred to as identical twins) and are *always* of the same sex, whilst dizygotic twins can look very similar too, but can be of different sexes. Therefore, $P(GG|M) = P(BB|M) = 0.5$, $P(GG|D) = P(BB|D) = 0.25$, and $P(GB|M) = 0$, $P(GB|D) = 0.5$.

(a) By extending the argument, express $P(GG)$ in terms of $p(M)$, the prior probability that a set of twins is monozygotic.

(b) Again in terms of $p(M)$, find the probability that if twins are both girls they are dizygotic, *i.e.* $P(D|GG)$.

(c) Find $P(D|GG)$ when $p(M) = 0.5$.

2.4. In a study of a drug, 20 out of 50 patients respond. (a) Find the maximum likelihood estimate for the response rate, and use a normal approximation for the likelihood for the log(odds) to find a 95% interval of values for the response rate which are supported by the data. A second study is performed, but due to time constraints only 20 patients are observed, of whom 8 respond. (b) For the second study, what is the most likely value for the response rate and an approximate 95% interval?

2.5. Gardner *et al.* (2000) report the results of a trial to investigate whether a progesterone emitting intra-uterine device (IUD) can reverse endometrial changes in women being treated for breast cancer with tamoxifen. At the end of the trial 5 out 56 women in the IUD group were discovered to have a submucous fibroid, whilst the corresponding number in the control group was 13 out of 53. Obtain a normal approximation to the likelihood for the log(odds ratio), and hence give a 95% interval for the odds ratio.

2.6. In the breast cancer trial of Exercise 2.5, women recruited had received tamoxifen for varying lengths of time, and the investigators felt that it was important to adjust for this and other possible confounders (including parity, menopausal status, body-mass index and age) in any analysis. They therefore used logistic regression to obtain an adjusted odds ratio of 0.23 with associated 95% confidence interval (CI) from 0.07 to 0.76. Obtain a normal approximation to the likelihood for the adjusted log(odds ratio).

2.7. Allen-Mersh *et al.* (1994) reported the results of a trial in which patients undergoing chemotherapy for liver metastases were randomised to receive it either systematically, as was standard, or via hepatic arterial infusion (HAI). Of 51 randomised to HAI 44 died, and of 49 randomised to systemic therapy 46 died.

(a) Obtain a rough normal approximation to the likelihood for the log(hazard ratio).

(b) The reported hazard ratio was 0.60 (95% CI from 0.40 to 0.95). Why might the approximation be so poor?

2.8. Shepherd *et al.* (2002) report the results of the PROSPER placebo-controlled RCT to evaluate the use of pravastin in elderly patients on a combined primary endpoint of death from coronary heart disease, non-fatal myocardial infarction, or stroke (fatal or non-fatal). Of 2891 patients randomised to pravastatin, 408 experienced the primary endpoint, whilst in the placebo group of 2913 patients 473 experienced it. (a) Obtain a rough estimate of the log(hazard ratio), assuming equal follow-up. The authors reported the results of a Cox proportional hazards regression model adjusting for a large number of baseline characteristics, which resulted in a 15% proportionate reduction in the hazard of the primary endpoint with 95% CI from 3% to 26%. (b) Obtain a normal approximation to the likelihood for the adjusted log(hazard ratio).

2.9. The PROSPER RCT in Exercise 2.8 also considered whether cancer incidence was higher in those patients receiving statin therapy. In the statin arm 245 cancers occurred out of 2891 patients, and in the placebo arm 199 cancers occurred in 2913 patients.

  (a) Obtain a normal approximation to the likelihood for the log(odds ratio).

  (b) Calculate a classical two-sided P-value.

  (c) Assess whether the data support a change in cancer incidence with statin use.

2.10. Suppose that 10% of patients taking anti-retroviral therapy currently experience a particular adverse event. Preliminary evidence suggests a new therapy might reduce this rate to 5%.

  (a) What is the hypothesised log(odds ratio)?

  (b) Estimate the number of events that would be required in an RCT in order to detect such a change, assuming a two-sided 5% level of statistical significance is to be used with a required power of 80%.

  (c) How many patients would be required in each arm of an RCT in order to observe this many events?

3

An Overview of the Bayesian Approach

In this chapter we shall introduce the core issues of Bayesian reasoning: these include subjectivity and context, the use of Bayes theorem, Bayes factors, interpretation of study results, prior distributions, predictions, decision-making, multiplicity, using historical data, and computation. This overview necessarily covers a wide range of material and ideas at an introductory level, and the issues will be further developed in subsequent chapters. A structure for reporting Bayesian analyses is proposed, which will provide a uniform style for the examples presented in this book. A number of starred sections can be omitted without loss of continuity.

3.1 SUBJECTIVITY AND CONTEXT

The standard interpretation of probability describes long-run properties of repeated random events (Section 2.1.1). This is known as the *frequency* interpretation of probability, and standard statistical methods are sometimes referred to as 'frequentist'. In contrast, the Bayesian approach rests on an essentially 'subjective' interpretation of probability, which is allowed to express generic uncertainty or 'degree of belief' about any unknown but potentially observable quantity, whether or not it is one of a number of repeatable experiments. For example, it is quite reasonable from a subjective perspective to think of a probability of the event 'Earth will be openly visited by aliens in the next ten years', whereas it may be difficult to interpret this potential event as part of a 'long-run' series. Methods of assessing subjective probabilities and probability distributions will be discussed in Section 5.2.

The rules of probability listed in Section 2.1.1 are generally taken as self-evident, based on comparison with simple chance situations such as rolling dice or drawing coloured balls out of urns. In these experiments there will be a

Bayesian Approaches to Clinical Trials and Health-Care Evaluation   D. J. Spiegelhalter, K. R. Abrams and J. P. Myles
© 2004 John Wiley & Sons, Ltd   ISBN: 0-471-49975-7

general consensus about the probabilities due to assumptions about physical symmetries: if a balanced coin is to be tossed, the probability of it coming up 'heads' will usually be assigned 0.5, whether this is taken as a subjective belief about the next toss or whether the next toss is thought of as part of a long series of tosses. However, as Lindley (2000) emphasises, the rules of probability do not need to be assumed as self-evident, but can be *derived* from 'deeper' axioms of reasonable behaviour of an individual (say, *You*) in the face of Your own uncertainty. This 'reasonable behaviour' features characteristics such as Your unwillingness to make a series of bets based on expressed probabilities, such that You are bound to lose (a so-called 'Dutch book'), or Your unwillingness to state probabilities that can always be improved upon in terms of their expected accuracy in predicting events. It is perhaps remarkable that from such conditions one can prove the three basic rules of probability (Lindley, 1985): as a simple example, if I state probabilities of 0.7 that it will rain tomorrow, and 0.4 that it will not rain, and I am willing to bet at these odds, then a good bookmaker can accept a series of bets from me such that I am bound to lose. (For example, assuming small stakes, I would consider it a good deal to bet 14 units of money for a return of 21 if it rained, since my expected profit is $0.7 \times 21 - 14 = 0.7$, and simultaneously I would bet 8 units of money for a return of 21 if it did *not* rain. Thus the bookmaker is certain to make a profit of 1 unit whatever happens.) Such probabilities are said not to 'cohere', and are assumed to be avoided by all rational individuals.

The vital point of the subjective interpretation is that Your probability for an event is a property of Your relationship to that event, and not an objective property of the event itself. This is why, pedantically speaking, one should always refer to probabilities *for* events rather than probabilities *of* events, and the conditioning context H used in Section 2.1.1 includes the observer and all their background knowledge and assumptions. The fact that the probability is a reflection of personal uncertainty rather than necessarily being based on future unknown events is illustrated (from personal experience) by a gambling game played in casinos in Macau. Two dice are thrown out of sight of the gamblers and immediately covered up: the participants then bet on different possible combinations. Thus, they are betting on an event that has already occurred, but about which they are personally ignorant. (Incidentally, their beliefs also do not appear to be governed by the assumed physical symmetries of the dice: although they have 2 minutes to bet, everyone remains totally still for at least 90 seconds, and then when the first bet is laid the crowd follow in a rush, apparently believing in the good fortune of the one confident individual.)

The subjective view of probability is not new, and in past epochs has been the standard ideology. Fienberg (1992) points out that Jakob Bernoulli in 1713 introduced 'the subjective notion that the probability is personal and varies with an individual's knowledge', and that Laplace and Gauss both worked with posterior distributions two hundred years ago, which became known as 'the inverse method'. However, from the mid-nineteenth century the frequency approach started to dominate, and controversy has sporadically continued.

Dempster (1998) quotes Edgeworth in 1884 as saying that the critics who 'heaped ridicule upon Bayes' theorem and the inverse method' were trying to elicit 'knowledge out of ignorance, something out of nothing'. Polemical opinions are still expressed in defence of the explicit introduction of subjective judgement into scientific research: 'it simply makes no sense to take seriously every apparent falsification of a plausible theory, any more than it makes sense to take seriously every new scientific idea' (Matthews, 1998).

Bayesian methods therefore explicitly allow for the possibility that the conclusions of an analysis may depend on who is conducting it and their available evidence and opinion, and therefore the context of the study is vital: 'Bayesian statistics treats subjectivity with respect by placing it in the open and under the control of the consumer of data' (Berger and Berry, 1988). Apart from methodological researchers, at least five different viewpoints might be identified for an evaluation of a health-care intervention:

- *sponsors*, e.g. the pharmaceutical industry, medical charities or granting agencies;
- *investigators*, *i.e.* those responsible for the conduct of a study, whether industry or publicly funded;
- *reviewers*, e.g. regulatory bodies;
- *policy makers*, e.g. agencies setting health policy;
- *consumers*, e.g. individual patients or clinicians acting on their behalf.

Each of these broad categories can be further subdivided. An analysis which might be carried out solely for the investigators, for example, may not be appropriate for presentation to reviewers or consumers: 'experimentalists tend to draw a sharp distinction between providing their opinions and assessments for the purposes of experimental design and in-house discussion, and having them incorporated into any form of externally disseminated report' (Racine *et al.*, 1996). The roles of these different stakeholders in decision-making is further explored in Chapter 9.

A characteristic of health-care evaluation is that the investigators who plan and conduct a study are generally not the same body as those who make decisions on the basis of the evidence provided in part by that study: such decision-makers may be regulatory authorities, policy-makers or health-care providers. This division is acknowledged in this book by separating Chapter 6 on the design and monitoring of trials from Chapter 9 on policy-making.

3.2 BAYES THEOREM FOR TWO HYPOTHESES

In Section 2.1.3 Bayes theorem was derived as a basic result in probability theory. We now begin to illustrate its use as a mechanism for learning about unknown quantities from data, a process which is sometimes known as 'prior to posterior' analysis. We start with the simplest possible situation. Consider two hypotheses H_0 and H_1 which are 'mutually exhaustive and exclusive', *i.e.* one and only one is true. Let the *prior* probability for each of the two hypotheses, before we have access to the evidence of interest, be $p(H_0)$ and $p(H_1)$; for the moment we will not concern ourselves with the source of those probabilities. Suppose we have observed some data y, such as the results of a test, and we know from past experience that the probability of observing y under each of the two hypotheses is $p(y|H_0)$ and $p(y|H_1)$, respectively: these are the *likelihoods*, with the vertical bar representing 'conditioning'.

Bayes theorem shows how to revise our prior probabilities in the light of the evidence in order to produce *posterior probabilities*. Specifically, by adapting (2.3) we have the identity

$$p(H_0|y) = \frac{p(y|H_0)}{p(y)} \times p(H_0), \qquad (3.1)$$

where $p(y) = p(y|H_0)p(H_0) + p(y|H_1)p(H_1)$ is the overall probability of y occurring.

Now $H_1 = $ 'not H_0' and so $p(H_0) = 1 - p(H_1)$ and $p(H_0|y) = 1 - p(H_1|y)$. In terms of odds rather than probabilities, Bayes theorem can then be re-expressed (see (2.5)) as

$$\frac{p(H_0|y)}{p(H_1|y)} = \frac{p(y|H_0)}{p(y|H_1)} \times \frac{p(H_0)}{p(H_1)}. \qquad (3.2)$$

Now $p(H_0)/p(H_1)$ is the 'prior odds', $p(H_0|y)/p(H_1|y)$ is the 'posterior odds', and $p(y|H_0)/p(y|H_1)$ is the ratio of the likelihoods, and so (3.2) can be expressed as

posterior odds = likelihood ratio × prior odds.

By taking logarithms we also note that

log (posterior odds) = log (likelihood ratio) + log (prior odds).

where the log(likelihood ratio) has also been termed the 'weight of evidence': this term was invented by Alan Turing when using these techniques for breaking the Enigma codes at Bletchley Park during the Second World War.

Example 3.1 shows how this formulation is commonly used in the evaluation of diagnostic tests, and reveals that our intuition is often poor when processing probabilistic evidence, and that we tend to forget the importance of the prior probability (Section 5.2).

Example 3.1   *Diagnosis: Bayes theorem in diagnostic testing*

Suppose a new home HIV test is claimed to have '95% sensitivity and 98% specificity', and is to be used in a population with an HIV prevalence of

1/1000. We can calculate the expected status of 100 000 individuals who are tested, and the results are shown in Table 3.1. Thus, for example, we expect 100 truly HIV positive individuals of whom 95% will test positive, and of the remaining 99 900 HIV negative individuals we expect 2% (1998) to test positive. Thus of the 2093 who test positive (*i.e.* have observation *y*), only 95 are truly HIV positive, giving a 'predictive value positive' of only $95/2093 = 4.5\%$.

Table 3.1 Expected status of 100 000 tested individuals in a population with an HIV prevalence of 1/1000.

	HIV−	HIV+	
Test −	97 902	5	97 907
Test +	1 998	95	2 093
	99 900	100	100 000

We can also do these calculations using Bayes theorem. Let H_0 be the hypothesis that the individual is truly HIV positive, and *y* be the observation that they test positive. The disease prevalence is the prior probability ($p(H_0) = 0.001$), and we are interested in the chance that someone who tests positive is truly HIV positive, *i.e.* the posterior probability $p(H_0|y)$.

Let H_1 be the hypothesis that they are truly HIV negative; '95% sensitivity' means that $p(y|H_0) = 0.95$, and '98% specificity' means that $p(y|H_1) = 0.02$. To use (3.2), we require two inputs: the prior odds $p(H_0)/p(H_1)$ which are 1/999, and the likelihood ratio $p(y|H_0)/p(y|H_1)$ which is $0.95/0.02 = 95/2$. Then from (3.2) the posterior odds are $(95/2) \times 1/999 = 95/1998$. These odds correspond to a posterior probability $p(H_0|y) = 95/(95 + 1998) = 0.045$, as found directly from the table.

Alternatively, we can use the form of Bayes theorem given by (3.1). Now $p(y) = p(y|H_0)p(H_0) + p(y|H_1)p(H_1) = 0.95 \times 0.001 + 0.02 \times 0.999 = 0.020\,93$. Thus (3.1) says that $p(H_0|y) = 0.95 \times 0.001/0.020\,93 = 0.045$.

The crucial finding is that over 95% of those testing positive will, in fact, not have HIV.

Figure 3.1 shows Bayes theorem for two hypotheses in either odds or probability form, for a range of likelihood ratios. The likelihood ratio from a positive result in Example 3.1 is $0.95/0.02 = 47.5$. From a rough inspection of Figure 3.1 we can see that such a likelihood ratio is sufficient to turn a moderately low prior probability, such as 0.2, into a reasonably high posterior probability of around 0.9; however, if the prior probability is as low as it is in Example 3.1 (*i.e.* 0.001), then the posterior probability is still somewhat small.

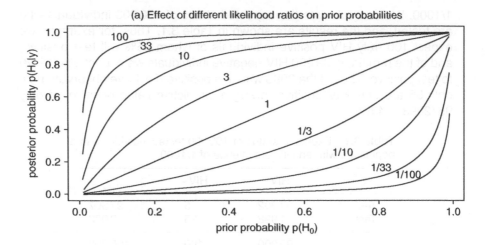

(a) Effect of different likelihood ratios on prior probabilities

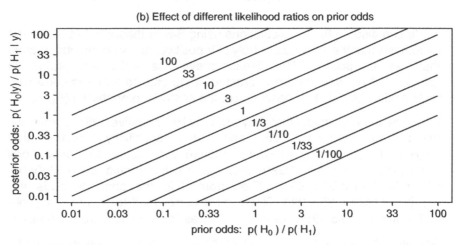

(b) Effect of different likelihood ratios on prior odds

Figure 3.1   Bayes theorem for two hypotheses H_0 and H_1 = 'not H_0' in (a) probability $p(H_0)$ and (b) odds $p(H_0)/p(H_1)$ form. By specifying the prior probability or odds, and the likelihood ratio $p(y|H_0)/p(y|H_1)$, the posterior probability or odds can be read off the graph. Note that (b) uses a logarithmic scaling, under which Bayes theorem gives a linear relationship.

3.3   COMPARING SIMPLE HYPOTHESES: LIKELIHOOD RATIOS AND BAYES FACTORS

In Section 3.2 we showed how data y influence the relative probabilities of two hypotheses H_0 and H_1 through the likelihood ratio $p(y|H_0)/p(y|H_1)$, and hence the likelihoods contain all the relevant evidence that can be extracted from the data: this is the likelihood principle, discussed in more detail in Section 4.3. This measure of the relative likelihood of two hypotheses is also known as the 'Bayes factor' (BF), although Cornfield (1976) also termed this the 'relative betting odds' between two hypotheses: see, for example, Goodman (1999b) for a detailed exposition. The Bayes factor can vary between 0 and ∞, with small values being considered as both evidence *against* H_0 and evidence *for* H_1. The scale in Table 3.2 was provided by the Bayesian physicist, Harold Jeffreys, and dates from 1939 (Jeffreys, 1961, p. 432).

The crucial idea is that the Bayes factor transforms prior to posterior odds: this uses expression (3.2), and the results can be read off Figure 3.1. In Example 3.1 we observed a Bayes factor (likelihood ratio) after a positive HIV test of $BF = 47.5$ in favour of being HIV positive (H_0). Table 3.2 labels this as 'very strong' evidence in itself in favour of H_0, but when combined with strong prior opinion against H_0 (prior odds of 1/999) does not lead to a very convincing result (posterior odds $\approx 1/21$).

Bayes factors can also be obtained for composite hypotheses that include unknown parameters: this is discussed in Section 4.4 and is a feature when using a prior distribution that puts a 'lump' of probability on a (null) hypothesis (Section 5.5.4). The relationship between Bayes factors and traditional ways of hypothesis testing has been the subject of considerable research and controversy, and is discussed further in Section 4.4.

The use of Bayes theorem in diagnostic testing is an established part of formal clinical reasoning. More controversial is the use of Bayes theorem in general statistical analyses, where a parameter θ is an unknown quantity such as the mean benefit of a treatment on a specified patient population, and its prior distribution $p(\theta)$ needs to be specified. This major step might be considered as a natural extension of the subjective interpretation of probability, but the following (starred) section provides a further argument for why a prior distribution on a parameter may be a reasonable assumption.

Table 3.2 Calibration of Bayes factor (likelihood ratio) provided by Jeffreys.

Bayes factor range	Strength of evidence in favour of H_0 and against H_1
> 100	Decisive
32 to 100	Very strong
10 to 32	Strong
3.2 to 10	Substantial
1 to 3.2	'Not worth more than a bare mention'
	Strength of evidence against H_0 and in favour of H_1
1 to 1/3.2	'Not worth more than a bare mention'
1/3.2 to 1/10	Substantial
1/10 to 1/32	Strong
1/32 to 1/100	Very strong
< 1/100	Decisive

3.4   EXCHANGEABILITY AND PARAMETRIC MODELLING*

In Section 2.2.3 we introduced the concept of independent and identically distributed (i.i.d.) variables $Y_1, \ldots, Y_n$ as a fundamental component of standard statistical modelling. However, just as we found in Section 3.1 that the rules of probability could themselves be derived from more basic ideas of rational behaviour, so we can derive the idea of i.i.d. variables and prior distributions of parameters from the more basic subjective judgement known as 'exchangeability'. Exchangeability is a formal expression of the idea that we find no systematic reason to distinguish the individual variables $Y_1, \ldots, Y_n$ – they are similar but not identical. Technically, we judge that $Y_1, \ldots, Y_n$ are exchangeable if the probability that we assign to any set of potential outcomes, $p(y_1, \ldots, y_n)$, is unaffected by permutations of the labels attached to the variables. For example, suppose Y_1, Y_2, Y_3 are the first three tosses of a (possibly biased) coin, where $Y_1 = 1$ indicates a head, and $Y_1 = 0$ indicates a tail. Then we would judge $p(Y_1 = 1, \ Y_2 = 0, \ Y_3 = 1) = p(Y_2 = 1, \ Y_1 = 0, \ Y_3 = 1) = p(Y_1 = 1, Y_3 = 0, Y_2 = 1)$, *i.e.* the probability of getting two heads and a tail is unaffected by the particular toss on which the tail comes. This is a natural judgement to make if we have no reason to think that one toss is systematically any different from another. Note that it does *not* mean we believe that $Y_1, \ldots, Y_n$ are independent: independence would imply $p(y_1, \ldots, y_n) = p(y_1) \times \ldots \times p(y_n)$ and hence the result of a series of tosses does not help us predict the next, whereas a long series of heads would tend to make us believe the coin was seriously biased and hence would lead us to predict a head as more likely.

An Italian actuary, Bruno de Finetti, published in 1930 a most extraordinary result (de Finetti, 1930). He showed that if a set of binary variables $Y_1, \ldots, Y_n$ were judged exchangeable, then it implied that

$$p(y_1, \ldots, y_n) = \int \prod_{i=1}^{n} p(y_i|\theta)p(\theta)d\theta. \qquad (3.3)$$

Now (3.3) is unremarkable if we argue from right to left: if $Y_1, \ldots, Y_n$ are i.i.d., each with distribution $p(y_i|\theta)$, their joint distribution (conditional on θ) is $p(y_1, \ldots, y_n|\theta) = \prod_{i=1}^{n} p(y_i|\theta)$ (2.16). Hence, their marginal distribution $p(y_1, \ldots, y_n)$ (2.7), given a distribution $p(\theta)$, is given by (3.3). However, de Finetti's remarkable achievement was to argue from left to right: exchangeable random quantities can be thought of as being i.i.d. variables drawn from some common distribution depending on an unknown parameter θ, which itself has a prior distribution $p(\theta)$. Thus, from a subjective judgement about observable quantities, one derives the whole apparatus of i.i.d. variables, conditional independence, parameters and prior distributions. This was an amazing achievement.

De Finetti's results have been extended to much more general situations (Bernardo and Smith, 1994), and the concept of exchangeability will continually recur throughout this book.

3.5   BAYES THEOREM FOR GENERAL QUANTITIES

This small section is the most important in this book.

Suppose θ is some quantity that is currently unknown, for example the true success rate of a new therapy, and let $p(\theta)$ denote the prior distribution of θ. As discussed in Section 3.1, this prior distribution should, strictly speaking, be denoted $p(\theta|H)$ to remind us that it represents Your judgement about θ conditional on a context H, where You are the person for whom the analysis is being performed (the client), and not the statistician who may be actually carrying out the analysis. The interpretation and source of such distributions are discussed in Section 3.9 and Chapter 5.

Suppose we have some observed evidence y, for example the results of a clinical trial, whose probability of occurrence is assumed to depend on θ. As we have seen, this dependence is formalised by $p(y|\theta)$, the (conditional) probability of y for each possible value of θ, and when considered as a function of θ is known as the likelihood. We would like to obtain the new, posterior, probability for different values of θ, taking account of the evidence y; this probability has the conditioning reversed and is denoted $p(\theta|y)$.

Bayes theorem applied to a general quantity θ was given in (2.6) and says that

$$p(\theta|y) = \frac{p(y|\theta)}{p(y)} \times p(\theta). \qquad (3.4)$$

Now $p(y)$ is just a normalising factor to ensure that $\int p(\theta|y)\, d\theta = 1$, and its value is not of interest (unless we are comparing alternative models). The essence of Bayes theorem only concerns the terms involving θ, and hence it is often written

$$p(\theta|y) \propto p(y|\theta) \times p(\theta), \qquad (3.5)$$

which says that the posterior distribution is proportional to (*i.e.* has the same shape as) the product of the likelihood and the prior. The deceptively simple expression (3.5) is the basis for the whole of the rest of this book, since it shows how to make inferences from a Bayesian perspective, both in terms of estimation and obtaining credible intervals and also making direct probability statements about the quantities in which we are interested.

3.6   BAYESIAN ANALYSIS WITH BINARY DATA

In Section 2.2.4 we considered a probability θ of an event occurring, and derived the form of the likelihood for θ having observed n cases in which r events occurred. Adopting a Bayesian approach to making inferences, we wish to combine this likelihood with initial evidence or opinion regarding θ, as expressed in a prior distribution $p(\theta)$.

3.6.1 Binary data with a discrete prior distribution

First, suppose only a limited set of hypotheses concerning the true proportion θ are being entertained, corresponding to a finite list denoted $\theta_1, \ldots, \theta_J$. Suppose in addition a prior probability $p(\theta_j)$ of each has been assessed, where $\sum_j p(\theta_j) = 1$. For a single Bernoulli trial with outcome 0 or 1, the likelihood for each possible value for θ is given by (2.15),

$$p(y|\theta_j) = \theta_j^y (1 - \theta_j)^{1-y}, \tag{3.6}$$

i.e. $p(y|\theta_j) = \theta_j$ if $y = 1$, and $p(y|\theta_j) = 1 - \theta_j$ if $y = 0$.

Having observed an outcome y, Bayes theorem (3.5) states that the posterior probabilities for the θ_j obey

$$p(\theta_j|y) \propto \theta_j^y (1 - \theta_j)^{1-y} \times p(\theta_j), \tag{3.7}$$

where the normalising factor that ensures that the posterior probabilities add to 1 is

$$p(y) = \sum_j \theta_j^y (1 - \theta_j)^{1-y} \times p(\theta_j).$$

After further observations have been made, say with the result that there have been r 'successes' out of n trials, the relevant posterior will obey

$$p(\theta_j|r) \propto \theta_j^r (1 - \theta_j)^{n-r} \times p(\theta_j). \tag{3.8}$$

A basic example of these calculations is given in Example 3.2.

Example 3.2 *Drug: Binary data and a discrete prior*

Suppose a drug has an unknown true response rate θ, and for simplicity we assume that θ can only take one of the values $\theta_1 = 0.2$, $\theta_2 = 0.4$, $\theta_3 = 0.6$ or $\theta_4 = 0.8$. Before experimentation we adopt the 'neutral' position of assuming each value θ_j is equally likely, so that $p(\theta_j) = 0.25$ for each $j = 1, 2, 3, 4$.

Suppose we test the drug on a single subject and we observed a positive response ($y = 1$). How should our belief in the possible values of θ be revised?

First, we note that the likelihood is simply $p(y|\theta_j) = \theta_j^y (1 - \theta)^{(1-y)} = \theta_j$. Table 3.3 displays the components of Bayes theorem (3.7): the 'Likelihood $\times$ prior' column, normalised by its sum $p(y)$, gives the posterior probabilities. It is perhaps initially surprising that a single positive response makes it four times as likely that the true response rate is 80% rather than 20%.

Table 3.3 Results after observing a single positive response, $y = 1$, for a drug given an initial uniform distribution over four possible response rates θ_j.

j	θ_j	Prior $p(\theta_j)$	Likelihood $p(y|\theta_j)$	Likelihood × prior $p(y|\theta_j)p(\theta_j)$	Posterior $p(\theta_j|y)$
1	0.2	0.25	0.2	0.05	0.10
2	0.4	0.25	0.4	0.10	0.20
3	0.6	0.25	0.6	0.15	0.30
4	0.8	0.25	0.8	0.20	0.40
	$\sum_j$	1.0		0.50	1.0

Suppose we now observe 15 positive responses out of 20 patients, how is our belief revised? Table 3.4 shows that any initial belief in $\theta_1 = 0.2$ is now completely overwhelmed by the data, and that the only remaining contenders are $\theta_3 = 0.6$ with about 30% of the posterior probability, and $\theta_4 = 0.8$ with about 70%.

We note that, had we given any non-zero probability to the extreme values of $\theta = 0$, 1, *i.e.* the drug either never or always worked, these would give a zero likelihood and hence zero posterior probability.

Table 3.4 Results after observing 15 positive responses, $y = 15$, for a drug out of 20 cases, given an initial uniform distribution over four possible response rates θ_j.

j	θ_j	Prior $p(\theta_j)$	Likelihood $\theta_j^{15}(1 - \theta_j)^5$ $(\times 10^{-7})$	Likelihood × prior $\theta_j^{15}(1 - \theta_j)^5\, p(\theta_j)$ $(\times 10^{-7})$	Posterior $p(\theta_j|X = 1)$
1	0.2	0.25	0.0	0.0	0.000
2	0.4	0.25	0.8	0.2	0.005
3	0.6	0.25	48.1	12.0	0.298
4	0.8	0.25	112.6	28.1	0.697
	$\sum_j$	1.0		40.3	1.0

3.6.2   Conjugate analysis for binary data

It is generally more realistic to consider θ a continuous parameter, and hence it needs to be given a continuous prior distribution. One possibility is that we think all possible values of θ are equally likely, in which case we could summarise this by a uniform distribution (Section 2.6.4) so that $p(\theta) = 1$ for $0 \leqslant \theta \leqslant 1$. Applying Bayes theorem (3.5) yields

$$p(\theta|y) \propto \theta^r(1 - \theta)^{n-r} \times 1, \qquad (3.9)$$

where r is the number of events observed and n is the total number of individuals.

We may recognise that the functional form of the posterior distribution in (3.9) is proportional to that of a beta distribution (Section 2.6.3). Rewriting the posterior distribution (3.9) as $\theta^{(r+1)-1}(1 - \theta)^{(n-r+1)-1}$, we can see that the posterior distribution is in fact Beta $[r + 1, \ n - r + 1]$. This immediately means that we can now summarise the posterior distribution in terms of its mean and variance, and make probability statements based on what we know about the beta distribution (for example, many common statistical packages will calculate tail area probabilities for the beta distribution).

Instead of a uniform prior distribution for θ we could take a Beta $[a, \ b]$ prior distribution and obtain the following analysis:

$$\text{Prior} \propto \theta^{a-1}(1 - \theta)^{b-1}$$
$$\text{Likelihood} \propto \theta^r(1 - \theta)^{n-r}$$
$$\text{Posterior} \propto \theta^{a-1}(1 - \theta)^{b-1}\theta^r(1 - \theta)^{n-r} \qquad (3.10)$$
$$\propto \theta^{a+r-1}(1 - \theta)^{b+n-r-1}$$
$$= \text{Beta}[a + r, \ b + n - r].$$

Thus we have specified a beta prior distribution for a parameter, observed data from a Bernoulli or binomial sampling distribution, worked through Bayes theorem, and ended up with a beta posterior distribution. This is a case of *conjugate analysis*. Conjugate models occur when the posterior distribution is of the same *family* as the prior distribution: other examples include the gamma distribution being conjugate with a Poisson likelihood, normal priors being conjugate with normal likelihoods (Section 3.7), and gamma priors for unknown precisions of normal likelihoods (Section 2.6.5).

Example 3.3   *Drug (continued): Binary data and a continuous prior*

Suppose that previous experience with similar compounds has suggested that response rates between 0.2 and 0.6 could be feasible, with an expectation around 0.4. We can translate this into a prior Beta[a, b] distribution as follows.

We first want to estimate the mean m and standard deviation s of the prior distribution. For normal distributions we know that $m \pm 2s$ includes just over 95% of the probability, so if we were assuming a normal prior we might estimate $m = 0.4$, $s = 0.1$. However, we know from Section 2.6.3 that beta distributions with reasonably high a and b have an approximately normal shape, so these estimates might also be used for a beta prior.

Next, from Section 2.6.3, we know that for a beta distribution

$$m = a/(a+b),\tag{3.11}$$

$$s^2 = m(1-m)/(a+b+1).\tag{3.12}$$

Expression (3.12) can be rearranged to give $a+b = m(1-m)/s^2 - 1$. Using the estimates $m = 0.4$, $s = 0.1$, we obtain $a+b = 23$. Then, from (3.11), we see that $a = m(a+b)$, and hence we finally obtain $a = 9.2$, $b = 13.8$: this can be considered a 'method of moments'. A Beta[9.2,13.8] distribution is shown in Figure 3.2(a), showing that it well represents the prior assumptions. It is convenient to think of this prior distribution as that which would have arisen had we started with a 'non-informative' prior Beta[0,0] and then observed $a = 9.2$ successes in $a+b = 23$ patients (however, this is only a heuristic argument as there is no agreed 'non-informative' beta prior, with Beta[0,0], Beta[$\frac{1}{2},\frac{1}{2}$], Beta[1,1] all having been suggested (Section 5.5.1)).

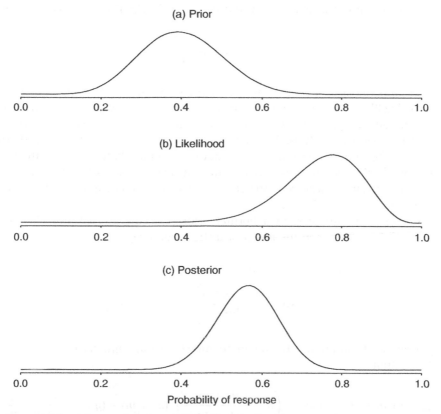

Figure 3.2 (a) is a Beta[9.2,13.8] prior distribution supporting response rates between 0.2 and 0.6, (b) is a likelihood arising from a binomial observation of 15 successes out of 20 cases, and (c) is the resulting Beta[24.2, 18.8] posterior from a conjugate beta-binomial analysis.

If we now observed $r = 15$ successes out of 20 trials, we know from (3.10) that the parameters of the beta distribution are updated to $[a + 15, b + 20 - 5] = [24.2, 18.8]$. The likelihood and posterior are shown in Figures 3.2(b) and 3.2(c): the posterior will have mean $24.2/(24.2 + 18.8) = 0.56$.

3.7   BAYESIAN ANALYSIS WITH NORMAL DISTRIBUTIONS

In Section 2.4 we saw that in many circumstances it is appropriate to consider a likelihood as having a normal shape, although this may involve working on somewhat uninituitive scales such as the logarithm of the hazard ratio. With a normal likelihood it is mathematically convenient, and often reasonably realistic, to make the assumption that the prior distribution $p(\theta)$ has the form

$$p(\theta) = \mathrm{N}\left[\theta \middle| \mu, \frac{\sigma^2}{n_0}\right], \qquad (3.13)$$

where μ is the prior mean. We note that the same standard deviation σ is used in the likelihood and the prior, but the prior is based on an 'implicit' sample size n_0. The advantage of this formulation becomes apparent when we carry out prior-to-posterior analysis. We note in passing that as n_0 tends to 0, the variance becomes larger and the distribution becomes 'flatter', and in the limit the distribution becomes essentially uniform over $(-\infty, \infty)$. A normal prior with a very large variance is sometimes used to represent a 'non-informative' distribution (Section 5.5.1).

Suppose we assume such a normal prior $\theta \sim \mathrm{N}[\mu, \sigma^2/n_0]$ and likelihood $y_m \sim \mathrm{N}[\theta, \sigma^2/m]$. Then the posterior distribution obeys

$$p(\theta|y_m) \propto p(y_m|\theta)p(\theta)$$

$$\propto \exp\left[-\frac{(y_m - \theta)^2 m}{2\sigma^2}\right] \times \exp\left[-\frac{(\theta - \mu)^2 n_0}{2\sigma^2}\right],$$

ignoring irrelevant terms that do not include θ. By matching terms in θ it can be shown that

$$(y_m - \theta)^2 m + (\theta - \theta_0)^2 n_0 = \left(\theta - \frac{n_0\theta_0 + my_m}{n_0 + m}\right)^2 (n_0 + m) + (y_m - \mu)^2\left(\frac{1}{m} + \frac{1}{n_0}\right),$$

and we can recognise that the term involving θ is exactly that arising from a posterior distribution

$$p(\theta|y_m) = \text{N}\left[\theta \left| \frac{n_0\mu + my_m}{n_0 + m} , \frac{\sigma^2}{n_0 + m} \right. \right].$$ (3.14)

Equation (3.14) is very important. It says that our posterior mean $(n_0\mu + my_m)/(n_0 + m)$ is a weighted average of the prior mean μ and parameter estimate y_m, weighted by their precisions, and therefore is always a compromise between the two. Our posterior variance (1/precision) is based on an implicit sample size equivalent to the sum of the prior 'sample size' n_0 and the sample size of the data m: thus, when combining sources of evidence from the prior and the likelihood, we *add precisions* and hence always decrease our uncertainty. As Senn (1997a, p. 46) claims, 'A Bayesian is one who, vaguely expecting a horse and catching a glimpse of a donkey, strongly concludes he has seen a mule'. Note that as $n_0 \to 0$, the prior tends towards a uniform distribution and the posterior tends to the same shape as the likelihood.

Suppose we do not adopt the convention for expressing prior and sampling variances as σ^2/n_0 and σ^2/m, and instead use the general notation $\theta \sim \text{N}[\mu, \tau^2]$ and likelihood $y_m \sim \text{N}[\theta, \sigma_m^2]$. Then it is straightforward to show that the posterior distribution (3.14) can be expressed as

$$p(\theta|y_m) = \text{N}\left[\theta \left| \frac{\frac{\mu}{\tau^2} + \frac{y_m}{\sigma_m^2}}{\frac{1}{\tau^2} + \frac{1}{\sigma_m^2}} , \frac{1}{\frac{1}{\tau^2} + \frac{1}{\sigma_m^2}} \right. \right].$$ (3.15)

We will sometimes find this general form useful, but will generally find (3.14) more intuitive.

Example 3.4 provides a simple example of Bayesian reasoning using normal distributions.

Example 3.4   *SBP: Bayesian analysis for normal data*

Suppose we are interested in the long-term systolic blood pressure (SBP) in mmHg of a particular 60-year-old female. We take two independent readings 6 weeks apart, and their mean is 130. We know that SBP is measured with a standard deviation $\sigma = 5$. What should we estimate her SBP to be?

Let her long-term SBP be denoted θ. A standard analysis would use the sample mean $y_m = 130$ as an estimate, with standard error $\sigma/\sqrt{m} = 5/\sqrt{2} = 3.5$: a 95% confidence interval is $y_m \pm 1.96 \times \sigma/\sqrt{m}$, *i.e.* 123.1 to 136.9.

However, we may have considerable additional information about SBPs which we can express as a prior distribution. Suppose that a survey in the same population revealed that females aged 60 had a mean long-term SBP of 120 with standard deviation 10. This population distribution can be considered as a prior distribution for the specific individual, and is shown in Figure 3.3(a): if we express the prior standard deviation as $\sigma/\sqrt{n_0}$ (*i.e.* variance σ^2/n_0), we can solve to find $n_0 = (\sigma/10)^2 = 0.25$.

Figure 3.3(b) shows the likelihood arising from the two observations on the woman. From (3.14) the posterior distribution of θ is normal with mean $(0.25 \times 120 + 2 \times 130)/(0.25 + 2) = 128.9$ and standard deviation $\sigma/\sqrt{n_0 + m} = 5/\sqrt{2.25} = 3.3$, giving a 95% interval of $128.9 \pm 1.96 \times 3.3 = (122.4, 135.4)$. Figure 3.3(c) displays this posterior distribution, revealing some 'shrinkage' towards the population mean, and a small increase in precision from not using the data alone.

Intuitively, we can say that the woman has somewhat higher measurements than we would expect for someone her age, and hence we slightly adjust our estimate to allow for the possibility that her two measures happened by chance to be on the high side. As additional measures are made, this possibility becomes less plausible and the prior knowledge will be systematically downgraded.

3.8   POINT ESTIMATION, INTERVAL ESTIMATION AND INTERVAL HYPOTHESES

Although it is most informative to plot an entire posterior distribution, there will generally be a need to produce summary statistics: we shall consider point estimates, intervals, and the probabilities of specified hypotheses.

Point estimates.   Traditional measures of location of distributions include the mean, median and mode, and – by imposing a particular penalty on error in estimation (Berger, 1985) – each can be given a theoretical justification as a point estimate derived from a posterior distribution. If the posterior distribution is symmetric and unimodal, as in Figure 3.3, then the mean, median and mode all coincide in a single value and there is no difficulty in making a choice. We shall find, however, that in some circumstances posterior distributions are considerably skewed and there are marked differences between, say, mean and median. We shall prefer to quote the median in such contexts as it is less sensitive to the tails of the distribution, although it is perhaps preferable to report all three summary measures when they show wide disparity.

Interval estimates.   Any interval containing, say, 95% probability may be termed a 'credible' interval to distinguish it from a Neyman–Pearson 'confidence interval', although we shall generally refer to them simply as posterior intervals. Three types of intervals can be distinguished – we assume a continuous parameter θ with range on $(-\infty, \infty)$ and a posterior conditional on generic data y:

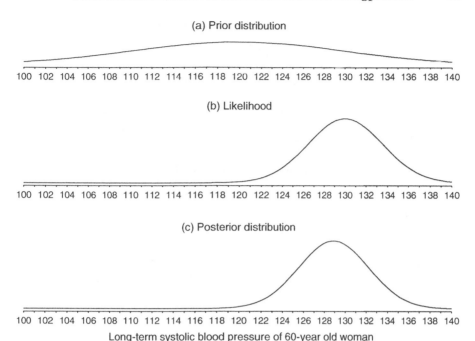

(a) Prior distribution

100 102 104 106 108 110 112 114 116 118 120 122 124 126 128 130 132 134 136 138 140

(b) Likelihood

100 102 104 106 108 110 112 114 116 118 120 122 124 126 128 130 132 134 136 138 140

(c) Posterior distribution

100 102 104 106 108 110 112 114 116 118 120 122 124 126 128 130 132 134 136 138 140

Long-term systolic blood pressure of 60-year old woman

Figure 3.3 Estimating the true long-term underlying systolic blood pressure of a 60-year-old woman: (a) the prior distribution is $N[120, 10^2]$ and expresses the distribution of true SBPs in the population; (b) the likelihood is proportional to $N[130, 3.5^2]$ and expresses the support for different values arising from the two measurements made on the woman; (c) the posterior distribution is $N[128.9, 3.3^2]$ and is proportional to the likelihood multiplied by the prior.

One-sided intervals. For example, a one-sided upper 95% interval would be (θ_L, ∞), where $p(\theta < \theta_L|y) = 0.05$.

Two-sided 'equi-tail-area' intervals. A two-sided 95% interval with equal probability in each tail area would comprise (θ_L, θ_U), where $p(\theta < \theta_L|y) = 0.025$, and $p(\theta > \theta_U|y) = 0.975$.

Highest posterior density (HPD) intervals. If the posterior distribution is skewed, then a two-sided interval with equal tail areas will generally contain some parameter values that have lower posterior probability than values outside the interval. An HPD interval does not have this property – it is adjusted so that the probability ordinates at each end of the interval are identical, and hence it is also the narrowest possible interval containing the required probability. Of course if the posterior distribution has more than one mode, then the HPD may be made up of a set of disjoint intervals.

These alternatives are illustrated in Figure 3.4, suggesting that HPD intervals would be preferable — unfortunately they are generally difficult to compute. For normal posterior distributions these intervals require only the use of tables or

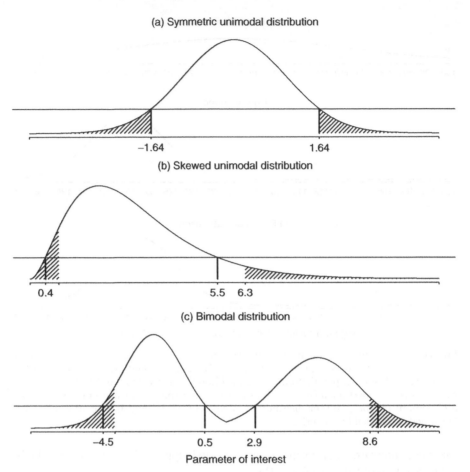

(a) Symmetric unimodal distribution

−1.64   1.64

(b) Skewed unimodal distribution

0.4   5.5   6.3

(c) Bimodal distribution

−4.5   0.5   2.9   8.6

Parameter of interest

Figure 3.4 (a) shows a symmetric unimodal distribution in which equi-tail-area and HPD intervals coincide at −1.64 to 1.64. (b) is a skewed unimodal distribution in which the equi-tail-area interval is 0.8 to 6.3, whereas the HPD of 0.4 to 5.5 is considerably shorter. (c) shows a bimodal distribution in which the equi-tail-area interval is −3.9 to 8.6, whereas the HPD appropriately consists of two segments.

programs giving tail areas of normal distributions (Sections 2.3 and 3.7). In more complex situation we shall generally be simulating values of θ and one- and two-sided intervals are constructed using the empirical distribution of simulated values (Section 3.19.3). It will not usually be possible to find HPD intervals when using simulation methods.

Traditional confidence intervals and Bayesian credible intervals differ in a number of ways.

1. Most important is their interpretation: we say there is a 95% probability that the true θ lies in a 95% credible interval, whereas this is certainly *not* the

interpretation of a 95% confidence interval. In a long series of 95% confidence intervals, 95% of them should contain the true parameter value – unlike the Bayesian interpretation, we cannot give a probability for whether a *particular* confidence interval contains the true value, it either does or does not and all we have to fall back on is the long-run properties of the procedure. Of course, the direct Bayesian interpretation is often wrongly ascribed to confidence intervals.

2. Credible intervals will generally be narrower due to the additional information provided by the prior: for an analysis assuming the normal distribution they will have width $2 \times 1.96 \times \sigma/\sqrt{n_0 + m}$, compared to $2 \times 1.96 \times \sigma/\sqrt{m}$ for the confidence interval.

3. Some care is required in terminology: while the width of classical confidence intervals is governed by the *standard error* of the estimator, the width of Bayesian credible intervals is dictated by the posterior *standard deviation*.

Interval hypotheses. Suppose a hypothesis of interest comprises an interval $H_0 : \theta_L < \theta < \theta_U$, for some prespecified θ_L, θ_U indicating, for example, a range of clinical equivalence. Then it is straightforward to report the posterior probability $p(H_0|y) = p(\theta_L < \theta < \theta_U|y)$, which may again be obtained using standard formulae or simulation methods.

Example 3.5   *SBP (continued): Interval estimation*

We extend Example 3.4 to encompass testing the hypothesis that the woman has a long-term SBP greater than 135, and the provision of 95% intervals.

The probability of the hypothesis H_0: $\theta_L < \theta < \infty$, $\theta_L = 135$, is

$$p(H_0|y) = p(\theta > \theta_L|y) = 1 - \Phi\left(\frac{\theta_l - \dfrac{n_0\mu + my_m}{n_0 + m}}{\sigma/\sqrt{n_0 + m}}\right)$$

and is shaded in Figure 3.5(a). Figure 3.5(b) displays a 95% posterior interval comprising the posterior mean $\pm 1.96 \times \sigma/\sqrt{n_0 + m}$. Table 3.5 provides the results for both prior and posterior.

We can contrast the Bayesian analysis with the classical conclusions drawn from the likelihood alone. This would comprise a 95% confidence interval $y_m \pm 1.96 \times \sigma/\sqrt{m}$, and a one-sided P-value

$$p(Y < y_m|H_0) = \Phi\left(\frac{y_m - \theta_L}{\sigma/\sqrt{m}}\right);$$

(a)

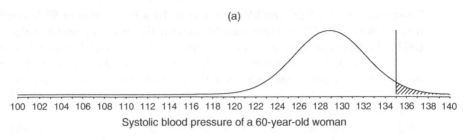

100 102 104 106 108 110 112 114 116 118 120 122 124 126 128 130 132 134 136 138 140
Systolic blood pressure of a 60-year-old woman

(b)

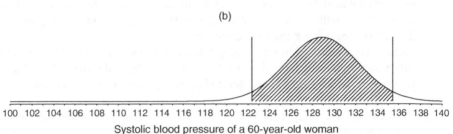

100 102 104 106 108 110 112 114 116 118 120 122 124 126 128 130 132 134 136 138 140
Systolic blood pressure of a 60-year-old woman

Figure 3.5 Inference from the posterior distribution of the true underlying systolic blood pressure of a 60-year-old woman: (a) shaded area is the probability 0.033 that $\theta > 135$; (b) a two-sided 95% interval (both equi-probability and HPD).

this is numerically identical to the tail area of the posterior with a uniform prior obtained by setting $n_0 = 0$.

We note from Table 3.5 that a traditional one-sided *p*-value for the hypothesis H_0: $\theta > 135$ is 0.08, while the Bayesian analysis has used the prior opinion to reduce this to 0.03.

Table 3.5 Bayesian and traditional intervals and tests of hypothesis H_0: $\theta > 135$.

	Mean	SD	95% credible interval	$p(H_0|y_m)$
Prior	120.0	10.0	100.4 to 139.6	0.067
Posterior	128.9	3.3	122.4 to 135.4	0.033
	Estimate	SE	95% CI	$p(Y < y_m|H_0)$
Classical	130.0	3.5	123.1 to 136.9	0.078

If we were to express the (rather odd) prior belief that all values of θ were equally likely, then $p(\theta)$ would be constant and (3.5) shows that the resulting posterior distribution is simply proportional to the likelihood: (3.14) shows this is equivalent to assuming $n_0 = 0$ in an analysis assuming a normal distribution. In many standard situations a traditional confidence interval is essentially equivalent to a credible interval based on the likelihood alone, and Bayesian and classical results may therefore be equivalent when using a uniform or 'flat' prior. Burton (1994) claims that 'it is already common practice in medical statistics to interpret a frequentist confidence interval as if it did represent a Bayesian posterior probability arising from a calculation invoking a prior density that is uniform on the fundamental scale of analysis'. In our examples we shall present the likelihood and often interpret it as a posterior distribution after having assumed a 'flat' prior: this can be termed a 'standardised likelihood', and some possible problems with this are discussed in Section 5.5.1.

Example 3.6 presents a Bayesian analysis of a published trial: it uses a highly structured format which will be discussed further in Section 3.21. We are aware of the potentially confusing discussion in terms of mortality rates, odds ratios, log(odds ratios) and risk reduction – this multiple terminology is unfortunately inevitable and it is best to confront it early on.

Example 3.6 *GREAT (continued): Bayesian analysis of a trial of early thrombolytic therapy*

Reference: Pocock and Spiegelhalter (1992).

Intervention: Thrombolytic therapy after myocardial infarction, given at home by general practitioners.

Aim of study: To compare anistreplase (a new drug treatment to be given at home as soon as possible after a myocardial infarction) and placebo (conventional treatment).

Study design: Randomised controlled trial.

Outcome measure: Thirty-day mortality rate under each treatment, with the benefit of the new treatment measured by the odds ratio, OR, *i.e.* the ratio of the odds of death following the new treatment to the odds of death on the conventional: OR $<$ 1 therefore favours the new treatment.

Statistical model: Approximate normal likelihood for the logarithm of the odds ratio (Section 2.4).

Prospective Bayesian analysis?: No, it was carried out after the trial reported its results.

Prior distribution: The prior distribution was based on the subjective judgement of a senior cardiologist, informed by empirical evidence derived from one unpublished and two published trials, who expressed belief that 'an expectation of 15–20% reduction in mortality is highly plausible, while the extremes of no benefit and a 40% relative reduction are both unlikely'. This has been translated to a normal distribution on the log(OR) scale, with a prior mean of $\mu_0 = -0.26$ (OR $= 0.78$) and symmetric 95% interval of -0.51 to 0.00 (OR 0.60 to 1.00), giving a standard deviation of 0.13. This prior is shown in Figure 3.6(a).

Loss function or demands: None specified.

Computation/software: Conjugate normal analysis (3.14).

Evidence from study: The 30-day mortality was 23/148 on control and 13/163 on new treatment.

We have already seen in Example 2.5 that the estimated log(OR) is $y_m = -0.74$ (OR $= 0.48$), with estimated standard error 0.36, giving a 95% classical confidence interval for log(OR) from -1.45 to -0.03 (OR from 0.24 to 0.97). The traditional standardised test statistic is therefore $-0.74/0.36 = 2.03$, and the null hypothesis of no effect is therefore rejected with a two-sided *P*-value of $2\Phi(-2.03) = 0.04$ (GREAT Group, 1992). Figure 3.6(b) shows the likelihood expressing reasonable support for values of θ representing a 40–60% reduction in odds of death. As explained in Example 2.5, it is convenient to express the variance of y_m as σ^2/m, and take $\sigma = 2$ and $m = 30.5$.

Bayesian interpretation: Figure 3.6(c) shows the posterior distribution, obtained by multiplying the prior and likelihood together and then making the total area under the curve equal to one (*i.e.* 'certainty'). The prior distribution has a standard deviation of 0.13, and expressing this as $\sigma/\sqrt{n_0}$ leads to an equivalent number of observations $n_0 = \sigma^2/0.13^2 = 236.7$. Thus the prior can be thought to have around $236.7/30.5 \approx 8$ times as much information as the likelihood, showing the strength of the subjective judgement in this example.

The equivalent number of observations in the posterior is then $n_0 + m = 236.7 + 30.5 = 267.2$, with a posterior mean equal to the weighted average $(n_0\mu + my_m)/(n_0 + m) = -0.31$ with standard deviation $\sigma/\sqrt{n_0 + m} = \sigma/\sqrt{267.2} = 0.12$. Thus, the estimated odds ratio is around $e^{-0.31} = 0.73$, or 27% risk reduction (half that observed in the trial). A 95% credible interval can be calculated on the log(OR) scale to be from -0.55 to -0.07, which corresponds to odds ratios from 0.58 to 0.93, or a 95% probability that the true risk reduction lies between 7% and 42%. The posterior probability that the reduction is at least 50% can be calculated by noting this is equivalent to a log(OR) of -0.69, which gives a probability of $\Phi((-0.69 + 0.31)/0.12) = \Phi(-3.11) = 0.001$. We can also calculate the posterior probability that there is any treatment effect as $p(\theta < 0|y_m) = \Phi((0 + 0.31)/0.12) = \Phi(2.54) = 0.995$ and so, adopting the prior provided by the 'expert', we can be 99.5% certain the new treatment is of benefit. Nevertheless, the evidence in the likelihood has been pulled back towards the prior distribution – a formal representation of the belief that the results were 'too good to be true'.

Sensitivity analysis: As an alternative prior formulation, we consider an observer who has no prior bias one way or another, but is more sceptical about large treatment effects than the current expert: this can be

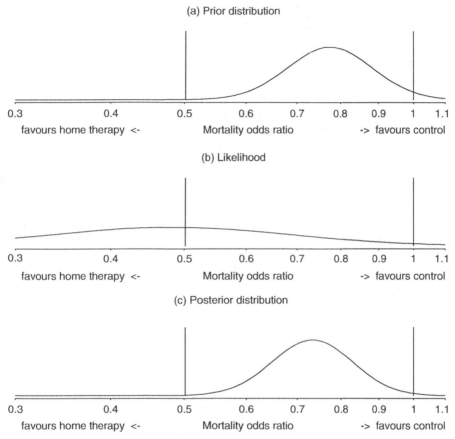

Figure 3.6 Prior, likelihood and posterior distributions arising from GREAT trial of home thrombolysis. These are all normal on the $\theta = \log(OR)$ scale.

represented by a normal prior centred on $\log(OR) = 0$ ($OR = 1$) and with a 95% interval that runs from a 50% reduction in odds of death ($OR = 0.5$, $\log(OR) = -0.69$), to a 100% increase ($OR = 2.0$, $\log(OR) = 0.69$). On a $\log(OR)$ scale, this prior has a 95% interval from -0.69 to 0.69, and so has a standard deviation $0.69/1.96 = 0.35$ and hence $m = 4/0.35^2 = 32.3$, approximately the same weight of evidence as the likelihood. The prior can therefore be thought of as providing equivalent evidence to that arising from an *imaginary* balanced trial, in which around 16 deaths were observed on each arm. This prior is shown in Figure 3.7, together with the likelihod and posterior distribution, which has mean -0.36 ($OR = 0.70$) and equivalent size $n_0 + m = 62.8$, leading to a standard deviation of 0.25. The probability that there is no benefit from the new treatment is now only $\Phi(-0.36/0.25) = \Phi(-1.42) = 0.08$, shown as the shaded area in Figure 3.7. This analysis suggests that a reasonably sceptical person may therefore not find the GREAT results convincing that there is a benefit: these ideas are formally explored in Section 3.11.

Comments: It is interesting to note that Morrison *et al.* (2000) conducted a meta-analysis of early thrombolytic therapy and estimated OR = 0.83 (95% interval from 0.70 to 0.98), far less impressive than the GREAT results and reasonably in line with the posterior distribution shown in Figure 3.6, which was calculated 8 years before publication of the meta-analysis. However, this finding should not be over-interpreted and two points should be kept in mind. First, Morrison *et al.* (2000) include some trials that contributed to the prior used by the expert in the above example, and so there is good reason why our posterior (which could be interpreted as a type of subjective meta-analysis) and the formal meta-analysis should correspond. Second, their primary outcome measure is in-hospital mortality, for which GREAT showed a non-significant (but still substantial) benefit of 11/163 vs. 17/148, with an estimated OR of 0.57.

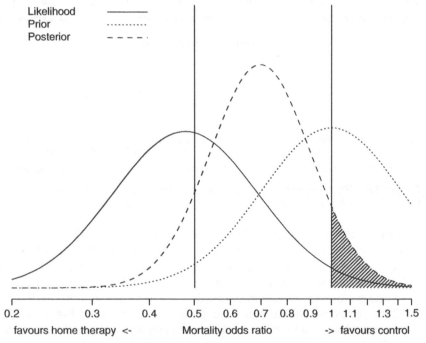

Figure 3.7 A prior distribution that expresses scepticism about large treatment effects would be centred on 0 and have, for example, a 95% interval for OR between 0.5 and 2.0. This is equivalent to a previous study in which 32.3 events occurred, divided equally between the two arms. Adopting this prior and updating it with the GREAT data leads to a posterior distribution as shown, with the shaded area representing a probability of 8% that the treatment is harmful.

3.9 THE PRIOR DISTRIBUTION

Bayesian analysis is driven by the prior distribution, and its source and use present many challenges. These will be covered in detail in Chapter 5, including elicitation from experts, derivation from historical data, the use of 'default' priors to represent archetypal positions of ignorance, scepticism and enthusiasm and, when multiple related studies are being simultaneously analysed, the assumption of a common prior that may be 'estimated'.

It is important to clarify a number of possible misconceptions that may arise. In particular, a prior is:

Not necessarily specified beforehand. Despite the name 'prior' suggesting a temporal relationship, it is quite feasible for a prior distribution to be decided *after* seeing the results of a study, since it is simply intended to summarise reasonable uncertainty given evidence external to the study in question. Cox (1999) states:

I was surprised to read that priors must be chosen before the data have been seen. Nothing in the formalism demands this. Prior does not refer to time, but to a situation, hypothetical when we have data, where we assess what our evidence would have been if we had had no data. This assessment may rationally be affected by having seen the data, although there are considerable dangers in this, rather similar to those in frequentist theory.

Naturally when making predictions or decisions one's prior distribution needs to be unambiguously specified, although even then it is reasonable to carry out analysis of sensitivity to alternative choices.

Not necessarily unique. There is no such thing as the 'correct' prior. Instead, researchers have suggested using a 'community' of prior distributions expressing a range of reasonable opinions. Thus a Bayesian analysis of evidence is best seen as providing a mapping from specified prior beliefs to appropriate posterior beliefs.

Not necessarily completely specified. When *multiple* related studies are being simultaneously analysed, it may be possible to have unknown parameters in the prior which are then 'estimated' – this is related to the use of hierarchical models (Section 3.17).

Not necessarily important. As the amount of data increases, the prior will, unless it is of a pathological nature, be overwhelmed by the likelihood and will exert negligible influence on the conclusions.

Of course, conclusions strongly based on beliefs that cannot be supported by concrete evidence are unlikely to be widely regarded as convincing, and so it is important to attempt to find consensus on reasonable sources of external evidence. As a true exemplification of the idea that the prior distribution should be under the control of the consumer of the evidence, Lehmann and Goodman (2000) describe ambitious interactive software which allows users to try their own prior distributions.

3.10   HOW TO USE BAYES THEOREM TO INTERPRET TRIAL RESULTS

There have been many connections made between the use of Bayes theorem in diagnostic testing (Example 3.1) and in general clinical research, pointing out that just as the prevalence of the condition (the prior probability) is required for the assessment of a diagnostic test, so the prior distribution on θ should supplement the usual information (P-values and confidence intervals) which summarises the likelihood. We need only think of the huge number of clinical trials that are carried out and the few clearly beneficial interventions found, to realise that the 'prevalence' of truly effective treatments is low. We should thus be cautious about accepting extreme results, such as observed in the GREAT trial, at face value; indeed, it has been suggested that a Bayesian approach provides 'a yardstick against which a surprising finding may be measured' (Grieve, 1994b). Example 3.7 illustrates this need for caution.

Example 3.7   *False positives: 'The epidemiology of clinical trials'*

Simon (1994b) considers the following (somewhat simplified) situation. Suppose 200 trials are performed, but only 10% are of truly effective treatments. Assume each trial is carried out with Type I error α of 5% (the chance of claiming an ineffective treatment is effective) and Type II error β of 20% (the chance of claiming an effective treatment is ineffective) – these are typical values adopted in practice. Table 3.6 displays the expected outcomes: of the 180 trials of truly ineffective treatments, 9 (5%) are expected to give a 'significant' result; similarly, of 20 trials of effective treatments, 4 (20%) are expected to be negative.

Table 3.6 shows that $9/25 = 36\%$ of trials with significant results are in fact of totally ineffective treatments: in diagnostic testing terms, the 'predictive

Table 3.6 The expected results when carrying out 200 clinical trials with $\alpha = 5\%, \beta = 20\%$, and of which only 10% of treatments are truly effective.

		Treatment		
		Truly ineffective	Truly effective	
Trial conclusion	Not significant	171	4	175
	Significant	9	16	25
		180	20	200

value positive' is only 64%. In terms of the odds formulation of Bayes theorem (3.2), when a 'significant result' is observed,

$$\frac{p(H_0|\text{'significant result'})}{p(H_1|\text{'significant result'})} = \frac{p(\text{'significant result'}|H_0)}{p(\text{'significant result'}|H_1)} \times \frac{p(H_0)}{p(H_1)}$$

$$= \frac{p(\text{Type I error})}{1 - p(\text{Type II error})} \times \frac{p(H_0)}{p(H_1)}.$$

Hence the prior odds 0.90/0.10 on the treatment being ineffective (H_0) are multiplied by the likelihood ratio $\alpha/(1 - \beta) = 0.05/0.80 = 1/16$ to give the posterior odds 9/16, corresponding to a probability of 9/25.

Qualitatively, this says that if truly effective treatments are relatively rare, then a 'statistically significant' result stands a good chance of being a false positive.

The analysis in Example 3.7 simplistically divides trial results into 'significant' or 'non-significant', the Bayes factor (likelihood ratio) for the null hypothesis is $\alpha/(1 - \beta)$: this might typically be $0.05/0.80 = 1/16$, categorised as 'strong' evidence against H_0 by Jeffreys (see Table 3.2). However, in Section 4.4.2 we describe how the relationship between Bayes factors and traditional hypothesis tests depends crucially on whether one knows the precise P-value or simply whether a result is 'significant'. We note that Lee and Zelen (2000) suggest selecting α so that the posterior probability of an effective treatment, having observed a significant result, is sufficiently high, say above 0.9. This is criticised by Simon (2000) and Bryant and Day (2000) as being based solely on whether the trial is 'significant' or not, rather than the actual observed data.

3.11 THE 'CREDIBILITY' OF SIGNIFICANT TRIAL RESULTS*

We have already seen in Example 3.6 how a 'sceptical' prior can be centred on 'no treatment difference' ($\theta = 0$) to represent doubts about large treatment effects. It is natural to extend this approach to ask how sceptical we would have to be *not* to find an apparently positive treatment effect convincing (Matthews, 2001). Specifically, suppose we have observed data y which is apparently 'significant' in the conventional sense, in that the classical 95% interval for θ on a normal likelihood lies wholly above or below 0. In addition, suppose our prior mean is 0, reflecting initial scepticism about treatment differences, with the variance of the prior expressing the degree of scepticism with which we view extreme treatment effects, either positive or negative. Matthews (2001) derives an expression for the critical prior distribution which would just lead to the corresponding posterior 95% interval including 0.

Suppose we observe $y_m < 0$. For a normal likelihood and prior with mean 0, (3.14) shows that

$$\theta \sim N\left[\frac{my_m}{n_0 + m}, \frac{\sigma^2}{n_0 + m}\right],$$

which means that the upper point u_m of the 95% posterior interval is

$$u_m = \frac{my_m}{n_0 + m} + 1.96\frac{\sigma}{\sqrt{n_0 + m}}.$$

The 95% interval will therefore overlap 0 if $u_m > 0$. Simple rearrangement shows this will happen provided

$$n_0 > \left(\frac{my_m}{1.96\sigma}\right)^2 - m = \frac{m^2}{1.96^2\sigma^2}\left(y_m^2 - \frac{1.96^2\sigma^2}{m}\right), \tag{3.16}$$

which provides a simple formula for determining the effective number of events in the sceptical prior that would just lead to a 95% posterior interval including 0.

Matthews (2001) shows that we can work directly in terms of the lower and upper points of a 95% interval based on the data alone, denoted l_D and u_D. Thus $l_D, u_D = y_m \pm 1.96\sigma/\sqrt{m}$. It follows that $(u_D - l_D)^2 = 4 \times 1.96^2\sigma^2/m$, and $u_D l_D = y_m^2 - 1.96^2\sigma^2/m$. Then from (3.16) the critical value of n_0 occurs when the lower point of the 95% prior interval, $l_0 = -1.96\sigma/\sqrt{n_0}$, obeys

$$l_0 = \frac{-1.96\sigma}{\sqrt{n_0}} = -\frac{(u_D - l_D)^2}{4\sqrt{u_D l_D}}.$$

Often we will be working, say, on a log(odds ratio) scale: if we let $l_0 = \log(L_0), l_D = \log(L_D), u_D = \log(U_D)$ then the corresponding expression is

$$L_0 = \exp\left(\frac{-\log^2(U_D/L_D)}{4\sqrt{\log(U_D)\log(L_D)}}\right). \tag{3.17}$$

L_0 is the critical value for the lower end of a 95% sceptical interval, such that the resulting posterior distribution has a 95% interval that just includes 1. Thus if one's prior belief lies wholly within $(L_0, 1/L_0)$ then one will not be convinced by the evidence, and Matthews suggests a significant trial result is not 'credible' unless prior experience indicates that odds ratios lying outside this critical prior interval are plausible. Figure 3.8 describes how this can be applied to assessment of 'significant' odds ratios.

Applying Figure 3.8 to the GREAT study, for which $L_D = 0.24, U_D = 0.97$, gives $L_0 = 0.10$. Hence, unless odds ratios more extreme than 0.1 can be considered as plausible, the results of the GREAT study should be treated with caution. Since such values do not seem plausible, we do not find the GREAT results 'credible'. This is easily seen to be a characteristic of any 'just significant' results such as those observed in the GREAT trial: just a minimal amount of prior scepticism is necessary to make the Bayesian analysis 'non-significant'. Examples of this approach to scepticism are given in Examples 3.8 and 3.13.

Example 3.8 *Credibility: Sumatriptan trial results*

Matthews (2001) considers the results of an early study of subcutaneous sumatriptan for migraine. This was a small study in which 79% of patients receiving sumatriptan reported an improvement compared to 25% with a placebo, with an estimated odds ratio in favour of sumatriptan of 11.4 and a wide 95% interval of 6.0 to 21.5: the likelihood is shown in Figure 3.9, and we note that odds ratios greater than 1 favour the new

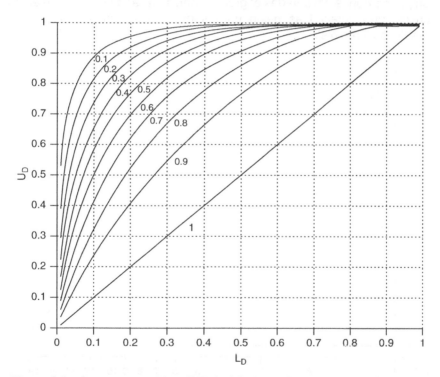

Figure 3.8 Assessment of 'credibility' of findings. Suppose one had observed a classical 95% interval (L_D, U_D) for an odds ratio. Then the value given in the graph is L_0, which is the lower end of a 95% prior interval centred on 1 expressing scepticism about large differences. L_0 is the critical value such that the resulting posterior distribution has a 95% interval that just includes 1, and hence does not produce 'convincing' evidence. Thus, unless values for the odds ratio more extreme than L_0 are judged plausible based on evidence external to the study, then the 'significant' conclusions should not be considered convincing.

treatment since in this application the events are 'positive'. It is reasonable to ask whether such extreme results are really 'too good to be true'. To use Figure 3.8 or (3.17) we first need to invert to odds ratios in favour of placebo, *i.e.* ORs less than 1: this leads to an estimated odds ratio of 0.088 with an interval (L_D, U_D) of (0.05, 0.17). Examination of Figure 3.8 reveals an approximate L_0 of 0.8: substitution in (3.17) gives an exact value of $L_0 = 0.84$. Transforming back to the original definition of the odds ratio gives a critical prior interval of $(1/L_0, L_0) = (0.84, 1/0.84) = (0.84, 1.19)$. Figure 3.9 shows this critical prior and the resulting posterior distribution whose 95% interval just includes OR = 1.

If 95% of our prior belief lies within this critical interval, then the posterior 95% interval would not exclude OR = 1 and we would not find the data convincing. However, it would seem unreasonable in this context to rule out on prior grounds advantages of greater than 19%, and hence we reject this critical prior interval as being unreasonably sceptical, and accept the results as 'credible'.

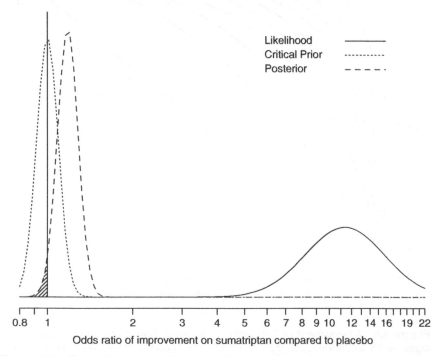

Figure 3.9 Sumatriptan example: the critical sceptical prior distribution (dotted) is centred on OR = 1 and is sufficiently sceptical to make the resulting posterior distribution have a 95% interval that just includes 1, *i.e.* the shaded area is 0.025. However, this degree of prior scepticism seems unreasonably extreme, and hence we might judge that the clinical trial findings are 'credible'.

3.12 SEQUENTIAL USE OF BAYES THEOREM*

Suppose we observe data in two or more segments, say y_m followed by y_n. Then after the first segment is observed our posterior distribution is given by (3.5):

$$p(\theta|y_m) \propto p(y_m|\theta)\, p(\theta). \qquad (3.18)$$

This posterior becomes the prior distribution for the next use of Bayes theorem, so after the next segment y_n is observed, the posterior conditioning on all the data, *i.e.* $p(\theta|y_n, y_m)$, obeys

$$p(\theta|y_n, y_m) \propto p(y_n|\theta, y_m)\, p(\theta|y_m). \qquad (3.19)$$

Combination of the two expressions (3.18) and (3.19) yields

$$p(\theta|y_n, y_m) \propto p(y_n|\theta, y_m)\, p(y_m|\theta)\, p(\theta);$$

this can also be derived by considering a single use of Bayes theorem with data y_n, y_m, but factorising the joint likelihood as $p(y_n, y_m|\theta) = p(y_n|\theta, y_m)p(y_m|\theta)$. In most situations the first term in (3.19) will not depend on y_m (*i.e.* Y_n is conditionally independent of Y_m given θ (Section 2.2.3)) and so $p(\theta|y_m)$ simply becomes the prior for a standard Bayesian update using the likelihood $p(y_n|\theta)$.

Example 3.9   *GREAT (continued): Sequential use of Bayes theorem*

Suppose the GREAT trial in Example 3.6 had a first analysis around half way through the trial with the results shown in Table 3.7(b). The estimated log(OR), its standard error and the effective number of events assuming $\sigma = 2$ are calculated as in Example 2.5, and are presented in Table 3.7 with the prior mean and effective number of events in the prior derived in Example 3.6. Bayes theorem assuming normal likelihoods leads to the posterior distribution shown in Table 3.7(c): as shown in (3.14), the effective number of events has been added to $236.7 + 18.1 = 254.8$, and the posterior mean is the weighted average of the prior and likelihood estimates $(236.7 \times -0.255) + (18.1 \times -0.654)/254.8 = -0.283$. The posterior standard deviation is obtained as $\sigma/\sqrt{254.8} = 0.125$.

The second half of the study then provided the data shown in Table 3.7(d), which made up the final totals of 23/144 under control and 13/163 under the new treatment. The sequential use of Bayes theorem means that the posterior following the first part of the study simply becomes the prior for the second, and the final posterior distribution arises in the same manner as described above.

Table 3.7 Possible results were the GREAT trial to have been analysed midway: the 'final' posterior is based on using the posterior from the first part of the trial as the prior for the second part, while the 'combined' posterior is based on pooling all the data into the likelihood. The results only differ through inadequacy of the normal approximation.

Stage	Control deaths/ cases	New treatment deaths/ cases	Estimated log(OR)	Effective no. events	Estimated SE
(a) Prior			−0.255	236.7	0.130
(b) Data – first half	13/74	8/82	−0.654	18.1	0.471
(c) Interim Posterior			−0.283	254.8	0.125
(d) Data – second half	10/74	5/81	−0.817	13.1	0.552
(e) 'Final' posterior			−0.309	267.9	0.122
(f) Combined data	23/144	13/163	−0.736	30.5	0.362
(g) 'Combined' posterior			−0.309	267.2	0.122

We note that the results obtained by carrying out the analysis in two stages (effective number of events 267.9) do not precisely match those obtained by using the total data shown in Table 3.7(g) (effective number of events 267.2). This is due to the quality of the normal approximation to the likelihood when such small numbers of events are observed.

3.13   PREDICTIONS

3.13.1   Predictions in the Bayesian framework

Making predictions is one of the fundamental objectives of statistical modelling, and a Bayesian approach can make this task reasonably straightforward. Suppose we wish to predict some future observations x on the basis of currently observed data y. Then the distribution we require is $p(x|y)$, and (2.8) shows we can extend the conversation to include unknown parameters θ by

$$p(x|y) = \int p(x|y, \theta)\, p(\theta|y)\, d\theta.$$

Now our current uncertainty concerning θ is expressed by the posterior distribution $p(\theta|y)$, and in many circumstances it will be reasonable to assume that x

and y are conditionally independent given θ, and hence $p(x|y, \theta) = p(x|\theta)$. The predictive distribution thus becomes

$$p(x|y) = \int p(x|\theta)\, p(\theta|y)\, d\theta,$$

the sampling distribution of x averaged over the current beliefs regarding the unknown θ. Provided we can do this integration, prediction becomes straightforward.

Such predictive distributions are useful in many contexts: Berry and Stangl (1996a) describe their use in design and power calculations, model checking, and in deciding whether to conduct a future trial, while Grieve (1988) provides examples in bioequivalence, trial monitoring and toxicology. Applications of predictions considered in this book include power calculations (Section 6.5), sequential analysis (Section 6.6.3), health policy-making (Section 9.8.4), and payback from research (Section 9.10).

3.13.2   Predictions for binary data*

Suppose θ is the true response rate for a set of Bernoulli trials, and that the current posterior distribution for θ has mean μ (note this might be a prior or posterior distribution, depending on whether data has yet been observed). We intend to observe a further n trials, and wish to predict Y_n, the number of successes. Then from the iterated expectation (2.13) given in Section 2.2.2 we know that

$$E(Y_n) = E_\theta[E(Y_n|\theta)] = E_\theta[n\theta] = n\mu, \tag{3.20}$$

which means, in particular, that the probability that the next observation ($n = 1$) is a success is equal to μ, the current posterior mean of θ. For example, after the single observation in Example 3.2, the probability that the next case shows a response is the current posterior mean of θ, *i.e.*

$$P(Y_1 = 1) = E(Y_1) = \sum_j \theta_j\, p(\theta_j|data)$$

$$= (0.2 \times 0.1) + (0.4 \times 0.2) + (0.6 \times 0.3) + (0.8 \times 0.4) = 0.6.$$

If our current distribution for θ is a conjugate Beta$[a, b]$, we can write down an expression for the exact predictive distribution for Y_n: this is known as the beta-binomial distribution and is given by

$$p(y_n) = \frac{\Gamma(a+b)}{\Gamma(a)\Gamma(b)} \binom{n}{y_n} \frac{\Gamma(a+y_n)\,\Gamma(b+n-y_n)}{\Gamma(a+b+n)}. \tag{3.21}$$

From (3.20) and the fact that $E(\theta) = a/(a+b)$, we immediately see that the mean of this distribution is

$$E(Y_n) = n\frac{a}{a+b}.$$

We can also obtain the variance by using the expression for the iterated variance (2.14) given in Section 2.2.2, to give

$$V(Y_n) = \frac{nab}{(a+b)^2}\frac{a+b+n}{(a+b+1)}. \tag{3.22}$$

We note two special cases of the beta-binomial distribution (3.21). First, when $a = b = 1$, the current posterior distribution is uniform and the predictive distribution for the number of successes in the next n trials is uniform over $0, 1, \ldots, n$. Second, when predicting the next single observation $(n = 1)$, (3.21) simplifies to a Bernoulli distribution with mean $a/(a+b)$.

Suppose, then, we start with a uniform prior for θ and then observe m trials, all of which turn out to be positive, so that our posterior distribution is now Beta$[m+1, 1]$ (Section 3.6.2). Then the probability that the event will occur at the next trial is $m/(m+1)$. This is known as 'Laplace's law of succession', and it means that even if an event has happened in every case so far (*e.g.* the sun rising every morning), we can still never be completely certain that it will happen at the next opportunity (that the sun will rise tomorrow).

Example 3.10 shows that the beta-binomial distribution can be used in designing experiments allowing for uncertainty in the true response rate.

Example 3.10   *Drug (continued): Making predictions for binary data*

In Example 3.3 we assumed an initial prior distribution for a drug's response rate that could be approximated by a Beta[9.2,13.8], and then observed 15/20 successes, leading to a posterior Beta[24.2,18.8] shown in Figure 3.10(a). The mean of this posterior distribution is 0.56, and hence from (3.20) this is the predictive probability that the next case responds successfully.

If we plan to treat 40 additional cases, then the predictive distribution of the total number of successes out of 40 is a beta-binomial distribution (3.21) which is shown in Figure 3.10(b), and has mean 22.5 and standard deviation 4.3.

Suppose we would consider continuing a development programme if the drug managed to achieve at least a further 25 successes out of these 40 future trials. The chance of achieving this number can be obtained by summing the probabilities in the right-hand tail of Figure 3.10(b), and comes to 0.329. In Example 3.15 we shall contrast this exact analysis with an approximation using simulation methods.

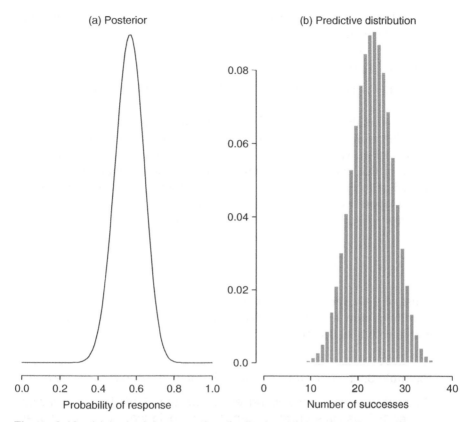

Figure 3.10 (a) is the beta posterior distribution after having observed 15 successes in 20 trials, (b) is the predictive beta-binomial distribution of the number of successes Y in the next 40 trials.

3.13.3   Predictions for normal data

Predictions are particularly easy when we are able to assume normal distributions. For example, suppose we assume a normal sampling distribution $Y_n \sim N[\theta, \sigma^2/n]$ for some future data Y_n, and a prior distribution $\theta \sim N[\mu, \sigma^2/n_0]$. We wish to make predictions concerning future values of Y_n, taking into account our uncertainty about its mean θ. We may write $Y_n = (Y_n - \theta) + \theta$, and so can consider Y_n as being the sum of two independent quantities: $Y_n - \theta \sim N[0, \sigma^2/n]$, and $\theta \sim N[\mu, \sigma^2/n_0]$. Now in Section 2.3 we observed that the sum of two independent normal quantities was normal with the sum of the means and the variances, and hence Y_n will therefore have a predictive distribution

$$Y_n \sim N\left[\mu, \sigma^2\left(\frac{1}{n}+\frac{1}{n_0}\right)\right]. \tag{3.23}$$

We could also derive (3.23) using the expressions for the iterated expectation (2.13) and variance (2.14) given in Section 2.2.2. Specifically,

$$E(Y_n) = E_\theta[E(Y_n|\theta)] = E_\theta[\theta] = \mu,$$
$$V(Y_n) = V_\theta[E(Y_n|\theta)] + E_\theta[V(Y_n|\theta)] = V_\theta[\theta] + E_\theta[\sigma^2/n] = \sigma^2(1/n_0 + 1/n).$$

Thus, when making predictions, we add variances and so *increase* our uncertainty. This is in direct contrast to combining sources of evidence using Bayes theorem, when we add precisions and *decrease* our uncertainty (Section 3.7). The use of this expression for comparison of prior distributions with data is described in Section 5.8, and for sample-size determination in Section 6.5.

Now suppose we had already observed data y_m and hence our distribution is $\theta \sim N[(n_0\mu + my_m)/(n_0 + m), \sigma^2/(n_0 + m)]$. Then

$$Y_n|y_m \sim N\left[\frac{n_0\mu + my_m}{n_0 + m}, \sigma^2\left(\frac{1}{n_0 + m} + \frac{1}{n}\right)\right]. \tag{3.24}$$

The use of this expression is illustrated in Example 3.11, and we shall see in Section 6.6.3 how to adapt these methods to predict the chance of a 'significant result' in a clinical trial setting.

Example 3.11 *GREAT (continued): Predictions of continuing the trial*

Suppose we were considering extending the GREAT trial to include a further 100 patients on each arm. What would we predict the observed OR in those future patients to be, with and without using the pre-trial prior information? It is important to remember that the precision with which the OR can be estimated does not depend on the actual number randomised (100 in each arm), but on the number of events (deaths) observed.

We assume the observed log(OR) in those future patients to be $Y_n \sim N[\theta, \sigma^2/n]$, where the future number of events is n and $\sigma = 2$: with 100 patients in each arm we can expect $n \approx 20$ events, given the current mortality rate of around 10%. From Example 3.6, the current posterior distribution is $\theta \sim N[-0.31, \sigma^2/(n_0 + m)]$ where $n_0 + m = 267.2$. Hence from (3.24) the predictive distribution of log(OR) has mean -0.31 and variance $\sigma^2(1/267.2 + 1/20.0) = \sigma^2/18.6 = 0.21 = 0.46^2$. This is shown in Figure 3.11: the great uncertainty in future observations is apparent.

Using the data from the trial alone is equivalent to setting $n_0 = 0$ and using a 'flat' prior, and hence the current posterior distribution is based on the likelihood alone, $\theta \sim N[-0.74, \sigma^2/m]$, where $m = 30.5$. Hence, ignoring the pre-trial prior based on the expert opinion, the predictive distribution of log(OR) has mean -0.74 and variance $\sigma^2(1/30.5 + 1/20.0) = \sigma^2/12.1$

$= 0.33 = 0.58^2$. Figure 3.11 shows that this predictive distribution is considerably flatter than when the prior is included.

We can use the predictive distributions to calculate the chance of any outcome of interest, say observing an OR of less than 0.50 in the future component of the trial. Using the fairly sceptical prior information, this probability is $p(Y_n < \log(0.50)|y_m) = \Phi((-0.69 + 0.31)/0.46) = \Phi(-0.83) = 0.21$, whereas if the prior distribution is ignored this rises to $\Phi((-0.69 + 0.74)/0.58) = \Phi(0.08) = 0.53$. So our prior opinion leads us to doubt that the current benefit will be observed in future patients if the trial is extended.

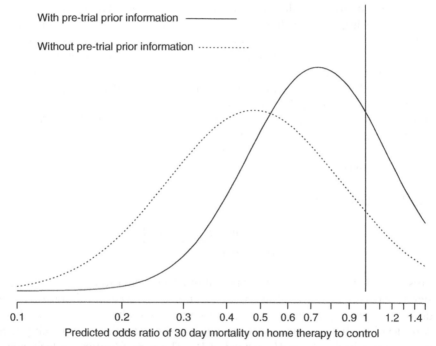

Figure 3.11 Predictive distributions for observed OR in a future 100 patients randomised to each arm in the GREAT trial, assuming around 20 events will be observed: with and without pre-trial prior information.

3.14 DECISION-MAKING

The appropriate role for formal decision theory in health-care evaluation is the subject of a long and continuing debate but is not the primary emphasis of this book. This section presents the basic ideas of which some are developed in later chapters, but for a full discussion we refer to classic texts such as DeGroot

(1970) and Lindley (1975), while Parmigiani (2002) provides a detailed exposition in a medical context.

Suppose we wish to make one of a set of decisions, and that we are willing to assess some value $u(d,\theta)$, known as a *utility*, of the consequences of taking each decision d when θ is the true unknown 'state of nature'. If we have observed some data y and our current probability distribution for θ is $p(\theta|y)$, then our expected utility of taking decision d is denoted

$$E(d) = \int u(d,\theta) \, p(\theta|y) \, d\theta,$$

where the integral is replaced by a sum if θ is discrete. The theory of optimal decision-making says we should choose the decision d^{opt} that maximises $E(d)$.

For example, suppose our unknown 'state of nature' comprises two hypotheses H_0 and H_1 with current posterior probabilities $p(H_0|y)$ and $p(H_1|y)$ respectively, and assume we face two possible decisions d_0 and d_1: we would choose d_0 if we believed H_0 to be true and d_1 if we believed H_1. Let $u(d_0,H_0)$ be the utility of taking decision d_0 when H_0 is true, and similarly define the other utilities. Then the theory of maximising expected utility states that we should take decision d_0 if $E(d_0) > E(d_1)$, which will occur if

$$u(d_0,H_0)p(H_0|y) + u(d_0,H_1)p(H_1|y) > u(d_1,H_0)p(H_0|y) + u(d_1,H_1)p(H_1|y),$$

which can be rearranged to give

$$\frac{p(H_0|y)}{p(H_1|y)} > \frac{u(d_1,H_1) - u(d_0,H_1)}{u(d_0,H_0) - u(d_1,H_0)}. \tag{3.25}$$

This inequality has an intuitive explanation. The numerator on the right-hand side is $u(d_1,H_1) - u(d_0,H_1)$, the additional utility involved in taking the correct decision when H_1 turns out to be the correct hypothesis – it could also be considered as the potential *regret*, in that it is the potential loss in utility when we erroneously decide on H_0 instead of H_1. The denominator similarly acts as the potential regret when H_0 is true. Hence (3.25) says we should only take decision d_0 if the posterior odds in favour of H_0 are sufficient to outweigh any extra potential regret associated with incorrectly rejecting H_1.

An alternative framework for using the principle of maximising expected utility occurs when our utility depends on future events, and our choice of action changes the probability of those events occurring. Suppose decision d_i can be taken at cost c_i, and leads to a probability p_i of an adverse event $Y = 0$ or 1 occurring with utility U_Y. Then the expected utility of taking decision i is

$$E(d_i) = p_i U_1 + (1 - p_i)U_0 - c_i,$$

and so, for example, d_0 will be preferred to d_1 if

$$p_0 U_1 + (1 - p_0)U_0 - c_0 > p_1 U_1 + (1 - p_1)U_0 - c_1.$$

Rearranging terms leads to a preference for d_0 if

$$p_1 - p_0 > \frac{c_0 - c_1}{U_0 - U_1} \qquad (3.26)$$

where the denominator $U_0 - U_1$ is positive since the event is considered undesirable. This is clearly obeyed if d_0 both costs less ($c_0 < c_1$) and reduces the risk of Y occurring ($p_0 < p_1$), since the right-hand side of (3.26) is negative and the left-hand side is positive. However, if d_0 costs more than d_1, then the right-hand side of (3.26) is positive, and d_0 will only be preferred if it reduces the risk by a sufficient quantity. We note that the decision depends on the risk difference $p_1 - p_0$, rather than a relative measure such as the odds ratio, and this led Ashby and Smith (2000) to show that (3.26) can be expressed as

$$\text{NNT} = \frac{1}{p_1 - p_0} < \frac{U_0 - U_1}{c_0 - c_1}. \qquad (3.27)$$

NNT denotes the 'number needed to treat' in order to prevent one adverse event (the expected number of events prevented when treating N individuals according to d_0 instead of d_1 is $N(p_1 - p_0)$, and hence one expects to prevent one event when treating $N = 1/(p_1 - p_0)$). So, if we are willing to assess the necessary costs and utilities to place in (3.27), we obtain a threshold for adopting a new treatment based on the NNT, without regard to any measure of 'significance'. Example 3.12 provides a somewhat stylised example.

Example 3.12 *Neural tube defects: Making personal decisions about preventative treatment*

Ashby and Smith (2000) consider a somewhat simplified example, but one that nevertheless illustrates the power (and the difficulties) of carrying out a formal decision analysis with utilities.

They consider a couple wishing to try and become pregnant but faced with the decision whether to take folic acid supplements to reduce the risk of a neural tube defect (NTD), such as spina bifida or anencephaly. Let d_0, d_1 denote respectively the decisions to take and not to take supplementation, with respective costs c_0, c_1, and let p_0, p_1 be the probabilities of a foetus having an NTD following each of the two decisions. Finally, let U_0, U_1 be the utilities of having a child without and with an NTD, respectively. The problem is structured as a decision tree in Figure 3.12.

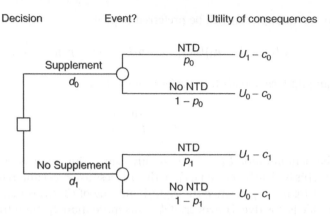

Decision   Event?   Utility of consequences

Figure 3.12 Decision tree for folic acid supplementation decision: the square node represents a decision, circular nodes represent chance events, and values at the end of branches represent utilities.

Inequality (3.26) can be rearranged to show that the couple should choose supplementation (d_1) if

$$U_0 - U_1 > \frac{c_0 - c_1}{p_1 - p_0}, \qquad (3.28)$$

and the issue becomes one of assigning reasonable values to these quantities. Estimates of p_0 and p_1 may be obtained from randomised trial and epidemiological evidence. Ashby and Smith (2000) provide the results of the sole available clinical trial of folic acid supplementation (carried out on couples who had already had a previous pregnancy resulting in an NTD): 21/602 randomised to placebo had pregnancies with an NTD, compared with 6/593 with supplementation. This corresponds to estimates of $p_0 = 0.010$, $p_1 = 0.035$, NNT $= 1/(p_1 - p_0) = 40.4$ and OR $= 0.30$. Suppose such a couple are deciding whether to take supplementation at a cost of $c_0 - c_1 = £c$; then (3.28) shows they should take the supplementation if the 'disutility' $U_0 - U_1$ of an NTD is greater than around $40c$. c may be costed in money terms if the couple will have to pay for a course of tablets, but Ashby and Smith (2000) suggest this may only be around £10, leading to a threshold of around £400. The problem lies in expressing the 'disutility' in £s.

This brings into focus the importance of identifying the appropriate decision-maker whose utilities are to be taken into account. If making public policy decisions regarding supplementation, it is reasonable that prevention of an NTD is worth more than around $40c$, even if the couple decide to terminate the pregnancy. However, from the couple's point of view, it may be best to think in terms of the utility U_0 of a 'healthy baby'. If this is of the order of £1 million, then they should take supplementation if the utility of an NTD is less than £999 600, which would suggest a fairly clear-cut decision. The crucial quantity is seen to be $S = c/U_0$, the cost of supplementation in terms of 'healthy baby' equivalents. Then the decision threshold (3.28) reduces to checking if

$$\frac{U_1}{U_0} < 1 - (S \times \text{NNT}).$$

Thus the previous analysis had $S \approx 0.000\,01$, NNT ≈ 40, and so supplementation is preferred if an NTD is valued at less than 0.9996 of a healthy baby.

Ashby and Smith (2000) also consider a couple with no previous history of an NTD, and they cite an incidence rate of 3.3 per 1000 pregnancies in a non-supplemented population. Taking this value as $p_0 = 0.0010$, and assuming the trial odds ratio applies to this group, leads to an estimate of $p_1 = 0.0033$, so that $p_1 - p_0 = 0.0023$, NNT $= 435$. We should therefore prefer supplementation if $U_1/U_0 < 1 - 0.000\,01 \times 435 \approx 0.996$. This threshold is again likely to be met, and the costs would need to become very substantial before the threshold was crossed into not preferring supplementation.

The use of Bayesian ideas in decision-making is a huge area of research and application, in which attention is more focused on the utility of consequences than the use of Bayesian methods to revise opinions. This activity blends naturally into cost-effectiveness analysis, but nevertheless the subjective interpretation of probability is essential, since the expressions of uncertainty required for a decision analysis can rarely be based purely on empirical data. There is a long history of attempts to apply this theory to medicine, and in particular there is a large literature on decision analysis, whether applied to the individual patient or for policy decisions. The journal *Medical Decision Making* contains an extensive collection of policy analyses based on maximising expected utility, some of which particularly stress the importance of Bayesian considerations. Any discussion of utility assessment must take careful account of the context in which the analysis is taking place, and our discussion is deferred until the chapter on cost-effectiveness and policy (Chapter 9).

There has been a long debate on the use of loss functions (defined as the negative of utility), in parallel to that concerning prior distributions, and some have continually argued that the design, monitoring and analysis of a study must explicitly take into account the consequences of eventual decisions (Berry, 1993). It is important to note that there is also a frequentist theory of decision-making that uses loss functions, but does not average with respect to prior or posterior distributions: the decision-making strategy is generally 'minimax' (DeGroot, 1970), where the loss is minimised whatever the true value of the parameter might be. This can be thought of as assuming the most pessimistic prior distribution. Thus 'ideological' approaches employing all combinations of the use of prior distributions and/or loss functions are possible: this is further discussed in Section 4.1 and, in the context of clinical trials, in Section 6.2.

It is particularly important to emphasise that the theory of optimal decision-making depends solely on the *expected* benefit, and hence any measures of uncertainty such as intervals or *P*-values are strictly speaking irrelevant, whether conducting clinical trials (Sections 6.2, 6.6.4 and 6.10) or policy-making (Chapter 9). An exception is when a decision can be made to obtain further information, and these ideas can be used for assessing the payback from research (Section 9.10).

3.15   DESIGN

Bayesian design of experiments can be considered as a natural combination of prediction and decision-making, in that the investigator is seeking to choose a design which they predict will achieve the desired goals. Nevertheless Bayesian design tends to be technically and computationally challenging (Chaloner and Verdinelli, 1995) except possibly in situations such as choosing the size of a clinical trial (Section 6.5).

Sequential designs present a particular problem known as 'backwards induction', in which one must work backwards from the end of the study, examine all the possible decision points that one might face, and optimise the decision allowing for all the possible circumstances in which one might find oneself. This can be computationally very demanding since one must consider what one would do in *all* possible future eventualities (Section 6.6.4), although approximations can be made such as considering only a single step ahead. A natural application is in dose-finding studies (Section 6.10). Early phases of clinical trials have tended to attract this approach: for example, Brunier and Whitehead (1994) consider the balancing of costs of experimentation and errors in treatment allocation (Section 6.12).

3.16   USE OF HISTORICAL DATA

Historical evidence has traditionally been used to help in the design of experiments and when pooling data in a meta-analysis, but Bayesian reasoning gives it a formal role in many aspects of evaluation. Here we introduce a brief taxonomy of ways in which historical data may be incorporated, which will be further developed in contexts such as the derivation of prior distributions

(Section 5.4), the use of historical controls in clinical trials (Section 6.9), the adjustment of observational studies for potential biases (Section 7.3) and the synthesis of multiple sources (Section 8.4).

We identify six broad relationships that historical data may have with current observations, ranging from being completely irrelevant to being of equal standing, with a number of possible means of 'downweighting' in between. There is an explicit reliance on judgement as to which is most appropriate in any situation.

(a) *Irrelevance*. The historical data provides no relevant information.
(b) *Exchangeable*. Current and past studies are 'similar' in the sense described in Section 3.17, and so their parameters can be considered exchangeable – this is a typical situation in a meta-analysis, and standard hierarchical modelling techniques can be adopted.
(c) *Potential biases*. Past studies are biased, either through lack of quality (internal bias) or because the setting is such that the studies are not precisely measuring the underlying quantity of interest (external bias), or both. The extent of the potential bias may be modelled and the historical results appropriately adjusted.
(d) *Equal but discounted*. Past studies may be assumed to be unbiased, but their precision is decreased in order to 'discount' past data.
(e) *Functional dependence*. The current parameter of interest is a logical function of parameters estimated in historical studies.
(f) *Equal*. Past studies are measuring precisely the parameters of interest and data can be directly pooled – this is equivalent to assuming exchangeability of individuals.

A fuller graphical and technical description of these stages is provided in Section 5.4.

3.17 MULTIPLICITY, EXCHANGEABILITY AND HIERARCHICAL MODELS

Evaluation of health-care interventions rarely concerns a single summary statistic. 'Multiplicity' is everywhere: clinical trials may present issues of 'multiple analyses of accumulating data, analyses of multiple endpoints, multiple subsets of patients, multiple treatment group contrasts and interpreting the results of multiple clinical trials' (Simon, 1994a). Observational data may feature multiple institutions, and meta analysis involves synthesis of multiple studies.

Suppose we are interested in making inferences on many parameters $\theta_1, \ldots, \theta_K$ measured on K 'units' which may, for example, be true treatment effects in subsets of patients, multiple institutions, or each of a series of trials. We can identify three different assumptions:

1. *Identical parameters.* All the θs are identical, in which case all the data can be pooled and the individual units ignored.
2. *Independent parameters.* All the θs are entirely unrelated, in which case the results from each unit can be analysed independently (e.g. using a fully specified prior distribution within each unit).
3. *Exchangeable parameters.* The θs are assumed to be 'similar' in the following sense. Suppose we were blinded as to which unit was which, and all we had was a label for each, say, A, B, C and so on. Suppose further that our prior opinion about any particular set of θs would not be affected by only knowing the labels rather than the actual identities, in that we have no reason to think specific units are systematically different. A set of random variables $Y_1, \ldots, Y_n$ with this property was termed 'exchangeable' in Section 3.4, equivalent, broadly speaking, to assuming the variables were independently drawn from some parametric distribution with a prior distribution on the parameter. The results of Section 3.4 can be equally applied to exchangeable parameters $\theta_1, \ldots, \theta_K$, and hence under broad conditions an assumption of exchangeable units is mathematically equivalent to assuming the θs are drawn at random from some population distribution, just as in a traditional random-effects model. This can be considered as a common prior for all units, but one with unknown parameters. Note that there does not need to be any actual sampling – perhaps these K units are the only ones that exist – since the probability structure is a consequence of the belief in exchangeability rather than a physical randomisation mechanism. Nor does the distribution have to be something traditional such as a normal (although we shall generally use that assumption in our examples): heavy-tailed or skewed distributions are possible, or 'partitions' that cluster units into groups that are equal or similar. We emphasise that an assumption of exchangeability is a *judgement* based on our knowledge of the context (Section 5.7).

If a prior assumption of exchangeability is considered reasonable, a Bayesian approach to multiplicity is thus to integrate all the units into a single model, in which it is assumed that $\theta_1, \ldots, \theta_K$ are drawn from some common prior distribution whose parameters are unknown: this is known as a hierarchical or multi-level model.

We illustrate these ideas assuming normal distributions. In each unit we shall observe a response Y_k assumed to have a normal likelihood

$$Y_k \sim N[\theta_k, s_k^2]. \tag{3.29}$$

The three situations outlined above are then treated as follows.

1. *Identical parameters (pooled effect).* We assume all the θ_k are identical and equal to a common treatment effect μ and, therefore, from (3.29),

$$Y_k \sim N[\mu, s_k^2].$$

Transforming to the notation $s_k^2 = \sigma^2/n_k$, assuming $\mu \sim N[0, \sigma^2/n_0]$ and sequential application of Bayes theorem, (3.14) gives a 'pooled' posterior distribution for μ (and hence each of the θ_k) of

$$\mu \sim N\left[\frac{\sum_k n_k y_k}{n_0 + \sum_k n_k}, \frac{\sigma^2}{n_0 + \sum_k n_k}\right]; \qquad (3.30)$$

the posterior mean for μ is equivalent to an overall sample mean, assuming the prior contributes n_0 'imaginary' observations of 0. As $n_0 \to 0$ the prior distribution on μ becomes uniform and the posterior for μ tends to

$$\mu \sim N\left[\frac{\sum_k n_k y_k}{\sum_k n_k}, \frac{\sigma^2}{\sum_k n_k}\right]. \qquad (3.31)$$

Reverting to the original notation $s_k^2 = \sigma^2/n_k$ reveals that

$$\mu \sim N\left[\frac{\sum_k y_k/s_k^2}{\sum_k 1/s_k^2}, \frac{1}{\sum_k 1/s_k^2}\right], \qquad (3.32)$$

where the posterior mean is simply the classical pooled estimate $\hat{\mu}$, which is the average of the individual estimates, each weighted inversely by its variance. A classical test for heterogeneity, *i.e.* whether it is reasonable to assume that all the trials are measuring the same quantity, is provided by

$$Q = \sum_k \frac{n_k}{\sigma^2}(y_k - \hat{\mu})^2, \qquad (3.33)$$

or equivalently $Q = \sum_k (y_k - \hat{\mu})^2/s_k^2$, which has a χ_{K-1}^2 distribution under the null hypothesis of homogeneity. It is well known that this is not a very powerful test (Whitehead, 2002), and so absence of a significant Q should not necessarily mean that the trial are homogenous.

2. *Independent parameters (fixed effects).* In this case each θ_k is estimated totally without regard for the others: assuming a uniform prior for each θ_k and the likelihood (3.29) gives the posterior distribution

$$\theta_k \sim N[y_k, s_k^2], \qquad (3.34)$$

which is simply the normalised likelihood.

3. *Exchangeable parameters (random effects).* The unit means θ_k are assumed to be exchangeable, and to have a normal distribution

$$\theta_k \sim N[\mu, \tau^2], \qquad (3.35)$$

where μ and τ^2 are 'hyperparameters' for the moment assumed known. After observing y_k, Bayes theorem (3.15) can be rearranged as

$$\theta_k|y_k \sim N[B_k\mu + (1 - B_k)y_k, \ (1 - B_k)s_k^2], \tag{3.36}$$

where $B_k = s_k^2/(s_k^2 + \tau^2)$ is the weight given to the prior mean. It can be seen that the pooled result (3.32) is a special case of (3.36) when $\tau^2 = 0$, and the independent result (3.34) a special case when $\tau^2 = \infty$.

An exchangeable model therefore leads to the inferences for each unit having *narrower* intervals than if they are assumed independent, but *shrunk* towards the prior mean response. This produces a degree of pooling, in which an individual study's results tend to be 'shrunk' by an amount depending on the variability between studies and the precision of the individual study. B_k controls the 'shrinkage' of the estimate towards μ, and the reduction in the width of the interval for θ_k. If we again use the notation $s_k^2 = \sigma^2/n_k$, $\tau^2 = \sigma^2/n_0$, then $B_k = n_0/(n_0 + n_k)$, clearly revealing how the degree of shrinkage increases with the relative information in the prior distribution compared to the likelihood.

The unknown hyperparameters μ and τ may be estimated directly from the data – this is known as the 'empirical Bayes' approach as it avoids specification of prior distributions for μ and τ. We shall not detail the variety of techniques available as they form part of classical random-effects meta-analysis (Sutton *et al.*, 2000; Whitehead, 2002). However, the simplest is the 'methods-of-moments' estimator (DerSimonian and Laird, 1986)

$$\hat{\tau}^2 = \frac{Q - (K - 1)}{N - \sum_k n_k^2/N}, \tag{3.37}$$

where Q is the test for heterogeneity given in (3.33), and $N = \sum_k n_k$; if $Q < (K - 1)$, then $\hat{\tau}^2$ is set to 0 and complete homogeneity is assumed. This estimator is used in Example 3.13 and in the Exercises, although we describe the use of 'profile-likelihood' in Section 3.18.

Alternatively, μ and τ^2 may be given a prior distribution (known as the 'full Bayes approach') and this is done later in the book, taking particular care in the choice of a prior distribution for the between-unit variation τ (Section 5.7.3). However, the results from either an empirical or full Bayes analysis will often be similar provided each unit is not too small and there are a reasonable number of units.

The use of hierarchical models is later discussed with respect to subset analysis (Section 6.8.1), N-of-1 studies (Section 6.11), institutional comparisons (Section 7.4) and meta-analysis (Section 8.2).

Example 3.13 *Magnesium: Meta-analysis using a sceptical prior*

Reference: Higgins and Spiegelhalter (2002).

Intervention: Epidemiology, animal models and biochemical studies suggested intravenous magnesium sulphate may have a protective effect after acute myocardial infarction (AMI), particularly through preventing serious arrhythmias. A series of small randomised trials culminated in a meta-analysis (Teo *et al.*, 1991) which showed a highly significant ($P < 0.001$) 55% reduction in odds of death. The authors concluded that 'further large scale trials to confirm (or refute) these findings are desirable', and the LIMIT-2 trial (Woods *et al.*, 1992) published results showing a 24% reduction in mortality in over 2000 patients. An editorial in *Circulation* subtitled 'An effective, safe, simple and inexpensive treatment' (Yusuf *et al.*, 1993) recommended further trials to obtain 'a more precise estimate of the mortality benefit'. Early results of the massive ISIS-4 trial pointed, however, to a lack of any benefit, and final publication of this trial on over 58 000 patients showed a non-significant adverse mortality effect of magnesium. ISIS-4 found no effect in any subgroups and concluded that 'Overall, there does not now seem to be any good clinical trial evidence for the routine use of magnesium in suspected acute MI' (Collins *et al.*, 1995).

Aim of study: To investigate how a Bayesian perspective might have influenced the interpretation of the published evidence on magnesium sulphate in AMI available in 1993. In particular, what degree of 'scepticism' would have been necessary in 1993 not to be convinced by the meta-analysis reported by Yusuf *et al.* (1993)?

Study design: Meta-analysis of randomised trials, allowing for prior distributions that express scepticism about large effects.

Outcome measure: Odds ratio for in-hospital mortality, with odds ratios less than 1 favouring magnesium.

Statistical model: All three approaches to modelling the multiple trials are investigated: (a) a 'pooled' analysis assuming identical underlying effects; (b) a fixed-effects analysis assuming independent, unrelated effects; and (c) a random-effects analysis assuming exchangeable treatment effects. For the last we assume a normal hierarchical model on the log(OR) scale, as given by (3.29) and (3.35). An empirical Bayes analysis is adopted using estimates of the overall mean μ and the between-study standard deviation τ, in order to use the normal posterior analysis given by (3.36).

Prospective analysis?: No.

Prior distribution: For the pooled- and fixed-effects analysis we assume a uniform prior for the unknown effects on the log(OR) scale. The empirical

Bayes analysis does not use any prior distributions on the parameters μ and τ (although the estimate for μ is equivalent to assuming a uniform prior on the log(OR) scale). Sensitivity analysis is conducted using 'sceptical' priors for μ centred on 'no effect'.

Loss function or demands: None.

Computation/software: Conjugate normal analysis.

Evidence from study: Table 3.8 gives the raw data and the estimated log-odds ratios y_k and their standard deviations s_k (Section 2.4.1). The classical test for heterogeneity Q (3.33) is not significant (9.35 on 7 degrees of freedom), and the method-of-moments estimate for τ is 0.29 (3.37). Figure 3.13 shows the profile log(likelihood) which summarises the support from the data for different values of τ, and is derived using the techniques described in Section 3.18.2: superimposed on this plot are the changing parameter estimates for different values of τ. The maximum likelihood estimate is $\hat{\tau} = 0$ although, from the discussion in Section 2.4.1, values for τ with a profile log(likelihood) above $-1.96^2/2 \approx -2$ might be considered as being reasonably supported by the data. $\hat{\tau} = 0$ would not appear to be a robust choice as an estimate since non-zero values of τ, which are well supported by the data, can have a strong influence on the conclusions. We shall assume, for illustration, the method-of-moments estimator $\hat{\tau} = 0.29$.

The results are shown in Figure 3.14. The standard pooled-effect analysis estimates an odds ratio OR $= 0.67$ (95% interval from 0.52 to 0.86). In the random-effects analysis the estimates of individual trials are 'shrunk' towards the overall mean by a factor given by B_k in Table 3.8, and individual trials have narrower intervals. The estimate of the 'average' effect is less precise, but still is 'significantly' less than 1: estimated odds ratio 0.58 (95% interval from 0.38 to 0.89).

Table 3.8 Summary data for magnesium meta-analysis, showing estimated odds ratios, log(odds ratios) (y_k), standard deviations for log(odds ratios) (s_k), the effective number of events assuming $\sigma = 2$ (n_k), and shrinkage coefficients $B_k = s_k^2/(s_k^2 + \hat{\tau}^2)$. $\hat{\tau}$ is taken to be 0.29.

Trial	Magnesium group		Control group		Estimated log(odds ratio) y_k	Estimated SD s_k	Effective no. events n_k	Shrinkage B_k
	Deaths	Patients	Deaths	Patients				
Morton	1	40	2	36	−0.65	1.06	3.6	0.93
Rasmussen	9	135	23	135	−1.02	0.41	24.3	0.65
Smith	2	200	7	200	−1.12	0.74	7.4	0.86
Abraham	1	48	1	46	−0.04	1.17	2.9	0.94
Feldstedt	10	150	8	148	0.21	0.48	17.6	0.72
Shechter	1	59	9	56	−2.05	0.90	4.9	0.90
Ceremuzynski	1	25	3	23	−1.03	1.02	3.8	0.92
LIMIT-2	90	1159	118	1157	−0.30	0.15	187.0	0.19

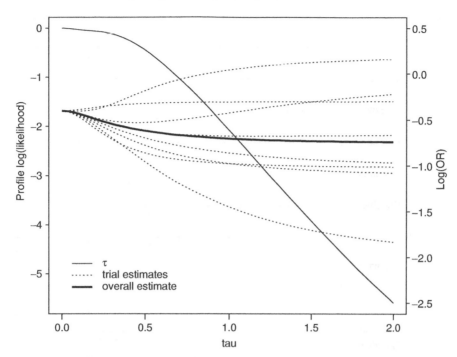

Figure 3.13 Profile log(likelihood) of τ, showing reasonable support for values of τ between 0 and 1. Also shown are individual and overall estimates of treatment effects for different values of τ: although $\tau = 0$ is the maximum likelihood estimate, plausible values of τ have substantial impact on the estimated treatment effects.

Bayesian interpretation: This random-effects analysis is not really a Bayesian technique, as it uses no prior distributions for parameters and conclusions are reported in the traditional way. One could, however, treat this as an approximate Bayesian analysis having assumed exchangeability between treatments and uniform priors on unknown parameters.

Sensitivity analysis: A meta-analysis using uniform prior distributions, whether a pooled- or random-effects analysis, finds a 'significant' benefit from magnesium. The apparent conflict between this finding and the results of the ISIS-4 mega-trial have led to a lengthy dispute, briefly summarised in Higgins and Spiegelhalter (2002). We shall return to this issue in Example 8.1, but for the moment we consider the robustness of the meta-analysis results to the choice of prior distribution. In particular, we use the credibility analysis described in Section 3.11 to check whether the findings are robust to a reasonable expression of prior scepticism concerning large benefits. We first consider the pooled analysis. From Figure 3.8, we can see that in order to find unconvincing the

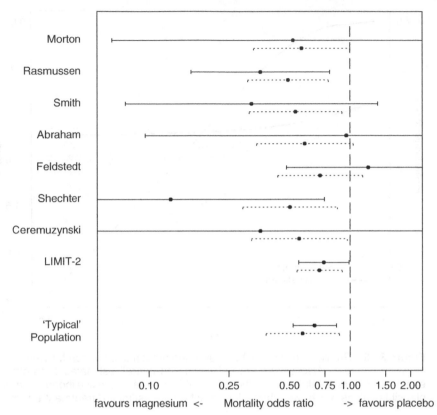

Figure 3.14 Fixed- (solid lines) and random-effects (dashed lines) meta-analysis of magnesium data assuming $\tau = 0.29$, leading to considerable shrinkage of the estimates towards a common value.

pooled analysis (95% interval from 0.52 to 0.86), a sceptical prior with a lower 95% point at around 0.80 would be necessary. Figure 3.15 displays the pooled likelihood, and the 'critical' sceptical prior distribution that leads to a posterior tail area of 0.025 above OR = 1. This prior is $N[0, 2^2/421]$, and hence is equivalent evidence to a trial in which 421 events have been observed, with exactly the same number in each arm. This seems a particularly extreme form of scepticism in that it essentially rules out all effects greater than around 20% on prior grounds. However, for the random-effects analysis (95% interval from 0.38 to 0.89), the lower end of the sceptical interval would need to be 0.6: the likelihood, 'critical' sceptical prior and posterior are shown in Figure 3.16. It might seem reasonable to find odds ratio below 0.6 extremely surprising, and

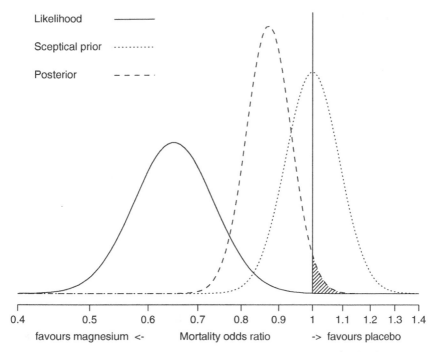

Figure 3.15 Critical sceptical prior for the pooled analysis, just sufficient to make posterior 95% interval include 1. This degree of scepticism seems unreasonably severe, as it equivalent to having already observed 421 events – 210.5 on each treatment.

hence a random-effects analysis and a reasonably sceptical prior render the meta-analysis somewhat unconvincing. This finding is reinforced by the comment by Yusuf (1997) that 'if one assumed that only moderate sized effects were possible, the apparent large effects observed in the meta-analysis of small trials with magnesium ... should perhaps have been tempered by this general judgment. If a result appears too good to be true, it probably is.'

Comments: One vital issue is that the maximum likelihood estimate of τ would lead to assuming a pooled estimate for the odds ratio, whereas there is reasonable evidence for considerable heterogeneity. A simplistic approach in which the maximum likelihood estimate is assumed to be true is therefore likely to substantially overstate the confidence in the conclusions. We note that we might question the exchangeability assumption of a large trial compared with many small ones, and this is further discussed in Higgins and Spiegelhalter (2002).

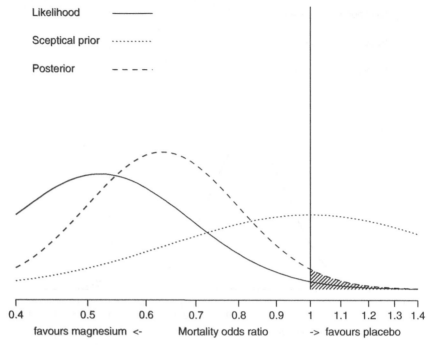

Likelihood   ———

Sceptical prior   ············

Posterior   – – – –

0.4   0.5   0.6   0.7   0.8   0.9   1   1.1   1.2   1.3 1.4

favours magnesium <-   Mortality odds ratio   -> favours placebo

Figure 3.16 Critical sceptical prior for random-effects analysis, just sufficient to make posterior 95% interval include 1. This degree of scepticism appears quite reasonable, corresponding to 58 events (29 in each arm) in a previous 'imaginary trial'.

3.18   DEALING WITH NUISANCE PARAMETERS*

3.18.1   Alternative methods for eliminating nuisance parameters*

In many studies we are focused on inferences on a single unknown quantity θ, such as the average treatment effect in a population of interest. However, there will almost always be additional unknown quantities which influence the data we observe but which are not of primary interest: these are known as 'nuisance' parameters and are a major issue in statistical modeling. Examples include the variance of continuous quantities, coefficients measuring the influence of background risk factors, baseline event rates in control groups, and so on.

Traditional statistical methods are primarily based on analysis of the likelihood for θ, and a number of methods have been developed to eliminate the nuisance parameters from this likelihood. These include the following:

1. Restricting attention to an estimator of θ whose likelihood (at least approximately) does not depend on the nuisance parameters. This technique is used extensively in this book in the form of approximate normal likelihoods for unknown odds ratios, hazard ratios and rate ratios (Section 2.4).
2. Estimating the nuisance parameters and 'plugging in' their maximum likelihood estimates into the likelihood for θ. This ignores the uncertainty concerning the nuisance parameters, and may be inappropriate if the number of nuisance parameters is large. In hierarchical modelling we might use this technique for the hyperparameters of the population distribution, and we saw in Section 3.17 that this is known as the empirical Bayes approach. Example 3.13 showed that conditioning on the maximum likelihood estimate might lead us to ignore an important source of uncertainty.
3. By conditioning on some aspect of the data that is taken to be uninformative about θ, forming a 'conditional likelihood' which depends only on θ.
4. Forming a 'profile likelihood' for θ, obtained by maximising over the nuisance parameters for each value of θ. This was used in Example 3.13 and is illustrated in Section 3.18.2, although here it is not applied to the parameter of primary interest.

Each of these techniques leads to a likelihood that depends only on θ, and which could then be combined with a prior in a Bayesian analysis.

However, a more 'pure' Bayesian approach would be as follows:

1. Place prior distributions over the nuisance parameters.
2. Form a joint posterior distribution over all the unknown quantities in the model.
3. Integrate out the nuisance parameters to obtain the marginal posterior distribution over θ.

This approach features in our examples when we do not assume normal approximations to likelihoods, such as modelling control group risks for binomial data in Examples 8.2 and 9.4, and control group rates for Poisson data in Example 8.3. We also consider full Bayesian modelling of sample variances for normal data in Examples 6.10 and 9.2. In other hierarchical modelling examples we shall generally adopt an approximation at the sampling level, but a full Bayesian analysis of the remaining nuisance parameter: the between-group standard deviation τ.

It is important to emphasise that sensitivity analysis of prior distributions placed on nuisance parameters is important, as apparently innocuous choices may exert unintended influence. For this reason it may be attractive to carry out a hybrid strategy of using traditional methods to eliminate nuisance parameters before carrying out a Bayesian analysis on θ alone, although we might wish to be assured that this was a good approximation to the full Bayesian approach.

3.18.2   Profile likelihood in a hierarchical model*

Consider the hierarchical model described in Section 3.17 and Example 3.13 in which

$$Y_k \sim N[\theta_k, s_k^2], \qquad \theta_k \sim N[\mu, \tau^2].$$

The hyperparameters μ and τ^2 will generally be unknown. From (3.24) the predictive distribution of Y_k, having integrated out θ_k, is

$$Y_k \sim N[\mu, s_k^2 + \tau^2].$$

Let the precision $w_k = 1/(s_k^2 + \tau^2)$ be the 'weight' associated with the kth study. Then the joint log(likelihood) for μ and τ is an arbitrary constant plus

$$L(\mu, \tau) = -\frac{1}{2} \sum_k [(y_k - \mu)^2 w_k - \log w_k]. \qquad (3.38)$$

By differentiating (3.38) with respect to μ and setting to 0, we find that, for fixed τ, the conditional maximum likelihood estimator of μ is

$$\hat{\mu}(\tau) = \sum_k y_k w_k / \sum_k w_k, \qquad (3.39)$$

with variance $1/\sum_k w_k$ (this is also the posterior mean and variance of μ when assuming a uniform prior distribution for μ). We can therefore substitute $\hat{\mu}(\tau)$ for μ in (3.38) and obtain the profile log(likelihood) for τ as

$$L(\tau) = -\frac{1}{2} \sum_k [(y_k - \hat{\mu}(\tau))^2 w_k - \log w_k]. \qquad (3.40)$$

This profile log(likelihood) may be plotted, as in Example 3.13, and maximised numerically to obtain the maximum likelihood estimate $\hat{\tau}$. This can then be substituted in (3.39) to obtain the maximum likelihood estimate of μ.

3.19   COMPUTATIONAL ISSUES

The Bayesian approach applies probability theory to a model derived from substantive knowledge and can, in theory, deal with realistically complex situations – the approach can also be termed 'full probability modelling'. It has to be acknowledged, however, that the computations may be difficult, with the specific problem being to carry out the integrations necessary to obtain the posterior distributions of quantities of interest in situations where non-standard prior distributions are used, or where there are additional 'nuisance parameters' in the model. These problems in integration for many years restricted Bayesian applications to rather simple examples. However, there has recently been enormous progress in methods for Bayesian computation, generally exploiting modern computer power to carry out simulations known as Markov chain Monte Carlo (MCMC) methods (Section 3.19.2).

In this book we shall downplay computational issues and many of our examples can be handled using simple algebra. In practice it is inevitable that MCMC methods will be required for many applications, and our later examples make extensive use of the WinBUGS software (Section 3.19.3).

3.19.1 Monte Carlo methods

Monte Carlo methods are a toolkit of techniques that all have the aim of evaluating integrals or sums by simulation rather than exact or approximate algebraic analysis. The basic idea of replacing algebra by simulation can be illustrated by the simple example given in Example 3.14.

Example 3.14 *Coins: A Monte Carlo approach to estimating tail areas of distributions*

Suppose we want to know the probability of getting 8 or more heads when we toss a fair coin 10 times. An *algebraic* approach would be to use the formula for the binomial distribution given in (2.39) to provide the probability of 8, 9 or 10 heads, which results in

$$
\begin{aligned}
P(\text{8 or more heads}) &= \binom{10}{8}\left(\frac{1}{2}\right)^{8}\left(\frac{1}{2}\right)^{2} + \binom{10}{9}\left(\frac{1}{2}\right)^{9}\left(\frac{1}{2}\right)^{1} + \binom{10}{10}\left(\frac{1}{2}\right)^{10}\left(\frac{1}{2}\right)^{0} \\
&= \frac{1}{2^{10}}(45 + 10 + 1) \\
&= \frac{56}{1024} \\
&= 0.0547.
\end{aligned}
$$

An alternative, *physical* approach would be to repeatedly throw a set of 10 coins and count the proportion of throws where there were 8 or more heads. Basic probability theory then says that eventually, after sufficient throws, this proportion will tend to the correct result of 0.0547. This rather exhausting procedure is best imitated by a *simulation* approach in which a computer program generates the throws according to a reliable random mechanism, say by generating a random number U between 0 and 1, and declaring a 'head' if $U \geq 0.5$. The results of 102 such simulated throws of 10 coins are shown in Figure 3.17(a): there were 4, 1 and 0 occurrences of 8, 9 and 10 heads respectively, an overall proportion of $5/102 = 0.0490$, compared to the true probability of 0.0547. Figure 3.17(b) shows the distribution of 10 240 throws, in which there were 428, 87 and 7 occurrences of 8, 9 and 10 heads respectively, instead of the expected counts of 450, 100, and 10. Overall we would therefore estimate the probability of 8 or more heads as $522/10\,240 = 0.0510$. After 10 240 000 simulated throws this empirical proportion is 0.05476, and can be made as close as required to the true value 0.0547 by simply running a longer simulation.

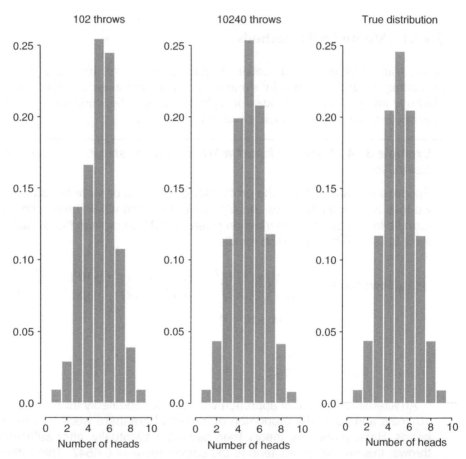

Figure 3.17   (a) Empirical distribution of the number of heads thrown in 102 tosses of 10 balanced coins, where the results of the tosses are obtained by a computer simulation. (b) Empirical distribution after 10 240 throws. (c) True distribution based on the binomial distribution.

The Monte Carlo method described in Example 3.14 is used extensively in risk modelling using software which allows sampling from a wide variety of distributions. The simulated quantities can then be passed into a standard spreadsheet, and the resulting distributions of the outputs of the spreadsheet will reflect the uncertainty about the inputs. This use of Monte Carlo methods can also be termed *probabilistic sensitivity analysis*, and we shall explore this in detail in the context of cost-effectiveness (Section 9.5).

Monte Carlo methods will be useful for Bayesian analysis provided the distribution of concern is a member of a known family – this distribution may be the prior (if no data are available) or current posterior. In conjugate Bayesian analysis it will be possible to derive such a posterior distribution algebraically as in Section 3.6.2 and hence to use Monte Carlo methods to find tail areas (although such tail areas may also be directly obtainable in software), or more usefully to find the distribution of complex functions of one or more unknown quantities as in the probabilistic sensitivity analysis mentioned above. An application of these ideas in power calculations is given in Example 6.5.

3.19.2 Markov chain Monte Carlo methods

Non-conjugate distributions or nuisance parameters (Section 3.18) will generally mean that in more complex Bayesian analysis it will not be possible to derive the posterior distribution in an algebraic form. Fortunately, Markov chain Monte Carlo methods have developed as a remarkably effective means of sampling from the posterior distribution of interest even when the form of that posterior has no known algebraic form. Only a brief overview of these methods can be given here: tutorial introductions are provided by Brooks (1998), Casella and George (1992) and Gilks *et al.* (1996).

The following form the essential components of MCMC methods:

- *Replacing analytic methods by simulation.* Suppose we observe some data y from which we want to make inferences about a parameter θ of interest, but the likelihood $p(y|\theta,\psi)$ also features a set of nuisance parameters (Section 3.18) ψ: for example, θ may be the average treatment effect in a meta-analysis, and ψ may be the control and treatment group response rates in the individual trials. The Bayesian approach is to assess a joint prior distribution $p(\theta,\psi)$, form the joint posterior $p(\theta,\psi|y) \propto p(y|\theta,\psi)p(\theta,\psi)$, and then integrate out the nuisance parameters in order to give the marginal posterior of interest, *i.e.*

$$p(\theta|y) = \int p(\theta, \psi|y)d\psi.$$

In most realistic situations this integral will not be a standard form and some approximation will be necessary. The idea behind MCMC is that we *sample* from the joint posterior $p(\theta,\psi|y)$, and save a large number of plausible values for θ and ψ: we can denote these sampled values as $(\theta^{(1)}, \psi^{(1)})$, $(\theta^{(2)}, \psi^{(2)})$, ..., $(\theta^{(j)}, \psi^{(j)})$, ... Then any inferences we wish to make about θ are derived from the sampled values $\theta^{(1)}$, $\theta^{(2)}$, ..., $\theta^{(j)}$, ...: for example, we use the sample mean of the $\theta^{(j)}$ as an estimate of the posterior mean $E(\theta|y)$. We can also create a smoothed histogram of all the sampled $\theta^{(j)}$ in order to estimate the shape of the posterior distribution $p(\theta|y)$. Hence we have replaced analytic integration by empirical summaries of sampled values.

- *Sampling from the posterior distribution.* There is a wealth of theoretical work on ways of sampling from a joint posterior distribution that is known to be proportional to a likelihood × prior, defined as $p(y|\theta,\psi)\,p(\theta,\psi)$, where the latter expression is of known form. These methods focus on producing a *Markov chain*, in which the distribution for the next simulated value $(\theta^{(j+1)}, \psi^{(j+1)})$ depends only on the current $(\theta^{(j)},\psi^{(j)})$. The theory of Markov chains states that, under broad conditions, the samples will eventually converge into an 'equilibrium distribution'. A set of algorithms are available that use the specified form of $p(y|\theta,\psi)p(\theta,\psi)$ to ensure that the equilibrium distribution is exactly the posterior of interest: popular techniques include Gibbs sampling and the Metropolis algorithm, but their details are beyond the scope of this book.

- *Starting the simulation.* The Markov chain must be started somewhere, and *initial values* are selected for the unknown parameters. In theory the choice of initial values will have no influence on the eventual samples from the Markov chain, but in practice convergence will be improved and numerical problems avoided if reasonable initial values can be chosen.

- *Checking convergence.* Checking whether a Markov chain, possibly with very many dimensions, has converged to its equilibrium distribution is not at all straightforward. *Lack* of convergence might be diagnosed simply by observing erratic behaviour of the sampled values, but the mere fact that a chain is moving along a steady trajectory does not necessarily mean that it is sampling from the correct posterior distribution: it might be stuck in a particular area due to the choice of initial values. For this reason it has become generally accepted that it is best to run multiple chains from a diverse set of initial values, and formal diagnostics exist to check whether these chains end up, to expected chance variability, coming from the same equilibrium distribution which is then assumed to be the posterior of interest. This technique is illustrated in Example 3.15, although in the remaining examples of this book we do not go into the details of convergence checking (in fact, our examples are generally well behaved and convergence is not a vital issue).

There are a vast number of published MCMC analyses, many of them using hand-tailored sampling programs. However, the WinBUGS software is widely used in a variety of applications and is essential for many of the examples in this book.

3.19.3   WinBUGS

WinBUGS is a piece of software designed to make MCMC analyses fairly straight-forward. Its advantages include a very flexible language for model specification, the capacity to automatically work out appropriate sampling methods, built-in graphics and convergence diagnostics, and a large range of examples and web presence that covers many different subject areas. It has two main disadvantages. The first is its current role as a 'stand-alone' program that is not integrated with a traditional statistical package for data manipulation, exploratory analyses and so on (although this is improving to some extent with the ability to call WinBUGS from other statistical packages). Secondly, it assumes that users are skilled at Bayesian analyses and hence can assess the impact of their chosen prior and likelihood, adequately check the fit of their model, check convergence and so on. It is therefore to be used with considerable care. WinBUGS may be obtained from www.mrc-bsu.cam.ac.uk/bugs/welcome.shtml (see also Section A.2).

A simple example of the model language was introduced in Example 3.14, which concerned the simulation repeated tosses of 10 'balanced coins'. This was carried out in WinBUGS using the program:

```
model{
Y ~ dbin (0.5, 10)
P8 <- step (Y-7.5)
}
```

where Y is binomial with probability 0.5 and sample size 10, and P8 is a step function which will take on the value 1 if $Y - 7.5$ is non-negative, *i.e.* if Y is 8 or more, 0 if 7 or less. There are only two connectives: The '$\sim$' indicates a distribution, '$< -$' indicates a logical identity. Running this simulation for 10 240 and 1 024 000 iterations, and then taking the empirical mean of P8, provided the estimated probabilities that Y will be 8 or more.

A more complex example is given in Example 3.15, which also illustrates the use of graphs to represent a model, and the use of scripts for running WinBUGS in the background.

Example 3.15 *Drug (continued): Using WinBUGS to implement Markov chain Monte Carlo methods*

In Example 3.10 we used the exact form of the beta-binomial distribution to obtain the predictive distribution of the number of successes in future Bernoulli trials, when the current uncertainty about the probability of success is expressed as a beta distribution. Here we use this example as a demonstration of the ability of the WinBUGS software to both carry out prior-to-posterior analysis and make predictions. In this instance we can compare the results with the exact results derived in Example 3.10; of course, the main use for WinBUGS is in carrying out analyses for which no algebraic solution is possible.

The basic components of the model being considered can be written as

θ ~ Beta[a, b] prior distribution

y ~ Bin[θ, m] sampling distribution

y_{pred} ~ Bin[θ, n] predictive distribution

P_{crit} = $P(y_{\text{pred}} \geq n_{\text{crit}})$ probability of exceeding critical threshold

which is expressed in the WinBUGS language as follows:

```
# WinBUGS analysis of Beta-Binomial 'drug' example
# Model description stored in file 'drug-model.txt'
model{
theta     ~ dbeta(a,b)         # prior distribution
y         ~ dbin(theta,m)      # sampling distribution
y.pred    ~ dbin(theta,n)      # predictive distribution
P.crit    <- step(y.pred-      # =1 if y.pred >= ncrit,
               ncrit+0.5)       # 0 otherwise
}
```

As mentioned in Section 3.19.3, the step function is used here as an indicator as to whether a quantity is greater than or equal to 0, so that the mean of P.crit over a large number of iterations will be the estimate of P_{crit}.

The model is also expressed graphically in Figure 3.18. The representation is described in the figure legend but should be fairly self-explanatory. The important point is that such a directed graph fully describes the joint distribution of all the unknown quantities, and in fact these graphs, known as *Doodles*, can be used by WinBUGS in place of the model syntax above. The part of WinBUGS that deals with the graphs, called DoodleBUGS, can interpret the graphs and either generate WinBUGS code or directly run the

name: y.pred   type: stochastic   density   dbin
proportion theta   order   n   lower bound   upper bound

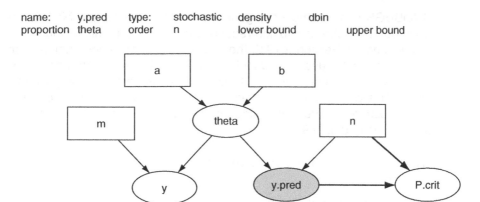

Figure 3.18 Doodle for drug example. The graphical model represents each quantity as a node in directed acyclic graph. Constants are placed in rectangles, random quantities in ovals. Stochastic dependence is represented by a single arrow, and a logical function as a double arrow. The resulting structure is much like a spreadsheet, but allowing uncertainty on the dependencies. WinBUGS allows models to be specified graphically and run directly from the graphical interface.

analysis from the Doodle. Graphical representations can be useful in explaining complex model structures without the distraction of equations; we use them in explaining alternative models for historical data (Section 5.4) and for evidence synthesis (Section 8.4 and Example 8.6).

The relevant values for the model are the parameters of the prior distribution, $a = 9.2$, $b = 13.8$; the number of trials carried out so far, $m = 20$; the number of successes so far, $y = 15$; the future number of trials, $n = 40$; and the critical value of future successes $n_{crit} = 25$. These values could have been placed in the model description, or alternatively can be written as a list using the format below. This list could be in a separate file or listed after the model description.

```
# data held in file 'data.txt'
# these values could alternatively have been given in model
description
list(
a = 9.2,        # parameters of prior distribution
b = 13.8,
y = 15,         # number of successes
m = 20,         # number of trials
n = 40,         # future number of trials
ncrit = 25)     # critical value of future successes
```

WinBUGS can automatically generate initial values for the MCMC analysis, but it is better to provide reasonable values in an initial-values list. As mentioned in Section 3.19.2, the best way to check convergence is to carry out multiple runs from widely dispersed starting points and check that, after a suitable 'burn-in', they give statistically indistinguishable chains. This example is simple enough not to require this level of care, but we illustrate the idea by setting up three initial-value files with starting points $\theta = 0.1$, 0.5, 0.9.

```
# initial values held in file 'drug-in1.txt'
list(theta=0.1)
# initial values held in file 'drug-in2.txt'
list(theta=0.5)
# initial values held in file 'drug-in3.txt'
list(theta=0.9)
```

It is possible to run WinBUGS from a 'point- and-click' interface, but once a program is working it is more convenient to use 'scripts' to carry out a simulation in the background. A script is shown below, checking the syntax of the model, reading in data and multiple initial values, carrying out the simulation and generating the results shown below.

```
# Script for running analysis
display('log')
check('c:/winbugs/drug-model.txt') # check syntax of model
data(' c:/winbugs/drug-dat.txt') # load data file
compile(3) # generate code for 3 simulations
inits(1, 'c:/winbugs/drug-in1.txt') # load initial values 1
for theta
inits(2, 'c:/winbugs/drug-in2.txt') # load initial values 2
for theta
inits(3, 'c:/winbugs' drug-in3.txt') # load initial values 3
for theta
gen.inits() # generate initial value for y.pred
set(theta) # monitor the true response rate
set(y.pred) # monitor the predicted number of successes
set(P.crit) # monitor whether 25 or more successes occur
update(11000) # perform 11000 simulations
```

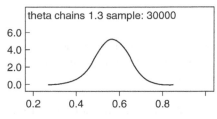
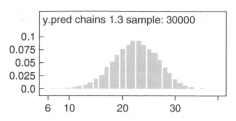

Figure 3.19 Some results based on 30 000 iterations. Convergence is rapidly achieved in such a simple model, and so the burn-in of 1000 iterations was hardly necessary.

```
gr(theta) # Gelman-Rubin diagnostic for convergence
beg(1001) # Discard first 1000 iterations as burn-in
stats(*) # Calculate summary statistics for all monitored
quantities
density(theta) # Plot distribution of theta
density(y.pred) # Plot distribution of y.pred
```

The statistics from the MCMC run are as follows:

```
node      mean     sd       MC error  2.5%   median  97.5%  start sample
P.crit   0.3273 0.4692   0.002631  0.0    0.0     1.0    1001  30000
theta    0.5633 0.07458  4.292E-4  0.4139 0.5647  0.7051 1001  30000
y.pred  22.52   4.278    0.02356   14.0   23.0    31.0   1001  30000
```

The exact answers are available from Example 3.10, and reveal that the posterior distribution has mean 0.563 and standard deviation 0.075, and the beta-binomial predictive distribution has mean 22.51 and standard deviation 4.31. The probability of observing 25 or more successes is 0.329. The MCMC results are within Monte Carlo error of the true values, and can achieve arbitrary accuracy by running the simulation for longer.

The flexibility of WinBUGS allows a variety of modelling issues to be dealt with in a straightforward manner: our examples include inference on complex functions of parameters (Examples 8.4, 8.7 and 9.3), alternative prior distributions (Examples 6.10 and 8.1), inference on ranks (Example 7.2), prediction of effects in new studies (Example 8.1), analysis of sensitivity to alternative likelihood assumptions (Example 8.2), and hierarchical models for both means and variances (Example 6.10).

3.20   SCHOOLS OF BAYESIANS

It is important to emphasise that there is no such thing as a single Bayesian approach, and that many ideological differences exist between researchers. Four broad levels of increasing 'purity' may be identified:

1. The **empirical** Bayes approach (Section 3.17), in which a prior distribution is estimated from multiple experiments. Analyses and reporting are in traditional terms, and justification is through improved sampling properties of procedures.
2. The **reference** Bayes approach, in which a Bayesian interpretation is given to conclusions expressed as posterior distributions, but an attempt is made to use 'objective' or 'reference' prior distributions. There have been a number of attempts to use Bayesian methods but with uniform priors, gaining the intuitive Bayesian interpretation while having essentially the classical results (see Section 5.5; see also Burton *et al.*, 1998; Gurrin *et al.*, 2000). For example, Shakespeare *et al.* (2001) use 'confidence levels' calculated from a normalised likelihood which is essentially a posterior distribution under a uniform prior – this type of activity has been termed an attempt to 'make the Bayesian omelette without breaking the Bayesian eggs'.
3. The **proper** Bayes approach, in which informative prior distributions are based on available evidence, but conclusions are summarised by posterior distributions without explicit incorporation of utility functions. Within this school there may be more or less focus on hypothesis testing using Bayes factors (Section 3.3): Bayes factor analyses essentially entertain the possibility of the precise truth of the null hypothesis (or at least values very close to the null), *i.e. either* θ is extremely close to 0, *or* we have almost no idea of regarding θ. Except in particular circumstances where such dichotomies may be feasible (perhaps in genetics), it might be considered more reasonable to express a 'smooth' sceptical prior: 'in most RCTs, estimation would be more appropriate than testing' (Kass and Greenhouse, 1989).
4. The **decision-theoretic** or 'full' Bayes approach, in which explicit utility functions are used to make decisions based on maximising expected utility. There has been long and vigorous debate on whether or not to incorporate an explicit loss function, and the extent to which a health-care evaluation should lead to an inference about a treatment effect or a decision as to future policy. Important objections to a decision-theoretic approach include the lack of a coherent theory for decision-making on behalf of multiple audiences with different utility functions, the difficulty of obtaining agreed utility values, and the fact that a strict decision-theoretic view would lead to future treatments being recommended on the basis of even marginal expected gains, without any concern as to the level of confidence with which such a recommendation is made (see Section 6.2 and Chapter 9).

Our personal leaning, and the focus in this book, is towards the third, *proper*, school of Bayesianism.

In spite of this apparent divergence in emphasis, the schools are united in their belief in the fundamental importance of three concepts that distinguish Bayesian from conventional methods: *coherence* of probability statements (Section 3.1), *exchangeability* (Section 3.17) and the *likelihood principle* (Section 4.3).

3.21   A BAYESIAN CHECKLIST

Bayesian methods tend to be inherently more complex than classical analyses, and thus there is an additional need for quality assurance. However, there are limited 'guidelines' available for reporting Bayesian analyses. Rudimentary guidance was provided by Lang and Secic (1997), who gave the following instructions:

1. Report the pre-trial probabilities and specify how they were determined.
2. Report the post-trial probabilities and their probability intervals.
3. Interpret the post-trial probabilities.

Similar advice is given in the *Annals of Internal Medicine*'s instructions to authors. The BaSiS (Bayesian Standards in Science) initiative (Section A.2) is seeking to establish guidelines for reporting.

In this section we present a checklist against which published accounts of Bayesian assessments of health-care interventions can be compared. We aim to ensure that an account which adequately contains all the points mentioned here would have the property that the analysis could be replicated by another investigator who has access to the full data. These guidelines should be seen as complementary to the CONSORT (Moher *et al.*, 2001) guidelines, in that they focus on those aspects crucial to an accountable Bayesian analysis, in addition to standard paragraphs concerning the intervention, the design and the results.

Our main examples attempt to use this structure, although it sets a high standard that we admit we do not always reach! In particular, it is often easier to present the evidence at the same time as the statistical model, particularly when there has been some iterative model construction. To avoid tedious repetition, the phrase 'should be clearly and concisely described' should be assumed to apply to each of the components below.

Background

- *The Intervention.* The intervention to be evaluated with regard to the population of interest and so on.
- *Aim of study.* It is important that a clear distinction is made between desired inferences on any quantity or quantities of interest, representing the

parameters to be estimated, and any decisions or recommendations for action to be made subsequent to the inferences. The former will require a prior distribution, while the latter will require explicit or implicit consideration of a loss or utility function.

Methods

- *Study design.* This is a standard requirement, but when synthesising evidence particular attention will be necessary to the similarity of studies in order to justify any assumptions of exchangeability.

- *Outcome measure.* The true underlying parameters of interest.

- *Statistical model.* The probabilistic relationship between the parameter(s) of interest and the observed data, either mathematically, or in such a way as to allow its mathematical form to be unambiguously obtained by a competent reader, including any model selection procedure, whether Bayesian or not.

- *Prospective Bayesian analysis?* It needs to be made clear whether the prior and any loss function were constructed preceding the data collection, and whether analysis was carried out during the study.

- *Prior distribution.* Explicit prior distributions for the parameters of interest should be given. If 'informative', then the derivation of the prior from an elicitation process or empirical evidence should be detailed. If claimed to be 'non-informative', then this claim should be justified. If it is intended to examine the effect of using different priors on the conclusion of the study, this should be stated and the alternative priors explicitly given.

- *Loss function or demands.* An explicit method of deducing scientific consequences is decided prior to the study. This will often be a range of equivalence (a range of values such that if the parameter of interest lies within it, two different technologies may be regarded as being of equal effectiveness), or a loss function whose expected value is to be minimised with respect to the posterior distribution of the parameter of interest. Any elicitation process from experts should be described.

- *Computation/software.* A mathematically competent reader should, if necessary, be able to repeat all the calculations and obtain the required results, and any mathematical software used to obtain the results should be described. If MCMC methods are being used the assumption of convergence should be justified.

Results

- *Evidence from study.* As much information about the observed data – sample sizes, measurements taken – as is compatible with brevity and data confidentiality should be given. It is also essential that the likelihood could be reconstructed, so that subsequent users can establish the contribution from the study to, say, a meta-analysis.

Interpretation

- *Bayesian interpretation.* The posterior distribution should be clearly summarised: in most cases, this should include a presentation of posterior credible intervals and a graphical presentation of the posterior distribution. If either a formal or informal loss function has been described, the results should be expressed in these terms.

  There should be a careful distinction between the report as a current summary for immediate action, in which case a synthesis of all relevant sources of evidence is appropriate, and the report as a contributor of information to a future evidence synthesis.

- *Sensitivity analysis.* The results of any alternative priors and/or expressions of the consequences of decisions.

- *Comments.* These should include an honest appraisal of the strengths and possible weaknesses of the analysis.

3.22 FURTHER READING

Historical references concerning Bayesian methods include Bayes (1763), Holland (1962), Fienberg (1992) and Dempster (1998). For general introductions, see the chapter by Berry and Stangl (1996a) in their textbook (Berry and Stangl, 1996b) which covers a whole range of modelling issues, including elicitation, model choice, computation, prediction and decision-making. Nontechnical tutorial articles include Lewis and Wears (1993), Bland and Altman (1998) and Lilford and Braunholtz (1996), while O'Hagan and Luce (2003) provide an excellent primer geared towards cost-effectiveness studies. Other authors emphasise different merits of Bayesian approaches in health-care evaluation: Eddy *et al.* (1990a) concentrate on the ability to deal with varieties of outcomes, designs and sources of bias, Breslow (1990) stresses the flexibility with which multiple similar studies can be handled, Etzioni and Kadane (1995) discuss general applications in the health sciences with an emphasis on decision-making, while Freedman (1996) and Lilford and Braunholtz (1996) concentrate on the ability to combine 'objective' evidence with clinical judgement. Stangl and Berry (1998) provide a recent review of biomedical applications.

There is a huge methodological statistical literature on general Bayesian methods, much of it quite mathematical. Cornfield (1969) provides a theoretical justification of the Bayesian approaches, in terms of ideas such as *coherence*. A rather old article (Edwards *et al.*, 1963) is still one of the best technical introductions to the Bayesian philosophy. Good tutorial introductions are provided by Lindley (1985) and Barnett (1982), while more recent books, roughly in order of increasing technical difficulty, include Berry (1996a), Lee (1997), O'Hagan (1994), Gelman *et al.* (1995), Carlin and Louis (2000), Berger (1985) and Bernardo and Smith (1994).

Recommended references for specific issues include DeGroot (1970) on decision theory, axiomatic approaches and backwards induction, Bernardo and Smith (1994) on exchangeability, and Kass and Raftery (1995) on Bayes factors. On computational issues, Carlin *et al.* (1993) and Etzioni and Kadane (1995) discuss a range of methods which may be used (normal approximations, Laplace approximations and numerical methods including MCMC), Gelman and Rubin (1996) review MCMC methods in biostatistics, and van Houwelingen (1997) provides a commentary on the importance of computational methods in the future of biostatistics.

With regard to hierarchical models Jerome Cornfield (1969, 1976) was an early proponent of the Bayesian approach to multiplicity (Section 6.8.1), while Breslow (1990) gives many examples of problems of multiplicity and reviews the use of empirical Bayes methods for longitudinal data, small-area mapping, estimation of a large number of relative risks in a case–control study, and multiple tumour sites in a toxicology experiment. Louis (1991) reviews the area and provides a detailed case study, while Greenland (2000) provides an excellent justification.

3.23   KEY POINTS

1. Bayesian methods are founded on the explicit use of judgement, formally expressed as prior beliefs and possibly loss functions. The analysis can therefore quite reasonably depend on the context and the audience. However, if the aim is to convince a wide range of opinion, subjective inputs must be strongly argued and be subject to sensitivity analysis.
2. Bayes theorem provides a natural means of revising opinions in the light of new evidence, and the Bayes factor or likelihood ratio provides a scale on which to assess the weight of evidence for or against specific hypotheses.
3. Bayesian methods are best seen as a transformation from initial to final opinion, rather than providing a single 'correct' inference.
4. Exchangeability is a vital judgement: exchangeable observations justify the use of parametric models and prior distributions, while exchangeable parameters lead to the use of hierarchical models.
5. Bayesian methods provide a flexible means of making predictions, and this is helped by MCMC methods.
6. Hierarchical models provide a flexible and widely applicable structure when wanting to simultaneously analyse multiple sources of evidence.
7. A decision-theoretic approach may be appropriate where the consequences of a study are considered reasonably predictable, but this is not the emphasis of this book.
8. Normal approximations can be used in many contexts, particularly when deriving likelihoods from standard analyses. This will generally entail transformation between different scales of measurement.

9. Standards for Bayesian reporting have not been established. The most important aspect is to provide details of each of the prior distributions, its justification and its influence assessed through sensitivity analysis.

EXERCISES

3.1. Altman (2001) considers the data in Table 3.9, showing the results of using a scan of the liver to detect abnormalities compared to classification at autopsy, biopsy or surgical inspection in 344 patients.
(a) Estimate the likelihood ratio for a positive scan.
(b) For the patients in Table 3.9 the prevalence of an abnormal pathology is 0.75. For this population estimate the posterior probability of an abnormal diagnosis after observing a positive scan result. What is the estimated posterior probability for a population in which the prevalence is 0.25?
3.2. Asked prior to a study of a new chemotherapy, an oncologist said that she would expect 90% of patients to respond, and that she thought it was unlikely to be less than 80%. (a) Use a 'method-of-moments' argument similar to that of Example 3.3 to summarise the oncologist's opinions in terms of a beta distribution, and plot this prior distribution. In 20 patients treated, 14 respond. (b) Plot the likelihood. (c) Update the beta parameters in the light of the data observed.
3.3. Show that if $y_1, \ldots, y_n$ are i.i.d. observations from a Poisson distribution with unknown mean θ, and that a gamma prior distribution with parameters α and β is specified for θ, the corresponding posterior distribution is also gamma, i.e. conjugate, with parameters $\alpha + \sum_{i=1}^{n} y_i$ and $\beta + n$.
3.4. Based on national statistics for a large number of similar hospitals, a manager believes that the mean number of patients attending a specialist clinic each week in his hospital should lie between 12 and 20. (a) Taking this range as approximately equivalent to a mean ± 2 standard deviations, use a 'method-of-moments' argument similar to that in Example 3.3 to summarise the manager's beliefs using a gamma distribution. The numbers of patients attending a specialist clinic each week for 5 weeks are 11, 15, 18, 13, 19, and are assumed to be independent observations

Table 3.9 Detection of abnormal liver pathology using scan compared to actual classification at autopsy, biopsy or surgical inspection in 344 patients.

Liver scan (Test)	Pathology (Truth)		Total
	Abnormal (+)	Normal (−)	
Abnormal (+)	231	32	263
Normal (−)	27	54	81
Total	258	86	344

from a Poisson distribution. (b) Obtain the posterior distribution for the mean number of patients per week based on the manager's prior beliefs. (c) Plot the prior and posterior densities. If your software permits it, calculate the prior and posterior probabilities that the mean is greater than 18.

3.5. Verify (3.14) algebraically, *i.e.* that a normal prior distribution is conjugate for the unknown mean of a normal likelihood.

3.6. Consider the GREAT trial of home thrombolytic therapy described in Example 3.6. Another cardiologist was more sceptical about the magnitude of benefit and thought that the relative reduction in odds of death was more likely to be around 10–15%, and that the extremes of a 25% relative reduction and a 2.5% increase were unlikely.

   (a) Fit a normal prior distribution for the log(odds ratio) to these opinions.
   (b) Obtain the posterior distribution for this cardiologist and compare it with the posterior distributions in Example 3.6.

3.7. Using the normal approximation to the likelihood derived in Exercise 2.5, assume a sceptical prior distribution, such that an odds ratio of 1 was most likely but with a 95% interval from 0.5 to 2.0. Obtain the posterior estimate for the log(odds ratio), odds ratio and associated 95% intervals.

3.8. Use the normal approximation to the likelihood derived in Exercise 2.8 and assume a sceptical prior distribution equivalent to the evidence in a balanced trial in which 50 events have occurred on each arm. Obtain the corresponding posterior distribution for the log(hazard ratio).

3.9. Using the methods of Section 3.11, consider the results seen in the PROSPER RCT in Exercise 2.8.

   (a) Find the sceptical prior distribution for the log(hazard ratio) with mean 0, such that the resulting posterior 95% interval for the hazard ratio just includes 1.
   (b) Do you think this degree of scepticism is reasonable, and hence are the trial results credible?

3.10. Baum *et al.* (1992) report the results of an RCT to investigate the use of tamoxifen compared to standard care for women treated for breast cancer, evaluated in terms of disease-free survival. In total, 2030 women were randomised and followed up for over 10 years. Overall, there were 484 events in the tamoxifen arm, whilst 419.6 were expected. (a) Assuming balanced randomisation and follow-up, estimate the number of events in the standard-care arm. During the first 5 years of the trial 387 events were observed compared to 320.2 expected, and in the second period of the trial 97 events were observed whilst 99.4 were expected. (b) Assuming a sceptical prior for the log(hazard ratio) centred at zero and with precision equivalent to having observed only 10 events, show that a sequential analysis of the accumulating trial data using the methods of Section 3.12 gives similar results to an analysis using all the trial data.

3.11. In Exercise 2.7 consider another 100 patients randomised between HAI and control.

(a) About how many deaths would we expect to observe?

(b) What would be the predictive distribution for the observed log(hazard ratio) using a sceptical prior distribution, *i.e.* centred at zero and equivalent to having observed 10 deaths?

(c) Repeat (b) for an optimistic prior that represented beliefs that there would be a 10% relative reduction in the risk of death associated with HAI with uncertainty equivalent to having observed 25 deaths.

3.12. Whitehead (2002) considers a meta-analysis of 9 RCTs to evaluate whether taking diuretics during pregnancy reduces the risk of pre-eclampsia and which is summarised in Table 3.10. For each study, (a) estimate the log(odds ratio) and its variance, and (b) obtain an estimate and 95% intervals for the pooled odds ratio. (c) Using the 'method of moments' (3.37), estimate the between-study variance τ^2. Hence obtain the posterior estimates and intervals for (d) the population odds ratio using random effects assuming the between-study variance is known, and (e) the odds ratios for each of the 13 studies assuming a random-effects model.

3.13. Cooper *et al.* (2002) report the results of an economic decision model to assess the cost-effectiveness of using prophylactic antibiotics in women undergoing Caesarean section. Evidence available includes the results of a Cochrane systematic review of 61 RCTs which evaluated the prophylactic use of antibiotics in women undergoing Caesarean section to prevent wound infection, which produces an estimated odds ratio of 0.40, where the baseline probability of wound infection without prophylactic use of antibiotics is estimated to be 0.08. Antibiotic treatment is assumed to cost £10. Women who have a Caesarean section and who do not develop an infection have a mean total cost of £1159 and are

Table 3.10 RCTs evaluating the use of diuretics during pregnancy to reduce risk of pre-eclampsia.

Study	Diuretic		Control	
	Cases	Total	Cases	Total
1	14	131	14	136
2	21	385	17	134
3	14	57	24	48
4	6	38	18	40
5	12	1011	35	760
6	138	1370	175	1336
7	15	506	20	524
8	6	108	2	103
9	65	153	40	102

assumed to have a utility in the subsequent year of 0.95 quality-adjusted life-years (QALYs), while women who have a Caesarean section and who develop an infection have mean total cost of £2320 and utility of 0.80 QALYs: it is assumed there is no difference between the groups after one year.

(a) Structure the decision as in Figure 3.12.

(b) Using the methods of Section 3.14, find the threshold for a policy decision-maker, in £ per QALY, at which the expected utility of using prophylactic antibiotics would exceed that of not using prophylactic antibiotics.

3.14. Use WinBUGS to repeat the analysis of the PROSPER RCT in Exercise 2.9, assuming a uniform prior (on a suitable wide range) for the log(odds ratio), and (a) the approximate normal likelihood, (b) exact binomial likelihoods.

3.15. Use WinBUGS to repeat the analysis in Exercise 3.4 of patients attending a specialist clinic.

4

Comparison of Alternative Approaches to Inference

4.1 A STRUCTURE FOR ALTERNATIVE APPROACHES

It would be misleading to dichotomise statistical methods as either 'classical' or 'Bayesian', since both terms cover a bewildering range of techniques. A rough taxonomy can be developed by distinguishing two characteristics: whether or not prior distributions are used for inferences, and whether the objective is estimation, hypothesis testing or a decision requiring a loss function of some form. All six combinations of these elements have been investigated in theory and, to some extent, in practice, and Table 4.1 assigns a label to each possible combination.

This categorisation can be made finer still, and in Section 3.20 an attempt was made to delineate the different schools of Bayesianism that exist. Empirical Bayes techniques can be considered as essentially Fisherian since there is no formal introduction of prior opinion, while reference Bayesian methods, based on attempts at 'objective' priors, fall somewhat between the Fisherian and proper Bayesian approaches. We acknowledge that many of the examples in

Table 4.1 A taxonomy of six possible 'philosophical' approaches to statistical inference, depending on the objective and the formal quantitative use of prior information.

		Objective		
		Inference (estimation)	*Hypothesis testing*	*Decision (loss function)*
Use of prior evidence	*Informal*	Fisherian	Neyman–Pearson	Classical decision theory
	Formal	Proper Bayesian	'Bayes factors'	Full decision-theoretic Bayesian

Bayesian Approaches to Clinical Trials and Health-Care Evaluation   D. J. Spiegelhalter, K. R. Abrams and J. P. Myles
© 2004 John Wiley & Sons, Ltd   ISBN: 0-471-49975-7

this book do not use informative prior distributions, and their results could be (approximately) obtained by a likelihood analysis.

With so many options the resulting arguments about their relative merits inevitably become somewhat complex, and in this chapter we can only highlight some major issues. The standard approach in the evaluation of medical interventions is a mixture of Fisherian and Neyman–Pearson philosophies and is briefly summarised in Section 4.2, although Neyman–Pearson ideas have attracted particularly strong criticism from both Fisherian and Bayesian perspectives (Section 4.3). P-values are critically compared with Bayes factors in Section 4.4.

In the midst of often polemical arguments, it has also been argued that it would be 'a great pity if differences of technical approach were exaggerated into differences about qualitative issues' (Cox and Farewell, 1997), while Armitage (1993) maintains it is not appropriate to polarise the argument as a choice between extremes. It also appears reasonable to suggest that the appropriate approach may depend crucially on context (Section 3.1): for example, both Koch (1991) and Whitehead (1993) claim that a proper Bayesian approach may be reasonable at early stages of a drug's development but is not acceptable in phase III trials.

4.2 CONVENTIONAL STATISTICAL METHODS USED IN HEALTH-CARE EVALUATION

Conventional approaches to inference can be divided into the two broad schools of Fisherian and Neyman–Pearson.

The *Fisherian* approach regarding inference on an unknown intervention effect θ is based on the likelihood function (Section 2.2.4), which expresses the relative support given to the different values of θ by the data. This gives rise to a maximum likelihood estimate comprising the most supported value for θ, and intervals based on ranges of values of θ with most likelihood. More controversially, Fisher suggested summarising the evidence against specified null hypotheses by P-values (the chance of getting a result as extreme as that observed were the null hypothesis true), although this was only intended as an informal guide to the strength of evidence in the specific experiment being reported (Goodman, 1999a). Hill *et al.* (2000) provide a good historical background, emphasising that the likelihood alone could be used for comparing hypotheses without calculation of P-values.

The *Neyman–Pearson* approach has a different perspective, rooted in an attempt at a theory of 'inductive behaviour', in seeking procedures for hypothesis testing and estimation that satisfy certain properties in long-run repeated use. Specifically, it focuses on the chances of making various types of error when making decisions on the basis of the data so that, for example, clinical trials are traditionally designed to have a fixed Type I error α (the chance of incorrectly rejecting the null hypothesis), usually taken as 5% or 1%, and fixed power (one minus the Type II error β, the chance of not detecting the alternative hypothesis), often 80% or 90%. Similarly, formulae for 95% confidence intervals are designed so that, in 95% of situations in which they are appropriately used, they will contain the true parameter value. The problem, as discussed in detail by Goodman (1999a), is that this restricts us in what we can say about the specific experiment being analysed.

In practice, a *combined* approach has developed, which is perhaps ironic in view of the enmity between the initial protagonists of the approaches (see below). Senn (1997b) points out that clinical trials are generally designed from a Neyman–Pearson standpoint, but analysed from a Fisherian perspective using P-values as measures of evidence. Methods used for observational methods and evidence synthesis tend to be more Fisherian, but Goodman (1999a) argues that the most common form of statistical analysis is to use P-values but, inappropriately, to interpret them as saying something about long-run properties.

Advantages of the conventional framework include its apparent separation of the evidence in the data from subjective factors, the general ease in computation, its wide acceptability and established criteria for 'significance', its relevance to the drug regulatory framework in which quality control of statistical submissions must be ensured, the availability of software, and the existence of robust non- and semi-parametric procedures.

Nevertheless, there has been continual criticism of these traditional approaches since their introduction in the 1920s and 1930s, and their development has been marked by considerable animosity and vituperative argument. When Neyman (1934) presented his theory of confidence intervals at a meeting of the Royal Statistical Society, Arthur Bowley, a strong advocate of the method of 'inverse probability' (the Bayesian approach), was given the task of proposing the vote of thanks. Towards the end of his remarks he said: 'I am not at all sure that the "confidence" is not a "confidence trick". He then went on to suggest a Bayesian approach was necessary: 'Does that really take us any further? . . . Does it really lead us towards what we need – the chance that in the universe which we are sampling the proportion is within . . . certain limits? I think it does not'. Fisher opened the discussion of Neyman (1935) on the attack: 'Were it not for the persistent efforts which Dr Neyman and Dr Pearson had made to treat what they speak of as problems of estimation, by means merely of tests of significance, he had no doubt that Dr Neyman would not have been in any danger of falling into the series of misunderstandings which his paper revealed'. Egon Pearson then came to Neyman's defence, saying that 'while he knew there was a widespread belief in Professor Fisher's infallibility, he must, in the first place, beg leave to question the wisdom of accusing a fellow-worker of incompetence without, at the same time, showing that he had succeeded in mastering the argument'.

In a strong attack on traditional methods, Cornfield (1976) claims that 'the paradox is that a solid structure of permanent value has, nevertheless, emerged, lacking only the firm logical foundation on which it was originally thought to have been built'. Generic criticisms include the failure of traditional methods to incorporate formally the inevitable background information that is available both at design and analysis, that they take no account of the consequences of the conclusions, and, from a more ideological perspective, that they disobey certain reasonable axioms of rational behaviour (Section 3.1). In addition, there is no doubt that classical inferences are often misinterpreted, in that P-values are mistaken for probabilities of null hypotheses being true, and 95% confidence intervals as meaning there is a 95% chance of their containing the true value. Our personal opinion is that the strongest argument against Neyman–Pearson methods and P-values is their disobedience of the likelihood principle: this crucial idea is now discussed within the context of sequential analysis.

4.3 THE LIKELIHOOD PRINCIPLE, SEQUENTIAL ANALYSIS AND TYPES OF ERROR

4.3.1 The likelihood principle

This principle (Berger and Wolpert, 1988) states that all the information that the data provide about the parameter is contained in the likelihood: we have already seen in Sections 3.2 and 3.3 how data only influence the *relative* plausibility of an alternative hypothesis through the relative likelihood and hence Bayesian inference automatically obeys this principle. This simple idea, however, has very strong consequences, as the following classic example demonstrates.

Example 4.1   *Stopping: The likelihood principle in action*

Goodman (1999a) considers the following classic problem. Suppose we hear that six people have each been given treatments A and B, and asked which they prefer. Five preferred A, and one preferred B. What evidence is this against the null hypothesis that A and B are preferred equally in the population?

Let θ be the true unknown proportion in the population preferring A, with $\theta = 0.5$ corresponding to the null hypothesis of 'no preference'. Then the likelihood arising from the experiment is proportional to $\theta^5(1 - \theta)$ (Section 2.2.4) and the likelihood principle states that all the evidence about θ to be derived from this experiment can be extracted from this function, using either likelihood or Bayesian methods.

In contrast, let us consider the P-value: the probability of observing a result at least as extreme as the data, given the null hypothesis $H_0: \theta = 0.5$. But what results are 'at least as extreme'? Suppose we are told that the experimenter decided in advance that six people were to be included, and the first five preferred A and the final one preferred B. The possible results of the experiment and their probabilities under H_0 are shown in

Table 4.2 under 'Design 1', with the 'at least as extreme as observed' outcomes highlighted in bold: these probabilities come from the binomial (0.5,6) distribution (Section 2.6.1). It is not clear how to handle the probability of the observation itself when defining what is 'as extreme' – here we adopt the standard convention of including half its probability so that the one-sided P-value is $\frac{1}{2}(6/64) + 1/64 = 0.0625$, with a two-sided P-value of 0.13; note that Goodman (1999a) considers the one-sided P-value including the whole contribution from the observed data, leading to $P = 0.11$. We may be disappointed that the result is not 'significant' at $P < 0.05$.

Table 4.2 Two different experimental designs: (1) ask six subjects whether they prefer A or B; (2) ask subjects sequentially until one prefers B and then stop. Observed data comprise 5 preferences for A and one for B. Highlighted values indicate potential data 'at least as extreme' as that observed under the null hypothesis H_0 of no overall preference in the population, i.e. the probability of either preference is 0.5.

Design 1		Design 2	
$Y_1 =$ No. subjects preferring A	Probability under H_0	$Y_2 =$ First subject preferring B	Probability under H_0
0	1/64	1	1/2
1	6/64	2	1/4
2	15/64	3	1/8
3	20/64	4	1/16
4	15/64	5	1/32
5	**6/64**	6	**1/64**
6	**1/64**	7	**1/128**
		8	**1/256**
		etc.	etc.

But then we hear that a mistake has been made in reporting the results, and that the experimenter in fact used a different (and admittedly rather strange) sampling procedure (Design 2): he had decided to carry on experimenting until he found someone who preferred B, and then stop. Table 4.2 again shows the possible results with those 'at least as extreme as observed' highlighted: the probabilities follow a 'geometric' distribution in which the chance of first getting a B preference on the nth trial is $1/2^n$. This time the P-value is $\frac{1}{2}(1/64) + 1/128 + 1/256 + \ldots = \frac{1}{2}(1/64) + 1/64 = 3/128 = 0.023$, with a two-sided P-value of 0.046, and we might now be delighted that it is 'significant' at $P < 0.05$.

A likelihood and Bayesian approach to this problem is described in Section 4.4.4.

In Example 4.1 the intention of the experimenter dictated the conclusions to be drawn from the results, and the P-values depended on what would have happened had something else been observed (Berry, 1987). The likelihood principle claims such behaviour is nonsensical, since only the observed data influence the conclusions and this is through the likelihood alone.

4.3.2 Sequential analysis

In a sequential experimental design the data are periodically analysed and the study stopped if sufficiently convincing results obtained. Such repeated analysis of the data can have a strong effect on the overall Type I error in the experiment, since there are many opportunities to obtain a false positive result. The traditional approach to sequential analysis identifies classes of 'stopping boundaries' with fixed overall Type I error α, and then chooses designs with minimum Type II error β (maximum power) for particular alternative hypotheses. At the end of a study P-values and confidence intervals should be adjusted for the sequential nature of the design (Whitehead, 1997a).

Sequential data fall naturally within the Bayesian framework, as the posterior distribution following each observation becomes the prior for the next (Section 3.12). As forcefully argued by Cornfield (1976), (3.25) shows that the evidence for taking alternative decisions depends only on the relative likelihood of alternative hypotheses (the Bayes factor), prior probabilities, and utilities, and hence provides a direct decision-theoretic justification for the likelihood principle within sequential trials. Sequential analysis therefore provides a primary focus for disagreement between frequentist and Bayesian approaches, since the likelihood principle means that concern about frequentist stopping rules retaining Type I error is entirely misplaced, and we can analyse trials at will. Criticism has been forceful: Anscombe (1963) baldly states that 'Sequential analysis is a hoax', and (1975) considers that 'provided the investigator has faithfully presented his methods and all of his results, it seems hard indeed to accept the notion that *I* should be influenced in my judgement by how frequently *he* peeked at the data while he was collecting it'.

We find the following argument particularly persuasive. If we were to assign weights to the relative importance of the two types of error that could be made, any resulting design would seek to minimise a linear combination of the Type I error rate α and Type II error rate β. Perhaps surprisingly, such a design would obey the likelihood principle, and this led Cornfield (1966) to point out that

the entire basis for sequential analysis depends upon nothing more profound than a preference for minimising β for given α rather than minimising their linear combination. Rarely has so mighty a structure, and one so surprising to scientific common sense, rested on so frail a distinction and so delicate a preference.

We shall return to this topic when discussing sequential clinical trials in Section 6.6.

4.3.3   Type I and Type II error

Neyman–Pearson theory has been strongly criticised from both a Bayesian and Fisherian perspective. Anscombe (1963) says 'the concept of error probabilities of the first and second kinds ... has no direct relevance to experimentation ... The formalism of opinions, decisions concerning further experimentation and other required actions, are not dictated in a simple prearranged way by the formal analysis of the experiment, but call for judgement and imagination'.

The selection of values for error rates in trials seems particularly arbitrary: Healy (1994) asks 'Why the invariable 5% for α? Conditional on this, why the larger 10% or even 20% for β? Is it really more important not to make a fool of yourself than it is to discover something new?' Sheiner (1991) provides a strong polemic against hypothesis testing and in favour of an approach in which 'we gather data to model and quantify nature'; shifting attention from hypothesis testing to confidence intervals does not really avoid the problem, since these are, essentially, just the set of hypotheses that cannot be rejected at a certain α level.

We have already identified the crucial issue that arises in any context in which simultaneous analysis of multiple studies, or multiple analyses of the same study, is required. The traditional approach warns that repeated hypothesis testing is bound to raise the chance of a Type I error (incorrectly rejecting a true null hypothesis), and so suggests some adjustment, such as Bonferroni, to try to retain a specified overall Type I error. This will typically give larger *P*-values and wider confidence intervals.

The problem lies in deciding the set in which to embed the particular analysis being carried out. Cornfield (1976) asks, with some irony: 'Do we want error control over a single trial, over all the independent trials on the same agent, on the same disease, over the lifetime of an investigator, etc.?' The need for any such adjustment, which necessarily depends on the number of hypotheses being tested, has been strongly questioned even from a non-Bayesian perspective, particularly in epidemiology; Cole (1979) states that 'in every study, every association should be evaluated on its own merits: its prior credibility and its features in the study at hand. The number of other variables is irrelevant'. Greenland and Robins (1991) are among the many who have argued that some adjustment *is* necessary, but rather than being based on Type I errors, it should be derived from an explicit model that reflects assumptions about variability, and hence leads naturally to the approach to multiplicity outlined in Section 3.17.

4.4   *P*-VALUES AND BAYES FACTORS*

4.4.1   Criticism of *P*-values

We noted in Section 4.3 that sequential trials present a particular problem for *P*-values. Other arguments against this procedure are that the null hypothesis may be neither plausible nor of great interest, the arbitrariness of the 0.05 and 0.01 level, and that P-values tend to create a false dichotomy between 'significant' and 'non-significant' which is inappropriate for consequent policy decisions. Furthermore, the definition of 'more extreme' and hence the value of P itself may be unclear even in some simple circumstances, such as testing association in a 2×2 table of counts, as well as requiring the choice between one- or two-sided tests.

The strongest criticism is, perhaps, that P-values focus on statistical rather than practical significance and hence their interpretation can be very dependent on sample size. This is illustrated in Example 4.2.

Example 4.2 *Preference: P-values as measures of evidence*

Freeman (1993) considered four hypothetical studies in which equal number of patients are given treatments A and B and asked which they prefer, with results shown in Table 4.3. Each results in an identical 'significant' two-sided P-value of 0.04. However, as Freeman states, the first trial

Table 4.3 Four theoretical studies all with the same two-sided P-value for the null hypothesis of equal preference in the population.

Number of patients receiving A and B	Numbers preferring A : B	% preferring A	two-sided P-value
20	15 : 5	75.00	0.04
200	115 : 86	57.50	0.04
2 000	1046 : 954	52.30	0.04
2 000 000	1 001 445 : 998 555	50.07	0.04

would be considered too small to permit reliable conclusions, while the last trial (with a preference proportion of 50.07%) would be considered as evidence *for* rather than *against* equivalence, since the preference rates are, from any practical perspective, equally balanced. Thus equal P-values can lead to very different conclusions depending on the sample size.

4.4.2 Bayes factors as an alternative to P-values: simple hypotheses

We have already seen (Section 3.3) that the Bayes factor or likelihood ratio is the natural way to compare the support for two alternative hypotheses: when these hypotheses are 'simple' (*i.e.* there are no unknown parameters), the Bayes factor is a measure of the evidence in the data alone and is not affected by any prior probabilities. In the rather unrealistic situation that data are only reported as being 'significant at the $100\alpha\%$ level', the Bayes factor is

$$\text{BF} = \frac{p(\text{'significant'}|H_0)}{p(\text{'significant'}|H_1)} = \frac{\alpha}{1-\beta} \tag{4.1}$$

where α and β are the standard Type I and Type II error rates (Example 3.7).

It is important to note the behaviour of (4.1) as the sample size increases but the alternative hypothesis H_1 remains fixed. In this case the power of the study increases, and hence β decreases and the Bayes factor decreases towards α: we are left with the conclusion of Peto *et al.* (1976) that a 'significant' result provides more evidence against the null hypothesis for larger sample sizes.

This finding can be contrasted with Lindley and Scott (1984), who preface their statistical tables with the claim that '*all significance tests are dubious because the interpretation to be placed on the phrase "significant at 5%" depends on the sample size: it is more indicative of the falsity of the null hypothesis with a small sample than with a large one*'. We therefore appear to have contradictory claims that both smaller *and* larger studies suggest increased evidence against the null hypothesis when reporting a 'significant' result.

For simple alternative hypotheses, Royall (1986) explains this apparent paradox by contrasting two situations: that we know a study was significant at the 5% level, and that we know the exact *P*-value was 5%. The first was covered by (4.1), while the second is now considered for normal distributions. Suppose

$$y_m \sim \text{N}[\theta, \sigma^2/m]$$

and we wish to compare two simple hypotheses $H_0: \theta = 0$ against $H_1: \theta = \theta_A > 0$. Then the Bayes factor is the likelihood ratio

$$\begin{aligned}
\text{BF} &= \frac{p(y_m|\theta = 0)}{p(y_m|\theta_A)} = \exp\left(-\frac{m}{2\sigma^2}[y_m^2 - (y_m - \theta_A)^2]\right) \\
&= \exp\left(-\frac{m\theta_A}{\sigma^2}\left[y_m - \frac{\theta_A}{2}\right]\right).
\end{aligned} \tag{4.2}$$

This reveals the intuitive behaviour that for $y_m < \theta_A/2$, the Bayes factor will exceed 1 and hence favour H_0, while if $y_m > \theta_A/2$ the Bayes factor will be less than 1 and favour H_1.

Equation (4.2) can also be written

$$\text{BF} = \exp\left(-\sqrt{m}z_m\delta + \frac{m\delta^2}{2}\right) \tag{4.3}$$

where $\delta = \theta_A/\sigma$ is a standardised version of the alternative hypothesis, and $z_m = y_m\sqrt{m}/\sigma$ is the standardised test statistic for H_0. The crucial observation is that, for fixed z_m and hence fixed *P*-value, the Bayes factor will *increase* with increasing sample size m, and hence support Lindley and Scott's observation that smaller sample sizes are more indicative of the falsity of the null hypothesis.

The apparent paradox for simple alternative hypotheses is seen to be resolved by being clearer by what we mean by a 'significant' result: when we only know a result achieved significance at a fixed level, the evidence against H_0 increases with sample size, while if we know the exact significance level, evidence against H_0 decreases with sample size. This reveals the complexity of comparing Bayes factors with P-values, and we shall now add to the potential confusion by considering composite alternative hypotheses, which are seen to obey *both* the behaviours contrasted above.

4.4.3   Bayes factors as an alternative to P-values: composite hypotheses

In most cases in which P-values are currently used H_1 will be 'composite', in that it encompasses a range of parameter values θ as alternatives to the single value specified by H_0, typically $\theta = 0$. We therefore need a method to obtain an overall likelihood $p(\text{data}|H_1)$ in order to obtain the Bayes factor, *i.e.* $p(\text{data}|H_0)/p(\text{data}|H_1)$.

A likelihood-based solution is to use the 'minimum' Bayes factors, $\text{BF}_{\min}$, under H_1 (Goodman, 1999b). For a general alternative hypothesis $H_1: \theta \neq 0$ in the normal model considered in (4.2), the minimum Bayes factor occurs when $\theta_A = y_m$, and from (4.3) is

$$\text{BF}_{\min} = \exp\left(-z_m^2/2\right), \tag{4.4}$$

where $z_m = y_m \sqrt{m}/\sigma$ is the standardised test statistic for H_0. This produces a direct mapping between one-sided P-values, given by $\Phi(z_m)$, and minimum Bayes factors that is displayed as part of Figure 4.1: using Jeffreys' descriptions contained in Table 3.2, a two-sided P-value (denoted $2P$) of 0.001 is 'decisive evidence', $2P = 0.01$ is on the border of 'strong' and 'very strong', and $2P = 0.05$ is 'substantial'. The minimum Bayes factor thus leads to conclusions that are qualitatively similar to P-values but obey the likelihood principle and so are unaffected by stopping rules. However, they still suffer from the criticism displayed in Example 4.2: all the four studies have significance corresponding (up to a normal approximation) to a z statistic of $z_{0.04/2} = -2.05$, and hence would have the same minimum Bayes factor of $\exp\left(-2.05^2/2\right) = 1/8.2$: 'substantial' evidence against H_0.

As an alternative to a likelihood-based approach, in a full Bayesian analysis we need to specify a prior $p(\theta|H_1)$ under the alternative hypothesis. If we assume

$$\theta|H_1 \sim \text{N}[0, \sigma^2/n_0],$$

then from (3.23) we have that

$$y_m|H_1 \sim N\left[0, \sigma^2 \left(\frac{1}{n_0} + \frac{1}{m}\right)\right],$$

and hence the Bayes factor is easily shown to be

$$BF = \sqrt{1 + \frac{m}{n_0}} \, \exp\left[\frac{-z_m^2}{2(1 + n_0/m)}\right]. \tag{4.5}$$

n_0 can approximately be interpreted as the number of 'imaginary' observations taking on the value of the null hypothesis $\theta = 0$, and hence reflects prior support under H_1 for parameter values 'near' (but not exactly) H_0. The problem then becomes that of assessing a reasonable value for n_0. This will be considered in Section 5.5.4 in which priors that explicitly consider the 'truth' of a (null) hypothesis are discussed, but we now note that Kass and Wasserman (1995) suggest that $n_0 = 1$ (a prior equivalent to a single observation) may be a reasonable choice in many circumstances.

Figure 4.1 displays the resulting relationship between two-sided P-values and Bayes factors for different choices of m/n_0, the ratio of data sample size to prior sample size under the alternative hypothesis. It is clear that Bayes factors can produce very different results from the standard measures of evidence, with a tendency towards preference for the null hypothesis: when m/n_0 is large we note that

$$BF \approx \sqrt{\frac{m}{n_0}} \, BF_{min}. \tag{4.6}$$

An alternative way of examining the relationship between Bayes factors and P-values is shown in Figure 4.2, in which the change in Bayes factor with increasing ratio m/n_0 is shown for fixed P-values. For example, evidence that is labelled as $2P = 0.001$ is considered only just 'strong' when the sample size is small relative to the prior precision, but becomes 'very strong' for moderate sample sizes, and then reduces to only 'substantial' for overwhelming large experiments. This non-monotonic relationship to sample size appears to match well the intuitive desire for measures of evidence brought out in Example 4.2.

As we have noted in Section 4.4.2, the importance of sample size and plausibility of benefits in interpreting P-values has often been stressed even within the non-Bayesian literature: for example, the ISIS-4 investigators state that 'when moderate benefits or negligibly small benefits are both much more plausible than extreme benefits, then a $2P = 0.001$ effect in a large trial or overview would provide much stronger evidence of benefit than the same significance level in a small trial, a small overview, or a small subgroup analysis' (Collins *et al.*, 1995). Examination of Figure 4.2 shows that their insight is again

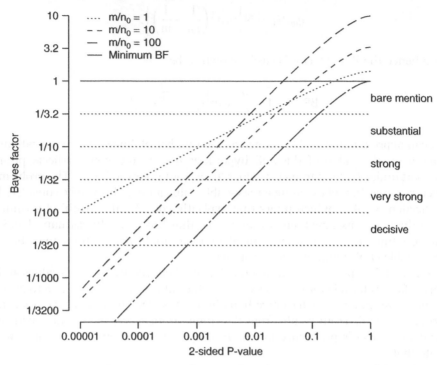

Figure 4.1 Bayes factors compared to *P*-values for composite normal hypotheses, showing bands corresponding to Jeffreys levels of evidence. The minimum Bayes factor is the Bayes factor against the maximum likelihood estimate for the parameter under H_1.

matched by the behaviour of the Bayes factor: smaller benefits being more plausible correspond to n_0 being relatively large, and hence m/n_0 lies in the 'dip' of Figure 4.2 in which stronger evidence is shown compared to smaller sample sizes. However, Figure 4.2 suggests a conclusion that is not mentioned by Collins *et al.* (1995) but seems quite appropriate: if the 'large trial or overview' becomes *extremely* large but still only significant at $2P = 0.001$, then the evidence for benefit will start to decline again.

For composite hypotheses it appears that neither of the views contrasted in Section 4.4.2 holds: there is no simple monotonic relationship between Bayes factors and *P*-values, and it is perhaps not surprising that so much apparent confusion has arisen.

Bayes factors can be obtained in the presence of nuisance parameters, but this makes the dependence on the prior distribution of even more concern. This is an area of substantial research and discussion (Kass and Raftery, 1995).

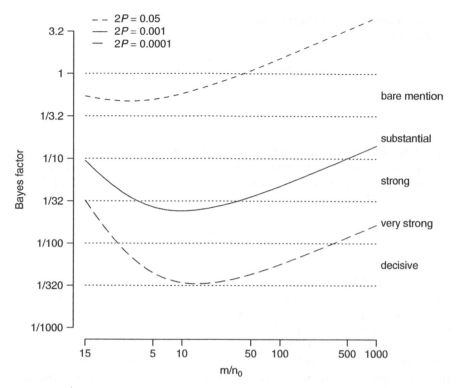

Figure 4.2 Bayes factors for composite normal hypotheses for fixed *P*-values and different m/n_0 ratios, *i.e.* ratio of observed to prior sample size, with areas delineated by Jeffreys' levels of evidence.

4.4.4 Bayes factors in preference studies

Consider the preference studies used in Examples 4.1 and 4.2, in which the underlying proportion of individuals preferring option *A* to *B* is assumed to be θ. Then the number of preferences *r* for option *A* out of *m* independent trials has a binomial distribution (Section 2.6.1)

$$p(r|\theta, m) = \binom{m}{r}\theta^r(1-\theta)^{m-r}.$$

The maximum likelihood estimator is $\hat{\theta} = r/m$, and so the minimum Bayes factor for the null hypothesis $H_0: \theta = 0.5$ is

$$\mathrm{BF_{min}} = \frac{p(r|\theta = 0.5)}{p(r|\theta = \hat{\theta})} = \frac{1}{2^m} \bigg/ \left(\frac{r}{m}\right)^r \left(1 - \frac{r}{m}\right)^{m-r}.$$

Assuming $p(\theta|H_1)$ is a uniform prior (as suggested by Jeffreys) gives the predictive distributions

$$p(r|m, H_0) = \binom{m}{r} \frac{1}{2^m}, \tag{4.7}$$

$$p(r|m, H_1) = \frac{1}{m+1}. \tag{4.8}$$

Equation (4.7) is simply the Binomial probability when $\theta = 0.5$, and (4.8) shows r has a uniform distribution over all its possible values $0, 1, 2, \ldots, m$, and is a special case of the beta-binomial distribution (Section 3.13.2) with $a = 1, b = 1$. Hence the exact Bayes factor is

$$BF = \binom{m}{r} \frac{m+1}{2^m}. \tag{4.9}$$

For both the likelihood and Bayesian approaches we can use approximations for large samples by calculating the P-value, obtaining a corresponding z-statistic, and substituting in (4.4) and (4.5). For the Bayesian approximation we do, however, need to specify a normal distribution for $p(\theta|H_1)$ instead of a uniform distribution, and the problem lies in choosing the normal variance. In 'interesting' situations the Bayes factor is driven by the ordinate of the $p(\theta|H_1)$ at the null hypothesis, and so we choose a normal distribution that has the same ordinate as a uniform distribution, namely 1. Were $\theta|H_1 \sim N[0.5, \sigma^2/n_0]$, then the ordinate at $\theta = 0.5$ would be $\sqrt{n_0/(2\pi\sigma^2)}$. σ^2 is the variance of a single obervation under H_0, and so $\sigma^2 = \theta(1-\theta) = \frac{1}{4}$ and equating the resulting ordinate $\sqrt{2n_0/\pi}$ to 1 gives $n_0 = \pi/2 = 1.57$, not far from the value of $n_0 = 1$ suggested by Kass and Wasserman (1995). Thus, for a preference study with a standardised test statistic of z_m, our approximate Bayes factors are

$$BF_{min} \approx \exp(-z_m^2/2), \tag{4.10}$$

$$BF \approx \sqrt{1 + \frac{m}{1.57}} \exp\left[\frac{-z_m^2}{2(1 + 1.57/m)}\right]. \tag{4.11}$$

The quality of these approximations is explored in Example 4.3.

We again emphasise that the Bayes factors, whether likelihood or Bayesian, are unaffected by whether the designs were sequential or fixed sample size.

Example 4.3 *Preference (continued): Bayes factors in preference studies*

Table 4.4 shows the quality of the approximate Bayes factors for the preference data, using the exact Bayes factors in (4.9), and approximations (4.10) and (4.11). The approximations for the Bayes factors appear reasonable, particularly for the minimum Bayes factor. For Example 4.1, both Bayes factors express minimal evidence against the null hypothesis, as would be expected from Figure 4.1. For the data in Example 4.2, the increasing sample size leaves the minimum Bayes factor constant at 'substantial' evidence against H_0, whereas the full Bayes factor changes from favouring H_1 to favouring H_0, and then steadily increases its support for H_0. This behaviour reflects the pattern shown in Figure 4.2 for increasing sample size and fixed P-value, following approximately the trajectory of $2P = 0.05$.

Table 4.4 Bayes factors for preference studies when m individuals asked whether they prefer A or B. The first row is from Example 4.1 and the other four rows from Example 4.2. z_m is a standardised test statistic that would give rise to the observed one-sided P-value. The approximate Bayes factor assumes $n_0 = 1.57$.

m	r prefer A	$\hat{\theta}$	One-sided P-value	z_m	Minimum Bayes factor		Bayes factor	
					Exact	Approx	Exact	Approx
6	5	0.83	0.063	1.53	0.23	0.31	0.65	0.86
20	15	0.75	0.02	2.05	0.07	0.12	0.31	0.53
200	115	0.575	0.02	2.05	0.10	0.12	1.20	1.41
2 000	1046	0.523	0.02	2.05	0.12	0.12	4.30	4.37
2 000 000	1 001 445	0.500 722 5	0.02	2.05		0.12	139.8	138.0

Rather than formulating these problems as hypothesis tests, it may be much more appropriate to assess a reasonable prior for θ and then report $p(\theta > 0.5|r, m)$ – the posterior probability that a majority of the population prefer A to B. Of course, such a measure suffers from exactly the same criticism of the P-values in Example 4.2: the posterior probability may be high even though the 'majority' that prefers A is negligible. In this case it may be more appropriate to assess an 'important majority' $\theta_S > 0.5$, and consider the $p(\theta > \theta_S|r, m)$. See Section 6.3 for applications of these ideas in clinical trials.

4.4.5   Lindley's paradox

Close examination of the top right-hand corner of Figure 4.1 reveals what might appear as odd behaviour: when the ratio m/n_0 is high, and the P-value is just marginally significant *against* H_0, the Bayes factor can be greater than 1 and hence *support* H_0. This somewhat surprising result is known as *Lindley's paradox*, after Lindley (1957). An informal explanation is as follows. First, for large sample sizes, a P-value can be small even if the data support values of θ very close to the null hypothesis, as shown for the large sample sizes in Example 4.2. Second, such data may indeed be unlikely under the null hypothesis, but are even more unlikely under an alternative that spreads the prior probability thinly over a wide range of potential values. Hence the Bayes factor can support H_0

when a significance test would reject it, essentially as the lesser of two evils. An example of this behaviour is shown in Example 4.4.

Example 4.4   *GREAT (continued): A Bayes factor approach*

From the 'Evidence from study' component of Example 3.6 we note that the standardised test statistic is $z = 2.03$: just significant evidence against H_0 at the traditional two-sided $P < 0.05$. The 'minimum' Bayes factor against H_0 is $BF_{min} = \exp(-z^2/2) = 0.13 = 1/7.8$, corresponding to 'substantial' evidence against H_0. Thus the classical and Bayesian approaches align to a reasonable extent if we allow the alternative hypothesis to be dictated by the data.

However, a fully Bayesian approach might place a prior on $\theta = \log(OR)$ under H_1, centred on 0 and with a large variance. For example, suppose we used a prior with $n_0 = 0.5$ which is essentially uniform over the log(OR) scale.

Since $m = 30.5$ we have a ratio of likelihood to prior precision of $m/n_0 = 61$. From (4.6) the Bayes factor is approximately $\sqrt{m/n_0}\, BF_{min} = 1.001$ (the exact value from substitution into (4.5) is 1.04), *i.e.* slight evidence *in favour* of H_0! This is an example of Lindley's paradox.

4.5   KEY POINTS

1. There is room for dispute over some of the fundamental principles of conventional statistical analysis.
2. The likelihood principle states that only the observed data should affect inferences: classical sequential analysis disobeys this.
3. The pragmatic interpretation of P-values strongly depends on sample size.
4. Minimum Bayes factors obey the likelihood principle, but have similar qualitative behaviour to P-values.
5. Proper Bayes factors can, for large sample sizes relative to the prior precision, support the null hypothesis when a classical analysis would lead to its rejection.

EXERCISES

4.1. Confirm the form of the Bayes factor given by (4.5).
4.2. Calculate the minimum Bayes factor corresponding to the three levels of significance considered in Figure 4.2. In what circumstances might the

minimum Bayes factor exaggerate the evidence against the null hypothesis, compared to a full Bayesian approach?

4.3. In the preference studies described in Section 4.4.4, suppose we observed data that were just 'significant', with a two-sided P-value of 0.05. Assume $n_0 = 1.57$.

(a) What sample size (approximately) would yield a Bayes factor of 1, *i.e.* indifference between the null and alternative hypotheses?

(b) What observed data would have given $2P = 0.05$ with this sample size?

4.4. For the PROSPER trial in Exercise 2.8 calculate the one-sided P-value, the minimum Bayes factor, and the Bayes factor corresponding to a sceptical prior distribution with an effective number of events $n_0 = 1$.

4.5. In Example 4.4, what would be the Bayes factor were we to adopt Kass and Wasserman's suggestion of $n_0 = 1$?

Prior Distributions

5.1 INTRODUCTION

There is no denying that quantifiable prior beliefs exist in medicine. For example, in the context of clinical trials, Peto and Baigent (1998) state that 'it is generally unrealistic to hope for large treatment effects' and that 'it might be reasonable to hope that a new treatment for acute stroke or acute myocardial infarction could reduce recurrent stroke or death rates in hospital from 10% to 9% or 8%, but not to hope that it could halve in-hospital mortality'. However, turning informally expressed opinions into a mathematical prior distribution is perhaps the most difficult aspect of Bayesian analysis. Five broad approaches are outlined below: elicitation of subjective opinion; summarising past evidence; default priors; 'robust' priors; and estimation of priors using hierarchical models. The discussion mainly focuses on priors for the primary treatment effects of interest, although we also consider the difficult issue of specifying a prior for the variance component in a hierarchical model. Finally, we consider the criticism of prior assessments, from both an empirical and a methodological perspective.

We should repeat the statements made in Section 3.9 concerning possible misconceptions about prior distributions: they are not necessarily prespecified, unique, known or important. Since there is no 'correct' prior, Bayesian analysis can be seen as a means of transforming prior into posterior opinions, rather than producing *the* posterior distribution. It is therefore vital to take into account the context and audience for the assessment (Section 3.1), and analysis of sensitivity to alternative assumptions should be considered essential. Kass and Greenhouse (1989) introduced the term 'community of priors' to describe the range of viewpoints that should be considered when interpreting evidence, and the suggestions in this chapter represent possible members of that community.

It is also important to keep in mind that, in certain circumstances, it may be quite reasonable for a prior to be elicited and used solely for design purposes, and excluded when publicly reporting a study. However, when wishing to convince an audience of the benefits of an intervention, it may be important

Bayesian Approaches to Clinical Trials and Health-Care Evaluation D. J. Spiegelhalter, K. R. Abrams and J. P. Myles
© 2004 John Wiley & Sons, Ltd   ISBN: 0-471-49975-7

to elicit their priors and possibly their utilities (Kadane and Wolfson, 1996).

From a mathematical and computational perspective, we have seen in Section 3.6.2 that it can be convenient if the prior distribution is a member of a family of distributions that is conjugate to the form of the likelihood, in the sense that they 'fit together' to produce a posterior distribution that is in the same family as the prior distribution. We also saw in Section 2.4 that in many circumstances likelihoods for treatment effects can be assumed to have an approximately normal shape, and thus in these circumstances it will be convenient to use a normal prior (the conjugate family), provided it approximately summarises the appropriate external evidence. Modern computing power is, however, reducing the need for conjugacy, and in this chapter we shall largely concentrate on the source and use of the prior rather than its precise mathematical form.

5.2   ELICITATION OF OPINION: A BRIEF REVIEW

5.2.1   Background to elicitation

A true subjectivist Bayesian approach requires only a prior distribution that expresses the personal opinions of an individual but, if the health-care intervention is to be generally accepted by a wider community, it would appear to be essential that the prior distributions have some evidential or at least consensus support. In some circumstances there may, however, be little 'objective' evidence available and summaries of expert opinion may be indispensable. We shall use the generic term 'clinical prior' for such expert assessments.

There is an extensive literature concerning the elicitation of subjective probability distributions from experts, with some good early references on statistical (Savage, 1971) and psychological aspects (Tversky, 1974), as well as on methods for pooling distributions obtained from multiple experts (Genest and Zidek, 1986). The fact that people are generally not good probability assessors is well known, and the variety of biases they suffer are summarised by Kadane and Wolfson (1997):

1. *Availability*. Easily recalled events are given higher probability, and vice versa.
2. *Adjustment and anchoring*. Initial assessments tend to exert an inertia, so that further elicited quantities tend to be insufficiently adjusted. For example, if a 'best guess' is elicited first, then subsequent judgements about an interval may be too close to the first assessment.
3. *Overconfidence*. Distributions are too tight.
4. *Conjunction fallacy*. A higher probability can be given to an event which is a subset of an event with a lower probability.
5. *Hindsight bias*. If the prior is assessed after seeing the data, the expert may be biased.

Nevertheless it has been shown that training can improve experts' ability to provide judgements that are 'well calibrated', in the sense that if a series of events are given a probability of, say, 0.6, then around 60% of these events will occur: see, for example, Murphy and Winkler (1977) with regard to weather forecasting.

Chaloner (1996) provides a thorough review of methods for prior elicitation in clinical trials, including interviews with clinicians, postal questionnaires, and the use of an interactive computer program to draw a prior distribution. She concludes that fairly simple methods are adequate, using interactive feedback with a scripted interview, providing experts with a systematic literature review, basing elicitation on 2.5th and 97.5th percentiles, and using as many experts as possible. Both Kadane and Wolfson (1996) and Berry and Stangl (1996a) emphasise the potential benefits of two approaches: eliciting predictive distributions of future events from which an implicit prior distribution can be derived, and asking additional questions as a consistency check.

5.2.2   Elicitation techniques

Methods used in practice can be divided into four main categories of increasing formality, which are listed here with some experience of their use:

1. *Informal discussion.* Prominent individuals can be informally interviewed for their opinion, as illustrated in Example 3.6. In a trial of paclitaxel in metastatic breast cancer, the study's principal clinical investigator expected the overall success rate to be 25% and had 50% belief that the true success rate lay between 15% and 35% (Rosner and Berry, 1995). Example 7.1 features priors obtained from two doctors for the relative risk of venous thrombosis associated with the use of oral contraceptives (Lilford and Braunholtz, 1996). There are clear difficulties in using such individual opinions in any formal context.
2. *Structured interviewing and formal pooling of opinion.* Freedman and Spiegelhalter (1983) describe an interviewing technique in which a set of experts were individually interviewed and hand-drawn plots of their prior distributions elicited, while deliberate efforts were made to prevent the opinions being overconfident (too 'tight'). The distributions were converted to histograms and averaged to produce a composite prior. This technique was also used for trials of thiotepa in superficial bladder cancer (Spiegelhalter and Freedman 1986) and osteosarcoma (Spiegelhalter *et al.*, 1993). Gore (1987) introduced the concept of 'trial roulette', in which 20 gaming chips, each representing 5% belief, could be distributed amongst the bins of a histogram: in a trial of artificial surfactant in premature babies, 12 collaborators were interviewed using this technique to obtain their opinion on the possible benefits of the treatment (Ten Centre Study Group, 1987). Using an elec-

tronic tool so that individuals in a group could respond without attribution, Lilford (1994) presented collaborators in a trial with a series of imaginary patients in order to elicit their opinions on the benefit of early delivery. The appropriate means of pooling such opinions is discussed in Section 5.2.3.

3. *Structured questionnaires.* The 'trial roulette' scheme described above was administered by post by Hughes (1991) for a trial in treatment of oesophageal varices and by Abrams *et al.* (1994) for a trial of neutron therapy. Parmar *et al.* (1994) elicited prior distributions for the effect of a new radiotherapy regime (CHART), in which the possible treatment effect was discretised into 5% bands and the form was sent by post to each of nine clinicians. Each provided a distribution over these bands and an arithmetic mean was then taken: see Example 5.1 for details. Tan *et al.* (2003) adapted this questionnaire, while Fayers *et al.* (2000) provide a similar questionnaire and document the variability between the elicited responses.

    Chaloner and Rhame (2001) provide a copy of the questionnaire they used to elicit opinions from 58 practising HIV clinicians concerning the baseline event rates and the potential benefit of two prophylactic treatments. This asks the minimum information comprising a point estimate and an estimated 95% interval. They used both post and telephone to carry out the elicitations.

4. *Computer-based elicitation.* Chaloner *et al.* (1993) provide a detailed case study of the use of a rather complex computer program that interactively elicited distributions from five clinicians for a trial of prophylactic therapy in AIDS. Kadane (1996) reports the results of an hour-long telephone interview with each of five clinicians, using software to estimate prior parameters from the results of a series of questions eliciting predictive probability distributions for responses of various patient types. When a second round of elicitation became necessary, the proposal was met by 'little enthusiasm'. Kadane and Wolfson (1996) provide an edited transcript of a computerised elicitation session in a non-trial context.

We agree with Chaloner (1996) that extremely detailed elicitation methods have not yet been shown to have any advantage over simple methods. However, it is feasible that complex policy problems, which necessarily may require substantial subjective input, would justify a more sophisticated approach. In any case, Chaloner and Rhame (2001) 'recommend documenting prior beliefs irrespective of whether a Bayesian or frequentist approach is taken to data analysis and formal statistical monitoring'.

5.2.3   Elicitation from multiple experts

Faced with varying prior distributions elicited from multiple experts, we could adopt one of a number of alternative strategies.

- *Elicit a consensus.* If the aim is to produce a single assessment expressing the belief of the group as a whole, then a range of techniques exist for bringing diverse opinions into consensus, including both informal and more formal Delphi-like methods. Care must of course be taken to avoid influence of dominant individuals.

- *Calculate a 'pooled' prior.* The choice of a method for pooling K multiple opinions is not clear cut, and Genest and Zidek (1986) provide a detailed annotated review of the issues. *Arithmetic pooling* simply takes the average of the height of the prior distributions for each parameter value θ, so that $p(\theta) = \sum_k p_k(\theta)/K$. This has the property that pooled probabilities for any event, such as tail areas, are also averages of the individually assessed tail areas. An alternative is *logarithmic pooling*, which takes the average of the logarithms of the density, equivalent to using a geometric mean of the original densities, so $p(\theta) \propto [\prod_k p_k(\theta)]^{1/K}$. This has the apparently attractive property that the same pooled posterior distribution is achieved, whether the pooling is done before or after the common likelihood is taken into account. With both proposals there is an opportunity to apply unequal weights to experts, dependent on their experience or past predictive ability. A further development is that of the *supra-Bayesian*, which takes the expressed opinions as data to manipulate using a statistical model.

- *Retain the individual priors.* The diversity of opinion might be just as important as the 'average' opinion, in that we may be interested in whether current evidence is sufficient to convince a full range of observers as to the benefits of a treatment, and hence to bring them into consensus. The extremes of opinion can be thought of as marking out the boundaries of the 'community of priors' mentioned in Section 5.1.

Our preference is to take a simple supra-Bayesian view, and treat the expressed heights of the prior distributions as data. Then, if we wish to assess the view of an 'average, well-informed participating clinician', it seems reasonable to simply use arithmetic pooling as in Example 5.1. Of course, we should not necessarily assume we have a random sample of clinicians, and so our estimate may be inevitably 'biased'.

Example 5.1   *CHART: Eliciting subjective judgements before a trial*

References: Parmar *et al.* (1994, 2001) and Spiegelhalter *et al.* (1994).

Intervention: In 1986 a new radiotherapy technique known as continuous hyperfractionated accelerated radio therapy (CHART) was introduced. The idea behind it was to give radiotherapy continuously (no weekend breaks), in many small fractions (three a day) and accelerated (the course completed in 12 days). There are clearly considerable logistical problems in efficiently delivering CHART.

Aim of studies: Promising non-randomised and pilot studies led the UK Medical Research Council to instigate two large randomised trials to compare CHART with conventional radiotherapy in both non-small-cell lung and head-and-neck cancer, and in particular to assess whether CHART provides a clinically important difference in survival that compensates for any additional toxicity and problems of delivering the treatment.

Study design: The trials began in 1990, randomised in the proportion 60:40 in favour of CHART, with planned annual meetings of the data monitoring committee (DMC) to review efficacy and toxicity data. No formal stopping procedure was specified in the protocol.

Outcome measure: Full data were to become available on survival (lung) or disease-free survival (head-and-neck), with results presented in terms of estimates of the hazard ratio, h, defined as the ratio of the hazard under CHART to the hazard under standard treatment. Hence, hazard ratios less than one indicate superiority of CHART.

Planned sample sizes: Lung: 600 patients were to be entered, with 470 expected deaths, with 90% power to detect at the 5% level a 10% improvement (15% to 25% survival). Using the methods described in Section 2.4.2, this can be seen to be equivalent to an alternative hypothesis of $h_A = \log(0.25)/\log(0.15) = 0.73$. Head-and-neck: 500 patients were to be entered, with 220 expected recurrences, with 90% power to detect at the 5% level a 15% improvement (45% to 60% disease-free survival), equivalent to an alternative hypothesis of $h_A = \log(0.60)/\log(0.45) = 0.64$.

Statistical model: Proportional hazards model, providing an approximate normal likelihood (Section 2.4.2) for the log(hazard ratio), $\delta = \log(h)$,

$$y_m \sim N\left[\theta, \frac{\sigma^2}{m}\right],$$

where y_m is the estimated log(hazard ratio), $\sigma = 2$ and m is the 'equivalent number of events' in a trial balanced in recruitment and follow-up.

Prospective analysis?: Yes, the prior elicitations were conducted before the start of the trials, and the Bayesian results presented to the DMC at each of their meetings.

Prior distribution: Although the participating clinicians were enthusiastic about CHART, there was considerable scepticism expressed by oncologists who declined to participate in the trial. Eleven opinions were elicited for the lung cancer trial and nine for the head-and-neck. The questionnaire used is described in detail in Parmar *et al.* (1994) and summarised in Figure 5.1.

	CHART worse than standard by %			CHART worse than standard by %						
	10 –15	5 –10	0 – 5	0 – 5	5 – 10	10 – 15	15 – 20	20 – 25	25+	TOTAL
Lung Study Your Entry										100
Head & Neck Study Your Entry										100
Hypothetical example	0	20	20	20	0	0	20	20	0	100

Figure 5.1   Part of the questionnaire used to elicit clinical opinions before the CHART trials. Participants were invited to distribute 100 points between the bins, indicating their 'weight of belief' in the true benefit from CHART. They were reminded to ignore the role of sampling variability – the hypothetical example was deliberately chosen to be a 'rather eccentric' radiotherapist so as not to provide an example that might inappropriately 'anchor' their opinions.

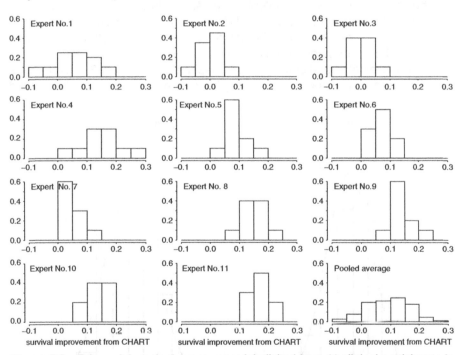

Figure 5.2   Prior opinions for lung cancer trial elicited from 11 clinical participants in the trial. The arithmetic average is used as the 'pooled' distribution.

Figure 5.2 shows the eleven lung cancer opinions as histograms. Note that subjects 7 and 11 have very different opinions and could be taken as extremes for a 'community' of priors. Here we use the arithmetic average of the distributions as a summary, since we wish to represent an 'average' clinician. The prior distribution expressed a median antici-pated 2-year survival benefit of 10%, and a 10% chance that CHART would offer no survival benefit at all. The histogram was then transformed to a log (hazard ratio) scale assuming a 15% baseline survival: for example, the 'bin' of the histogram with range 5% to 10% was transformed to one with upper limit $\log[\log(0.20)/\log(0.15)] = -0.16$ and lower limit $\log[\log(0.25)/\log(0.15)] = -0.31$. This subjective prior distribution had a mean of -0.28 and standard deviation of 0.232 (corresponding to an estimated hazard ratio of 0.76 with 95% interval from 0.48 to 1.19). A normal $N[\mu, \sigma^2/n_0]$ distribution with these characteristics was fitted, with $\mu = -0.28$, $\sigma = 2$, $\sigma/\sqrt{n_0} = 0.23$, which implies $n_0 = 74.3$. From Section 2.4.2, this prior could also be thought of as a posterior having observed a log-rank statistic $(L = O - E)$ such that $4L/n_0 = -0.28$, and so $L = -5.5$. The expected E under the null hypothesis is $n_0/2 = 37.2$ and so the observed O under CHART is $37.2 - 5.5 = 31.7$. Thus the prior can be interpreted as being approximately equivalent to a balanced 'imaginary' trial in which 74 deaths had occurred (32 under CHART, 42 under standard).

For the head-and-neck trial, the fitted prior mean log(hazard ratio) is $\mu = -0.33$ with standard deviation 0.26, equivalent to $n_0 = 61.0$.

The clinical prior distributions are displayed in Figure 5.3, which shows the average transformed onto a log(hazard-ratio) scale for both lung and

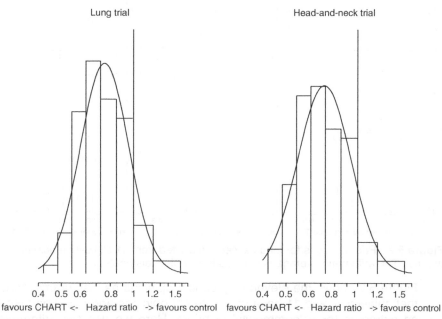

Lung trial   Head-and-neck trial

0.4 0.5 0.6   0.8   1 1.2 1.5   0.4   0.5 0.6 0.8   1   1.2 1.5
favours CHART <- Hazard ratio -> favours control   favours CHART <- Hazard ratio -> favours control

Figure 5.3 Average opinion for lung cancer and head-and-neck CHART trials with normal distributions fitted with matching mean and variance.

head-and-neck trials. The fit of the normal distribution is quite reasonable, and the similarity between the two sets of opinions is clear, each supporting around a 25% reduction in hazard, but associated with considerable uncertainty.

5.3   CRITIQUE OF PRIOR ELICITATION

There have been many criticisms of the process of eliciting subjective prior distributions in the context of health-care evaluation, and claims include the following:

1. *Subjects are biased in their opinions.* Gilbert *et al.* (1977) state that 'innovations brought to the stage of randomised trials are usually expected by the innovators to be sure winners', while the very fact that clinicians are participating in a trial is likely to suggest they expect the new therapy to be of benefit (Hughes, 1991) – we shall see that this appears to be borne out in the results to be shown in Table 5.3. Altman (1994) warns that investigators may even begin to exaggerate their prior beliefs in order to make their prospective trial appear more attractive (although we could claim this already happens both in public and industry-funded studies). Fisher (1996) believes the effort put into elicitation is misplaced, since the measured beliefs are likely to be based more on emotion than on scientific evidence.
2. *The choice of subject biases results.* The biases discussed in Section 5.2 mean that the choice of subject for elicitation is likely to influence the results. If we wish to know the distribution of opinions among well-informed clinicians, then trial investigators are not a random sample and may give biased conclusions. Fayers *et al.* (2000) provide a detailed case study in which there is clear over-optimism of investigators (see Example 6.4). Lewis (1994) says statisticians reviewing the literature may well provide much better prior distributions than clinicians, while Chalmers (1997) suggests even lay people are biased towards believing new therapies will be advances, and therefore we need empirical evidence on which to base the prior probability of superiority. Pocock (1994) states that the 'hardened sceptical trialist, the hopeful clinician and the optimistic pharmaceutical company will inevitably have grossly different priors'. An extreme view is that uncertainty as to whose prior to use militates against any use of Bayesian methods (Fisher, 1996).
3. *Timing of elicitation has an influence.* Senn (1997a) objects to any retrospective elicitation of priors as 'present remembrance of priors past is not the same as a true prior', while Hughes (1991) points out that opinions are likely to be biased by what evidence has recently been presented and by whom.

These concerns have led to a call for the evidential basis for priors to be made explicit, and for effort to go into identifying reasons for disagreement and attempting to resolve these (Fisher, 1996). Even advocates of Bayesian methods have suggested that the biases in clinical priors suggest more attention should be paid to empirical evidence from past trials, possibly represented as priors expressing a degree of scepticism concerning large effects: Fayers (1994) asks, given the long experience of negative trials, 'should we not be using priors strongly centred around 0, irrespective of initial opinions, beliefs and hopes of clinicians?'. Our view is similar: elicited priors from investigators show predictable positive bias and should be supplemented, if not replaced, by priors that are either based on evidence or reflect archetypal views of 'scepticism' or 'enthusiasm'. Taking context into account (Section 3.1) means that it is quite reasonable to allow for differing perspectives, and in many cases substantial effort in careful elicitation from representative clinicians may not be worthwhile.

5.4 SUMMARY OF EXTERNAL EVIDENCE*

If the results of previous similar studies are available, it is clear they may be used as the basis for a prior distribution. Suppose, for example, we have historical data $y_1, \ldots, y_H$ each assumed to have a normal likelihood

$$y_h \sim \mathrm{N}[\theta_h, \sigma_h^2],$$

where each of these estimates could itself be based on a pooled set of studies. Numerous options are available for specifying the relationship between $\theta_h, h = 1, \ldots, H$, and θ, the parameter of interest, and we shall expand on the list given in Section 3.16. Each option is represented graphically in Figure 5.4 using a similar convention to that in Section 3.19.3: these approaches for handling historical data are also considered when considering historical controls in randomised trials (Section 6.9), modelling the potential biases in observational studies (Section 7.3), and in pooling data from many sources in an evidence synthesis (Section 8.2).

(a) *Irrelevance.* Each θ_h is of no relevance to θ, and the prior will need to be formulated without reference to previous studies.
(b) *Exchangeable.* We might be willing to assume $\theta_h, h = 1, \ldots, H$, and θ are exchangeable so that, for example,

$$\theta_h, \theta \sim \mathrm{N}[\mu, \tau^2].$$

This leads to a direct use of a meta-analysis of many previous studies.

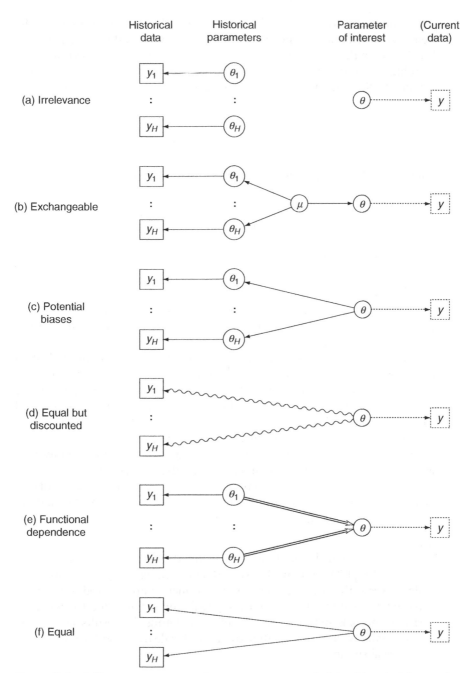

Figure 5.4   Different assumptions relating parameters underlying historical data to the parameter of current interest: single arrows represent a distribution, double arrows represent logical functions, and wavy arrows represent discounting.

It is important to note that the appropriate prior distribution for θ is the predictive distribution of the effect θ in a new study, and not the posterior distribution of the 'average' effect μ. In particular, assuming τ is known and adopting a uniform prior for μ before the historical studies, we have from Section 3.18.2 that the posterior distribution for μ given the historical studies is

$$\mu \mid y_1, \ldots, y_H \sim \text{N}\left[\frac{\sum_h y_h w_h}{\sum_h w_h}, \frac{1}{\sum_h w_h}\right],$$

where $w_h = 1/(\sigma_h^2 + \tau^2)$. Hence the prior distribution for θ is

$$\theta \mid y_1, \ldots, y_H \sim \text{N}\left[\frac{\sum_h y_h w_h}{\sum_h w_h}, \frac{1}{\sum_h w_h} + \tau^2\right].$$

If there is just a single historical study h, then

$$\theta \mid y_h \sim \text{N}[y_h, \ 2\tau^2 + \sigma_h^2].$$

In general τ will be unknown and need to be estimated, although with few historical studies it will need to be assumed known or be given an informative prior distribution.

Exchangeability is quite a strong assumption, but if this is reasonable then it is possible to use databases to provide prior distributions (Gilbert *et al.*, 1977). Lau *et al.* (1995) point out that cumulative meta-analysis can be given a Bayesian interpretation in which the prior for each trial is obtained from the meta-analysis of preceding studies, while DerSimonian (1996) derives priors for a trial of the effectiveness of calcium supplementation in the prevention of pre-eclampsia in pregnant women by a meta-analysis of previous trials using both random-effects and fixed-effects models.

(c) *Potential biases.* We could assume that θ_h, $h = 1, \ldots, H$, are functions of θ. A common choice is the existence of a bias δ_h so that $\theta_h = \theta + \delta_h$. Possibilities then include making the following assumptions:

1. δ_h is known.
2. δ_h has a known distribution with mean 0, say $\delta_h \sim \text{N}(0, \sigma_{\delta h}^2)$, and so $\theta_h \sim \text{N}(\theta, \sigma_{\delta h}^2)$. This is now almost identical to the exchangeability assumption, except that the previous study parameters are centred around the parameter of interest θ and not the population mean μ and the potential site of the bias may be study-specific. Adapting the results for the exchangeability case reveals that the posterior distribution for θ given the historical studies is

$$\theta \mid y_1, \ldots, y_H \sim \text{N}\left[\frac{\sum_h y_h w_h'}{\sum_h w_h'}, \frac{1}{\sum_h w_h'}\right],$$

where $w'_h = 1/(\sigma_h^2 + \sigma_{\delta h}^2)$, which follows by noting the predictive distribution $y_h \sim N[\theta, \sigma_h^2 + \sigma_{\delta h}^2]$. If there is just a single historical study h, then

$$\theta | y_h \sim N[y_h, \sigma_h^2 + \sigma_{\delta h}^2];$$

again, with only one historical study σ_δ^2 will need to be assumed known or have a strong prior distribution.

3. If we suspect systematic bias in one direction, we might take δ_h to have a known distribution with non-zero mean, say $\delta_h \sim N[\mu_\delta, \sigma_{\delta h}^2]$. We then obtain a prior distribution, for a single historical study,

$$\theta \sim N[y_h + \mu_\delta, \sigma_h^2 + \sigma_{\delta h}^2].$$

(d) *Equal but discounted.* Previous studies may not be directly related to the one in question, and we may wish to discount their influence: for example, in the context of control groups, Kass and Greenhouse (1989) state that 'we wish to use this information, but we do not wish to use it as if the historical controls were simply a previous sample from the same population as the experimental controls'. Ibrahim and Chen (2000) suggest the 'power' prior, in which we assume $\theta_h = \theta$, but discount the historical evidence by taking its likelihood $p(y_h|\theta_h)$ to a power α. For normal historical likelihoods this corresponds to adopting a prior distribution for θ, given the historical studies, of

$$\theta | y_1, \ldots, y_H \sim N\left[\frac{\sum_h y_h w''_h}{\sum_h w''_h}, \frac{1}{\alpha \sum_h w''_h}\right]$$

where $w''_h = 1/\sigma_h^2$; α varies between 0 (totally discount past evidence) to 1 (include past evidence in its totality and at 'face value'). If there is just a single historical study h, then

$$\theta | y_h \sim N[y_h, \sigma_h^2/\alpha].$$

For example, Greenhouse and Wasserman (1995) downweight a previous trial with 176 subjects to be equivalent to only 10 subjects, and Tan *et al.* (2002) take $\alpha = 0.25$ in basing a prior on a previous phase III study; see Example 5.2 for a detailed illustration of using such a 'power' prior. We note, however, that Eddy *et al.* (1992) are very strong in their criticism of this method, claiming it has no operational interpretation and hence no means of assessing a suitable value for α.

(e) *Functional dependence.* It is possible that the parameter of interest may be logically expressed as a function of parameters from historical studies. For example, suppose θ_1 were the treatment effect in men derived from a male-only study, and θ_2 were the treatment effect in women derived from a female-only study. Then the expected treatment effect in a study to be carried out in a population with proportion p males would be

$$\theta = p\theta_1 + (1 - p)\theta_2,$$

and a prior for θ could be derived from evidence on θ_1 and θ_2.

(f) *Equal*. This assumes the past studies have all been measuring identical parameters: if θ is a property of a single patient group rather than a treatment effect, this assumption is essentially equivalent to direct pooling of the past data with those in the current study, and hence is based on the very strong assumption of exchangeability of individual patients. In our normal model we would assume $\theta_h = \theta$ and individuals are exchangeable, and so completely pool the data to obtain a prior

$$\theta | y_1, \ldots, y_H \sim N\left[\frac{\sum_h y_h w_h''}{\sum_h w_h''}, \frac{1}{\sum_h w_h''}\right]$$

where $w_h'' = 1/\sigma_h^2$. If there is just a single historical study h, then

$$\theta | y_h \sim N[y_h, \sigma_h^2].$$

Such a strong assumption may be more acceptable if a prior is to be used in the design and not the analysis, and Brown *et al.* (1987) provide such an example using data from a pilot trial.

We note that, for the Normal model, exchangeability (b), bias (c) and discounting (d) could under certain circumstances all lead to the same prior distribution for θ, provided there is only one historical study. If there are multiple studies then these three approaches will generally all lead to different priors for θ.

Various combinations of these techniques are possible. For example, Berry and Stangl (1996a) assume a fixed probability p that each historical patient is exchangeable with those in the current study, *i.e.* either option (f) (complete pooling) with probability p, or option (a) (complete irrelevance) with probability $1 - p$. Example 9.3 illustrates the combination of an exchangeable and a bias model: a past parameter θ_h is assumed to have distribution $\theta_h \sim N[\mu + \delta_h, \tau^2]$, where the additional bias term has distribution $\delta_h \sim N(0, \sigma_{\delta h}^2)$. Hence the overall likelihood contribution from the past study is $\theta_h \sim N[\mu, \tau^2 + \sigma_{\delta h}^2]$; the variance can also be expressed as τ^2/q_h, where $q_h = \tau^2/(\tau^2 + \sigma_{\delta h}^2)$ can be considered as a 'quality weight' of the past study. Values of q_h near 1 mean little bias, near 0 mean substantial bias. This model formally justifies the use of 'quality-weights' in random-effects meta-analysis.

Example 5.2   *GUSTO: Using previous results as a basis for prior opinion*

References: Brophy and Joseph (1995), Fryback *et al.* (2001b), Harrell and Shih (2001), Brophy and Joseph (2000) and Ibrahim and Chen (2000).

Intervention: Streptokinase (SK) compared to tissue plasminogen activator (tPA) to dissolve clots in occluded coronary arteries following a myocardial infarction. tPA is considerably more expensive than SK.

Aim of study: Two previous trials of SK versus tPA (GISSI-2 and ISIS-3) showed minimal difference, although the stroke rate was consistently higher under tPA.

Study design: Parallel-group unblinded RCT, with two SK arms with different administrations of heparin (later pooled), tPA arm and an arm with both SK and tPA (ignored in this analysis).

Outcome measure: Odds ratio (OR) of stroke and/or death, with OR < 1 favouring tPA.

Planned sample size: The sample size of the GUSTO trial was calculated on the basis of having 80% power to detect a 15% relative reduction in the risk of death or a 1% absolute decrease at the 5% significance level.

Statistical model: A normal likelihood was assumed based on the estimated log(odds ratio) (Section 2.4.1); σ has been taken as 2.

Prospective analysis?: No.

Prior distribution: It is natural to base, to some extent, a prior distribution on the two preceding trials, whose results are shown in Table 5.1, using data presented by Brophy and Joseph (1995). Taking the previous trials at full weight, the pooled previous trials give rise to a prior for GUSTO with mean 0.0002 and standard deviation $\sigma/\sqrt{4604} = 0.03$: a very sceptical prior indeed, with a 95% interval for the OR from 0.94 to 1.06.

Table 5.1   Historical and observed data for GUSTO study. The *m*s are the 'effective number of events' in a balanced trial, obtained from setting the estimated variances of the log(odds ratios) to σ^2/m: the *m*s do not exactly match the actual number of events, particularly in GUSTO, due to imbalance in allocation. The 'pooled' results are obtained by adding the *m*s and weighting the log(odds ratios) by their respective *m*s: this pooled *m* can be relabelled n_0 if it is used as the basis for a prior distribution for GUSTO.

Trial	SK events/cases	%	tPA events/cases	%	OR	log(OR)	m (when $\sigma = 2$)
GISSI-2	985/10 396	9.5%	1067/10 372	10.3%	1.09	0.09	1847
ISIS-3	1596/13 780	11.6%	1513/13 746	11.0%	0.94	−0.06	2757
Pooled						0.0002	$n_0 = 4604$
GUSTO	1574/20 173	7.8%	714/10 343	6.9%	0.88	−0.13	1825

However, Brophy and Joseph (2000) emphasise important differences between the studies: the GUSTO study featured an 'accelerated' tPA protocol, more aggressive use of intravenous heparin, increased revascularisation in the tPA arm, and possible increased tPA benefit in US patients. This suggests downweighting the prior evidence in some way, and different authors have subsequently used almost all the approaches outlined in Section 5.4. We shall focus on simple discounting (method (d)), but other methods are mentioned under 'Comments'. Brophy and Joseph (1995) 'discounted' the previous trials, essentially implementing the power prior distributions of Ibrahim and Chen (2000), which is equivalent to adjusting the prior 'number of events' from n_0 to αn_0. They considered α to be 0, 0.1, 0.5 and 1.0, equivalent to taking the prior 'number of events' to be 0, 460.4, 2302 and 4604. Taking $\alpha = 0$ is equivalent to treating the previous trials as irrelevant (option (a)) and hence selecting a uniform prior on the log(odds ratio), while taking $\alpha = 1$ is equivalent to assuming the trials are measuring equal parameters (option (f)) – note that this is not equivalent to pooling the patients on each arm, but is equivalent to pooling the estimated treatment effects.

Loss function or demands: The GUSTO trial was designed around a 15% reduction in mortality, so we might take an odds ratio of 0.85 to reflect a clinically important difference.

Computation/software: Conjugate normal model.

Evidence from study: This is provided in Table 5.1. The standardised test statistic based on the data alone is $z_m = y_m\sqrt{m}/\sigma = -0.13\sqrt{1825}/2 = -2.78$, providing a two-sided P-value of 0.005.

Bayesian interpretation: Figure 5.5 shows plots of prior, likelihood and posterior under different assumptions concerning α, superimposed on a clinically important difference of 0.85. The probability that tPA is inferior to SK is very low unless the prior trials are considered at almost full weight. However, it is clear that although GUSTO may show 'statistical significance' in that the posterior probability that OR < 1 is high, there is not strong evidence of 'practical significance', in that the posterior probability that OR < 0.85 is moderate even when the prior evidence is totally ignored.

Sensitivity analysis: Figure 5.6 shows changing conclusions as α ranges from 0 (ignore historical evidence) to 1 (completely pool with historical evidence). This clearly shows evidence for benefit unless the past data are quite strongly weighted, but even slight inclusion of past data serves to exclude a clinically important difference of 15%.

Comments: We can fit previous approaches to this problem within the structure outlined in Section 5.4.

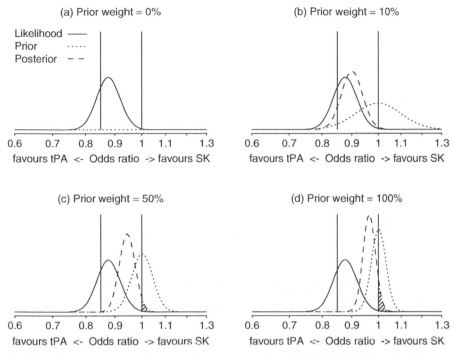

Figure 5.5 Posterior estimate of the odds ratio for the GUSTO trial under different prior assumptions: weighting the previous trial results by a factor (a) 0% (*i.e.* the reference prior in which the posterior is proportional to the likelihood), (b) 10%, (c) 50% and (d) 100% (*i.e.* full pooling with the past data). The shaded area represents the posterior probability that OR > 1 and hence favours SK, and is very low unless very high weight is given to the previous trials. However, the chance of an odds ratio less than 0.85 is only moderate even when using the trial data alone, and drops severely for even 10% weighting of the past trial data.

(a) *Irrelevance.* Harrell and Shih (2001) consider that the previous trials are entirely irrelevant to GUSTO due to the revised tPA protocol, and so only consider a 'reference' and 'sceptical' prior (Section 5.5): the reference prior is uniform on the log(OR) scale and hence the posterior distribution is the same shape as the likelihood, while the sceptical prior was centred on the null hypothesis of OR = 1, and expressed 95% belief that the true OR lay within the bounds 0.75–1.33, *i.e.* it is unlikely that there is more than a 25% relative change between the treatments: this prior is even more diffuse than that shown in Figure 5.5(b).

(b) *Exchangeable.* One of the models considered by Brophy and Joseph (2000) assumes the treatment effects in the three trials are exchangeable, and places a normal population distribution on the three log(odds ratios) – they use 'diffuse' priors on the parameters of mean and variance of the normal population. However, both the exchangeability

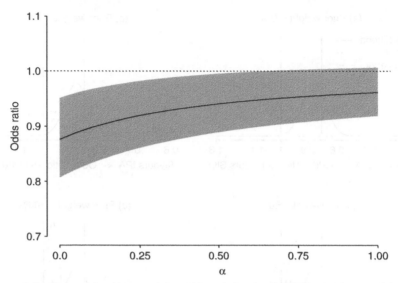

Figure 5.6   Posterior estimate of the odds ratio for the GUSTO trial downweighting previous trial results by varying amounts ($\alpha = 0$ implies total discounting, whilst $\alpha = 1$ implies acceptance of previous evidence at 'face-value').

assumption, and the attempt to estimate population parameters from just three trials (regardless of their size), make this prior formulation somewhat doubtful.

(c) *Potential biases*. Acknowledging the possible systematic differences between the trials, Brophy and Joseph (2000) also consider two possible sources of bias: differences in revascularisation rates in GUSTO, and differences in tPA administration between GUSTO and the previous trials. These are applied to the hierarchical model described under (b).

(d) *Equal but discounted*. In a different application of the discounting approach, Fryback *et al.* (2001b) suggests the SK arm in GUSTO is reasonably compatible with the SK arm in previous trials, and so adopt $\alpha_C = 1/3$ for SK. However, they severely discount the tPA arm from a sample size of around 24 000 to one of 50, so that $\alpha_T \approx 1/500$ for tPA.

Now $V(\log (OR)) = V(\log O_C) + V(\log O_T)$, where O_C, O_T are the odds on death under SK and tPA, respectively. With no discounting, $V(\log O_C) \approx V(\log O_T) = V$. With differential discounting,

$$V(\log (OR)) = \frac{V(\log O_C)}{\alpha_C} + \frac{V(\log O_T)}{\alpha_T} \approx V\left(\frac{1}{\alpha_C} + \frac{1}{\alpha_T}\right).$$

Thus the overall discount factor, relative to the undiscounted variance of 2V, is $\alpha = 2/(\alpha_C^{-1} + \alpha_T^{-1})$ which is the 'harmonic mean' of the individual discounts. Fryback *et al.*'s assumptions therefore lead to an overall discount factor of $2/(3 + 500) \approx 1/250$, which means the prior will have little impact on the likelihood.

(f) *Equal.* As an extreme of the discounting procedure, if we assume $\alpha = 1$ we are led to completely pool the results of the three trials.

5.5   DEFAULT PRIORS

It would clearly be attractive to have prior distributions that could be taken 'off the shelf', rather than having to consider all available evidence external to the study in their construction: such priors can, at a minimum, be considered as 'baselines' against which to measure the impact of past evidence or subjective opinion. Four main suggestions can be identified.

5.5.1   'Non-informative' or 'reference' priors

There has been a huge volume of research into so-called *non-informative* or *reference* priors, that are intended to provide a kind of default or 'objective' Bayesian analysis free from subjectivity. Kass and Wasserman (1996) review the literature, but emphasise the continuing difficulties in defining what is meant by 'non-informative', and the lack of agreed reference priors in all but simple situations.

In many situations we might adopt a uniform distribution over the range of interest, possibly on a suitably transformed scale of the parameter (Box and Tiao, 1973). Formally, a uniform distribution means the posterior distribution has the same shape as the likelihood function, which in turn means that the resulting Bayesian intervals and estimates will essentially match the traditional results. Results with reference priors are generally quoted as one part of a Bayesian analysis, and may even form the main basis for inferences. For example, Burton (1994) suggests that most doctors interpret frequentist confidence intervals as credible intervals, and also that information external to a study tends to be vague, and that therefore results from a study should be presented by performing a Bayesian analysis with a non-informative prior and quoting posterior probabilities for the parameter of interest being in various regions. The fact that a reference prior may produce essentially identical conclusions to a classical analysis, and yet allow more flexible and intuitive presentations, has led to the use of what are essentially Bayesian methods but under names such as 'confidence levels' (Shakespeare *et al.*, 2001).

Invariance arguments may be used as a basis for reference priors (Jeffreys, 1961): for example, if we feel a reference prior on an odds ratio OR should be the same whichever treatment is taken in the numerator of the odds ratio, then it means that the same prior should hold for OR and 1/OR, which means that we must be uniform on the log(OR) scale. Similar arguments can be used to justify a uniform prior on $\log(\sigma^2)$ for a sampling variance σ^2, since this prior is also equivalent to a uniform prior on $\log(\sigma)$ (or indeed any power of σ), and hence is invariant to whether one is working on the standard deviation or variance scale. This prior is equivalent to assuming $p(\sigma^2) \propto \sigma^{-2}$, or $p(\sigma) \propto \sigma^{-1}$. A standard result (DeGroot, 1970; Lee, 1997) is that, for normal likelihoods, this prior, combined with an independent uniform prior on the mean, gives rise to the familiar classical tail areas based on a t distribution.

The real problem with 'uniform' priors is that they are no longer uniform if the parameter is transformed, which is well illustrated by the problem of assigning a reference prior to the probability θ of an event. The classic solution, dating back to Bayes and Laplace in the eighteenth century, is to give a uniform prior for θ, equivalent to a Beta[1,1]. From the beta-binomial distribution (Section 3.13.2) we can show this leads to a uniform distribution over the number 0, 1, ..., n of occurrences in n Bernoulli trials, which might seem a reasonable justification for its claim to be 'non-informative'. However in many of our examples we place a uniform distribution over a log(odds) scale, *i.e.* $\log[p/(1-p)]$ has a uniform distribution. It can be shown that this is equivalent to a Beta[0,0] distribution for p – an improper distribution that strongly favours values of p near 0 or 1. As an intermediate suggestion, invariance arguments (Box and Tiao, 1973) have led to the use of a Beta[0.5,0.5] prior, which is proper but still favours extreme values of p (Section 2.6.3). Of course, all these priors will give essentially the same result with a large enough set of data, but could have some influence with rare events. Even when one has chosen a suitable scale for a uniform prior, it may be inappropriate to term it 'non-informative': Fisher (1996) points out that 'there is no such thing as a "noninformative" prior. Even improper priors give information: all possible values are equally likely'. There is a particular difficulty in assigning such a 'reference' prior to random-effect variances in hierarchical models, and we shall consider this issue in Section 5.7.

5.5.2   'Sceptical' priors

Informative priors that express scepticism about large treatment effects have been put forward both as a reasonable expression of doubt, and as a way of controlling early stopping of trials on the basis of fortuitously positive results (Section 6.6.2). Kass and Greenhouse (1989) suggest that a 'cautious reasonable sceptic will recommend action only on the basis of fairly firm knowledge', but that these sceptical 'beliefs we specify need not be our own, nor need they be the beliefs of any actual person we happen to know, nor derived in some way from any group of "experts"'.

Mathematically speaking, a sceptical prior about a treatment effect will have a mean of zero and a shape chosen to include plausible treatment differences which determine the degree of scepticism. Spiegelhalter *et al.* (1994) argue that a reasonable degree of scepticism may be feeling that the trial has been designed around an alternative hypothesis that is *optimistic*, formalised by a prior with only a small probability γ (say, 5%) that the treatment effect is as large as the alternative hypothesis θ_A (see Figure 5.7).

Assuming a prior distribution $\theta \sim N[0, \ \sigma^2/n_0]$ and such that $p(\theta > \theta_A)$ is a small value γ implies $\gamma = 1 - \Phi(\theta_A\sqrt{n_0}/\sigma)$ and so

$$-\sigma\frac{z_\gamma}{\sqrt{n_0}} = \theta_A, \tag{5.1}$$

where $\Phi(z_\gamma) = \gamma$. Now suppose the trial has been designed with size α and power $1 - \beta$ to detect an alternative hypothesis θ_A. Then we have the standard relation (2.38)

$$\sigma^2\frac{(z_{\alpha/2} + z_\beta)^2}{\theta_A^2} = n \tag{5.2}$$

between the proposed sample size n and θ_A. Equating θ_A in (5.1) and (5.2) gives

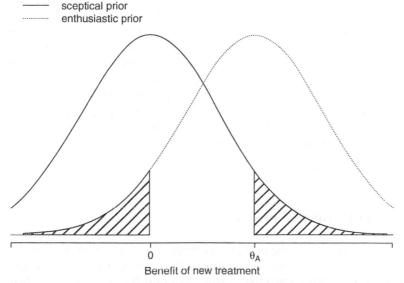

Figure 5.7 Sceptical and enthusiastic priors for a trial with alternative hypothesis θ_A. The sceptics' probability that the true difference is greater than θ_A is γ (shown shaded). This value has also been chosen for the enthusiasts' probability that the true difference is less than 0.

$$\frac{n_0}{n} = \left[\frac{z_\gamma}{z_{\alpha/2} + z_\beta}\right]^2.$$

Reasonable values might be $\alpha = 0.05$, $\beta = 0.1$ and $\gamma = 0.05$, which gives $n_0/n = 0.257$.

Thus in a trial designed with 5% size and 90% power, such a sceptical prior corresponds to adding a 'handicap' equivalent to already having run a 'pseudo-trial' with no observed treatment difference, and which contains around 26% of the proposed sample size.

This approach has been used in a number of case studies (Freedman *et al.*, 1994; Parmar *et al.*, 1994) and has been suggested as a basis for monitoring trials (Section 6.6) and when considering whether or not a confirmatory study is justified (Section 6.7). Other applications of sceptical priors include Fletcher *et al.* (1993), DerSimonian (1996), and Heitjan (1997) in the context of phase II studies, while a senior FDA biostatistician (O'Neill, 1994) has stated that he 'would like to see [sceptical priors] applied in more routine fashion to provide insight into our decision making'.

Example 5.3   *CHART (continued): Sceptical priors*

References: Parmar *et al.* (1994, 2001) and Spiegelhalter *et al.* (1994).

Prior distribution: A *sceptical prior* was derived using the ideas in Section
   5.5.2: the prior mean is 0 and the precision is such that the prior
   probability that the true benefit exceeds the alternative hypothesis is
   low (5% in this case). Thus a prior with mean 0 and standard deviation
   $\sigma/\sqrt{n_0}$ will show a 5% chance of being less than δ_A if $n_0 = (1.65\sigma/\theta_A)^2$ by
   (5.1). For the lung trial, the alternative hypothesis on the log(hazard
   ratio) scale is $\theta_A = \log(0.73) = -0.31$. Assuming $\sigma = 2$ gives $n_0 = 110$.
   For the head-and-neck trial, the alternative hypothesis is
   $\theta_A = \log(0.64) = -0.45$, which gives a sceptical prior with $n_0 = 54$.
      The sceptical prior distributions are displayed in Figure 5.8, with the
   clinical priors derived in Example 5.1.

5.5.3   'Enthusiastic' priors

As a counterbalance to the pessimism expressed by the sceptical prior, Spiegel-halter *et al.* (1994) suggest an 'enthusiastic' prior centred on the alternative hypothesis and with a low chance (say, 5%) that the true treatment benefit is negative. Use of such a prior has been reported in case studies (Freedman *et al.*, 1994; Heitjan, 1997; Vail *et al.*, 2001; Tan *et al.*, 2002) and as a basis for conservatism in the face of early negative results (Fayers *et al.*, 1997); see

Lung trial   Head-and-neck trial

Clinical prior	——		Clinical prior	——
CHART superior survival	0.857		CHART superior survival	0.891
Control superior survival	0.143		Control superior survival	0.109
Sceptical prior	·········		Sceptical prior	·········
CHART superior survival	0.5		CHART superior survival	0.5
Control superior survival	0.5		Control superior survival	0.5

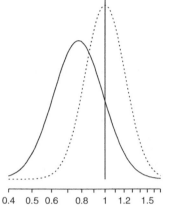

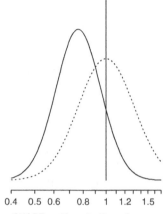

favours CHART <- Hazard ratio -> favours control   favours CHART <- Hazard ratio -> favours control

Figure 5.8 Sceptical and clinical priors for both lung and head-and-neck CHART trials, showing prior probabilities that CHART has superior survival. The sceptical priors express a 5% prior probability that the true benefit will be more extreme than the alternative hypotheses of HR = 0.73 for the lung trial and HR = 0.64 for the head-and-neck trial.

Section 6.6.2. Dignam *et al.* (1998) provide an example of such a prior but call it 'optimistic' (Example 6.7). Such a prior is intended to represent the opinion of an archetypal enthusiast and does not represent the opinion of an identifiable individual.

Other options for default priors are possible: for example, Cronin *et al.* (1999) adopt an 'indifference' prior that lies half-way between 'sceptical' and 'enthusiastic'.

5.5.4 Priors with a point mass at the null hypothesis ('lump-and-smear' priors)*

The traditional statistical approach expresses a qualitative distinction between the role of a null hypothesis, generally of no treatment effect, and alternative hypotheses. A prior distribution that retains this distinction would place a 'lump' of probability on the null hypothesis, and 'smear' the remaining probability over the whole range of alternatives; for example Cornfield (1969) uses a normal distribution centred on the null hypothesis, while Hughes (1993) uses a uniform prior over a suitably restricted range. The resulting posterior distribution retains this structure, giving rise to a posterior probability of the truth of the null hypothesis; this is apparently analogous to a P-value but is neither numerically nor conceptually equivalent.

A specific assumption used in our examples is the following:

$$H_0 : \theta = \theta_0 \quad \text{with probability } p,$$

$$H_A : \theta \sim N\left[\theta_0, \frac{\sigma^2}{n_0}\right] \text{ with probability } 1 - p,$$

where we label the 'lump' and the 'smear' as null and alternative hypotheses, respectively.

Cornfield repeatedly argued for this approach, which naturally gives rise to the 'relative betting odds' or Bayes factor (Section 3.3) as a sequential monitoring tool, defined as the ratio of the likelihood of the data under the null hypothesis to the average likelihood (with respect to the prior) under the alternative. If we assume a normal likelihood $y_m \sim N[\theta, \sigma^2/m]$, then we have shown in Section 4.4.3 that the Bayes factor is

$$\text{BF} = \frac{p(y_m|H_0)}{p(y_m|H_A)} = \sqrt{1 + \frac{m}{n_0}} \, \exp\left[\frac{-z_m^2}{2(1 + n_0/m)}\right]. \tag{5.3}$$

Since

$$\frac{p(H_0|y_m)}{p(H_A|y_m)} = \text{BF} \frac{p}{1 - p},$$

we can obtain the posterior probability $p(H_0|y_m)$.

The relative betting odds are independent of the 'lump' of prior probability placed on the null (while depending on the shape of the 'smear' over the alternatives), and do not suffer from the problem of 'sampling to a foregone conclusion' (Section 6.6.5). Cornfield suggests a 'default' prior under the alternative as a normal distribution centred on the null hypothesis and with expectation (conditional on the effect being positive) equal to the alternative hypothesis θ_A. Then from the properties of the half-normal distribution (Section 2.6.7) it follows that

$$E(\theta|\theta > 0) = \sqrt{\frac{2\sigma^2}{\pi n_0}}. \tag{5.4}$$

Equating this to θ_A leads to assuming a prior standard deviation under the alternative hypothesis of $\sqrt{\pi/2}\theta_A$. This is similar to the formulation of a sceptical prior described in Section 5.5.2, but with probability of exceeding the alternative hypothesis of $\gamma = \Phi(-\sqrt{2/\pi}) = 0.21$ – this is larger than the value of 5% often used for sceptical priors, but the lump of probability on the null hypothesis is already expressing considerable scepticism. Values for these prior distributions for 11 outcome measures are reported for the Urokinase Pulmonary Embolism Trial (Sasahara *et al.*, 1973, p. 27), and Example 5.4 considers one of these outcomes. This method was used in a number of major studies alongside more standard approaches (Coronary Drug Project Research Group, 1970; University Group Diabetes Program, 1970), although relative betting odds were later dropped from the analysis (Coronary Drug Project Research Group, 1975). A mass of probability on the null hypothesis has also been used in a cancer trial (Freedman and Spiegelhalter, 1992) and for sensitivity analysis in trial reporting (Hughes, 1993).

Although such an analysis provides an explicit probability that the null hypothesis is true, and so appears to answer a question of interest, the prior might be somewhat more realistic were the lump to be placed on a small range of values representing the more plausible null hypothesis of 'no clinically effective difference'. Lachin (1981) has extended the approach to this situation where the null hypothesis forms an interval, although Cornfield (1969) points out that the 'lump' is in any case just a mathematical approximation to such a prior.

Example 5.4 *Urokinase: 'lump and smear' prior distributions*

Reference: Sasahara *et al.* (1973).

Intervention: Urokinase treatment for pulmonary embolism.

Aim of study: To compare thrombolytic capability in urokinase (new) with heparin (standard).

Study design: RCT entering 160 patients between 1968 and 1970. There was no prespecified sample size or stopping rule, although data were examined four times yearly by an advisory committee but not released to the investigators.

Outcome measure: Eleven endpoints based on continuous measures from angiograms, lung scans and haemodynamics.

Statistical model: Normal likelihoods assumed for an estimate y_m of treatment effect θ based on m pairs of randomised patients.

Prospective analysis?: Yes, the prior elicitations were conducted before the start of the trials, and the Bayesian results presented to the advisory committee at each of their meetings.

Prior distribution: A 'lump-and-smear' prior was assessed for each outcome (Section 5.5.4). To select n_0, Cornfield (1969) suggests setting the
expectation, given there is a positive effect, to the alternative hypothesis,
so from (5.4) the prior standard deviation $\sigma/\sqrt{n_0}$ is $\sqrt{\pi/2}\theta_A$, and hence
$n_0 = 2\sigma^2/(\pi\theta_A^2)$. Alternative hypotheses were assessed by members of
the advisory committee 'based on what appeared reasonable from previous experience with thrombolytics'.

For the outcome 'Absolute improvement in resolution on lung scan',
we take σ to be the value observed in the study, 9.35 (see below). The
alternative hypothesis was selected to be $\theta = 8$, slightly less than a
1 standard deviation effect, giving rise to $n_0 = 0.87$. Thus the prior
under the alternative hypothesis is approximately equivalent to having
observed a single pair of patients, each with the same response. This is
a weak prior, but remarkably corresponds almost precisely to that recommended in recent theoretical work on Bayes factors (Kass and Wasserman, 1995); see Section 4.4.3.

Loss function or demands: None specified.

Computation/software: Conjugate normal analysis.

Evidence from study: For 'Absolute improvement in resolution on
24-hour lung scan', outcomes were available on 72 patients treated
with urokinase and 70 with heparin. The difference in mean responses
was $y_m = 3.61$, with standard error 1.11. Assuming $m = 71$ pairs, we
have $\sigma = 1.11\sqrt{m} = 9.35$, as mentioned above. Using (5.3) the 'relative
betting odds' (Bayes factor) can be calculated to be 0.052 – from Table
3.2 this corresponds to 'strong' evidence against the null hypothesis.
Setting $p = 0.5$ to represent equal prior belief in the null and alternative
hypotheses, this leads to a probability $0.052/(1 + 0.052) = 0.049$ that
the null hypothesis is true.

Bayesian interpretation: Figure 5.9 shows the size of the 'lump' dropping
dramatically from its prior level. The result is highly significant classically:
$z = 3.61/1.11 = 3.25$, with a two-sided P-value of 0.001; Sasahara *et al.*
(1973) report that due to many outcome measures and sequential analysis, only $z > 3$ would be taken as 'significant'. Note that the Bayesian
posterior on the null is only 0.047, and so is not as extreme as the
P-value (Section 4.4.3).

Comments: In this application, $m/n_0 = 71/0.87 = 82$; Figure 4.2 shows
that for such results with a classical two-sided P-value of 0.001,
the Bayes factor only provides 'strong' evidence against the null hypothesis. The prior drawn in Figure 5.9(a) provides a clue as to the difference
between the two approaches: although the data observed are unlikely
under the null hypothesis, the prior under the alternative is so diffuse

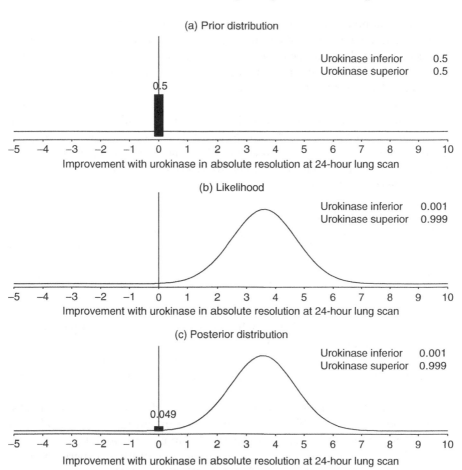

Figure 5.9 Results from the Urokinase trial analysed by Cornfield using 'relative betting odds' (Bayes factors). Data which are classically 'highly significant' ($z = 3.25$, two-sided *P*-value 0.001) only provide 'strong' evidence against the null hypothesis (Bayes factor $\approx 1/20$).

that it gives little weight to the parameter values suggested by the data. Hence the data are not strongly supported by either hypothesis, although the alternative receives the benefit of the doubt.

5.6   SENSITIVITY ANALYSIS AND 'ROBUST' PRIORS

An integral part of any good statistical report is a sensitivity analysis of assumptions concerning the form of the model (the likelihood). Bayesian approaches have the additional concern of sensitivity to the prior distribution, both in view of its controversial nature and because it is by definition a subjective assumption that is open to valid disagreement. We reiterate that this fits naturally into the idea of a 'community of priors' (Kass and Greenhouse, 1989).

A natural development when carrying out a Bayesian post-hoc analysis, rather than a full Bayesian pre-study design, is to avoid all prespecification of priors and simply report the impact of the data on a suitable range of opinion: O'Rourke (1996) emphasises that posterior probabilities 'should be clearly and primarily stressed as being a "function" of the prior probabilities and not *the* probability of treatment effects'. We can therefore take the following steps after having observed the data:

1. Select a suitably flexible class of priors.
2. Examine how the conclusions depend on the choice of prior.
3. Identify the subsets of priors that, if seriously held, would lead to posterior conclusions of specific interest (say, the clinical superiority of an intervention).
4. Report the results and hence allow the audience to judge whether their own prior lies in the identified 'critical' subsets.

This is known as the 'robust' approach, and is also known as 'prior partitioning' (Carlin and Sargent, 1996; Sargent and Carlin, 1996). See Section 6.6.2 for further discussion of this approach to monitoring clinical trials.

Three increasingly complex 'communities' of priors have been considered:

1. *Discrete set.* Many case studies carry out analysis of sensitivity to a limited list of possible priors, possibly embodying scepticism, enthusiasm, clinical opinion and 'ignorance'; see, for example, Examples 6.6 and 6.7. It is also possible to consider sensitivity to the opinions of multiple experts, perhaps summarised by their extremes of opinion (Section 5.2.3).
2. *Parametric family.* If the community of priors can be described by one varying parameter, then it is possible to graphically display the dependence of the main conclusion on that parameter. Hughes (1991) suggested examining sensitivity of conclusions to priors based on previous trial results and that reflecting investigators' opinions, and later Hughes (1993) gives an example which features a point-mass prior on zero, and an explicit plot of the posterior probability against the prior probability of this null hypothesis. Example 5.2 carries out a similar analysis in which the 'discount' parameter is continuously varied, and the 'credibility' analysis described in Section 3.11 provides such a tool for the class of normal sceptical priors.
3. *Non-parametric family.* The 'robust' Bayesian approach has been further explored by allowing the community of priors to be a non-parametric family in the neighbourhood of an initial prior. For example, Gustafson (1989), considers the ECMO study (Example 6.9) with a community centred around a 'non-informative' prior but 20% 'contaminated' with a prior with minimal

restrictions, such as being unimodal. The maximum and minimum posterior probability of the treatment's superiority within such a class can be plotted, providing a sensitivity analysis. A similar approach has also been taken by Greenhouse and Wasserman (1995) and Carlin and Sargent (1996).

One should, however, beware of carrying out too restricted a sensitivity analysis. Stangl and Berry (1998) emphasise the need for a fairly broad community, taking into account not just the spread of the prior but also its location. They also stress that sensitivity to exchangeability and independence assumptions should be examined and that, while sensitivity analysis is important, it should not serve as a substitute for careful thought about the form of the prior distribution.

There is limited experience of reporting such analyses in the medical literature, and it has been suggested (Koch, 1991; Hughes, 1991; Spiegelhalter *et al.*, 1994) that a separate 'interpretation' section is required to display how the data in a study would add to a range of currently held opinions (Section 3.21). It would be attractive for people to be able to carry out their own sensitivity analysis of their own prior opinion; Lehmann and Goodman (2000) describe a computing architecture for this, and available software and web pages are described in Section A.2.

5.7 HIERARCHICAL PRIORS

The essence of hierarchical models was summarised in Section 3.17: by assuming that multiple parameters of interest are drawn from some common prior distribution, *i.e.* they are exchangeable, we can 'borrow strength' between multiple substudies and improve the precision for each parameter. These models form an essential component of much of Bayesian analysis, but their added power does not come without cost. The three essential assumptions are: exchangeability of parameters θ_k, a form for the random-effects distribution of the θ_k, and a 'hyperprior' distribution for the parameters of the random-effects distribution of the θ_k. All these assumptions can be important, and none can be made lightly.

5.7.1 The judgement of exchangeability

An assumption of exchangeability underlies any random-effects analysis, whether Bayesian or classical. Nevertheless, Tukey (1977) says that 'to treat the true improvements for the classes concerned as a sample from a nicely behaved population ... does not seem to me to be near enough the real world to be a satisfactory and trustworthy basis for the careful assessment of strength of evidence'. But, as noted in Section 3.4, there does not need to be any actual population from which units are sampled, and the very fact that we are carrying out simultaneous analysis on a number of units suggests some relationship between them. In addition, if there are known reasons to suspect that specific units are systematically different, then those reasons might be modelled by including relevant covariates and then the residual variability more plausibly reflects exchangeability; for example, Dixon and Simon (1991) discuss the reasonableness of exchangeability assumptions in the context of subset analysis (Section 6.8.1), and observe that any subsets of prior interest should be considered separately.

5.7.2   The form for the random-effects distribution

This is generally taken to be normal until evidence shows otherwise: if there is no reason to suspect systematic difference between units, a central limit theorem argument could be used to justify normality as arising from the sum of many small unobserved differences between units. Normality is computationally helpful, although with the advent of MCMC methods it has less importance, and 'heavier-tailed' distributions such as the Student's t can be adopted (Smith *et al.*, 1995).

Unlike other prior assumptions, the form of the random-effects distribution can be empirically checked from the data, although strategies for this are outside the scope of this book; see, for example, Lange and Ryan (1989), Christiansen and Morris (1996) and Hardy and Thompson (1998).

5.7.3   The prior for the standard deviation of the random effects*

In a hierarchical model $\theta \sim N[\mu,\tau^2]$, the random-effects standard deviation τ plays an important role, and its value can be very influential in assessing the uncertainty concerning μ or in predicting future θs. However, there may be limited information in the data to provide a precise estimate of τ due either to there being few units, or to each unit providing little information, or both. This can make the prior for τ particularly important, and yet neither is there any generally accepted reference prior for τ, nor are there formally established techniques for assessing a subjective prior distribution.

Three strategies have been adopted which broadly follow the ideas for parameters of primary interest described earlier: elicitation (Section 5.2), summary of evidence (Section 5.4), and reference priors (Section 5.5).

Elicitation of opinion. In order to be able to make judgements about their relative plausibility, we need to have a clear interpretation of what different values of τ signify. We can first note that 95% of values of θ will lie in the interval $\mu \pm 1.96\tau$, and hence the 97.5% and 2.5% values of θ are $2 \times 1.96 \times \tau$ apart. θ will often be measured on a logarithmic scale, for example as a log(odds ratio), and hence the ratio of the 97.5% odds ratio to the 2.5% odds ratio is $\exp(3.92\tau)$, roughly representing the 'range' of odds ratios. For example, in the context of meta-analysis, Smith *et al.* (1995) thought that it was unlikely that the between-study odds ratios would vary by more than an order of magnitude, and hence considered $\exp(3.92\tau) = 10$, or $\tau = \log(10)/3.92 = 0.59$ to represent a 'high' value of the standard deviation τ.

An alternative approach is to imagine two randomly chosen θs drawn from the random-effects distribution, whose difference will have distribution $\theta_1 - \theta_2 \sim N[0, 2\tau^2]$ by (2.26). Their absolute difference $|\theta_1 - \theta_2|$ therefore has a normal distribution constrained to be greater than 0, which is a half-normal distribution $HN[2\tau^2]$ (Section 2.6.7). This distribution has median $\Phi^{-1}(0.75) \times \sqrt{2}\tau = 1.09\tau$, which is therefore the median difference between the maximum and minimum of a random pair of θs (Larsen *et al.*, 2000). If θ is, for example, a log(odds ratio), then $\exp(1.09\tau)$ is the median ratio of the maximum to the minimum of any random pair of odds ratios drawn from the distribution.

Table 5.2 illustrates these two interpretations for a range of values of τ when θ represents a log(odds ratio). It is apparent that $\tau = 1$ corresponds to a substantial heterogeneity, with a random pair having a median ratio of 3, for example one trial showing no effect and another showing an odds ratio of 3. $\tau = 2$ means the trials are effectively independent.

Table 5.2 Possible interpretations of τ, the standard deviation of the log(odds ratio) in a hierarchical model $\theta \sim N[\mu, \tau^2]$. The 'range' $\exp(3.92\tau)$ is actually the ratio of the 97.5% to the 2.5% point of the distribution of odds ratios, while $\exp(1.09\tau)$ is the median ratio of the maximum to minimum odds ratio in a random pair of θs drawn from the distribution.

τ	$\exp(3.92\tau)$: 'range' of odds ratios	$\exp(1.09\tau)$: median ratio of random pair
0.0	1.00	1.00
0.1	1.48	1.11
0.2	2.19	1.24
0.3	3.24	1.39
0.4	4.80	1.55
0.5	7.10	1.72
0.6	10.51	1.92
0.7	15.55	2.14
0.8	23.01	2.39
0.9	34.06	2.67
1.0	50.40	2.97
1.5	357.81	5.13
2.0	2540.20	8.84

In conclusion, values of τ from 0.1 to 0.5 may appear reasonable in many contexts, from 0.5 to 1.0 might be considered as fairly high, and above 1.0 would represent fairly extreme heterogeneity.

When assessing a subjective prior distribution for τ, we first need to consider whether $\tau = 0$ is a plausible value, representing no variability between θs. At the other extreme, we should think of an 'upper' value for τ which we shall label τ_u; Table 5.2 may be useful for this. A possible prior distribution is then a half-normal distribution $\text{HN}[(\tau_u/1.96)^2]$ (Pauler and Wakefield, 2000). This will have its mode at 0 and be steadily declining in τ, with an upper 95% point at τ_u. Its median will be $\Phi^{-1}(0.75) \times \tau_u/1.96 = 0.39\tau_u$. This is illustrated in Figure 5.10(a) for $\tau_u = 1$, which may be a reasonable prior in many situations; see Example 8.5.

Summary of evidence. It is natural to construct a prior distribution for τ from an analysis of past hierarchical models in the context being considered, in order to determine reasonable values of τ experienced in practice. Thus we could, for example, study the typical variability between subgroups, between institutions in their clinical performance, or between centres in multi-centre clinical trials. In the field of meta-analysis, Higgins and Whitehead (1996) and Smith *et al.* (1996) both consider empirical distributions of past τs: essentially they are carrying out a meta-analysis of meta-analyses. Higgins and Whitehead (1996) go on to formally construct an additional level in the hierarchical model in which τ is a random effect with a distribution. They restrict attention to gamma distributions for τ^{-2}, and estimate that a τ^{-2} for a new meta-analysis has a Gamma[1.0, 0.35] distribution. Transforming this onto the τ scale using standard theory for probability distributions yields a root-inverse-gamma distribution RIG[1, 0.35] (Section 2.6.6). This has its mode at $\tau = 0.48$, mean $\sqrt{0.35\pi} = 1.05$ and a standard deviation of ∞. Figure 5.10(b) reveals it to rule out low values of τ.

Default 'non-informative' priors. A number of suggestions have been made for placing a 'default' prior distribution on τ or, equivalently, τ^2. The standard reference prior for a sampling variance, $p(\sigma^2) \propto \sigma^{-2}$ (Section 5.5.1), is inappropriate at the random-effects level as it gives an *improper* posterior distribution (Berger, 1985). Five of the main contenders are listed below.

(a) *A 'just proper' prior.* An inverse gamma distribution such as

$$\tau^{-2} \sim \text{Gamma}[0.001, 0.001]$$

is proper and close to being uniform on $\log(\tau)$. Figure 5.10(c) shows that it gives a high weight near $\tau = 0$ and so, if the likelihood supports low values of τ, it could show a preference for a low variance. This may be reasonable behaviour but should be acknowledged.

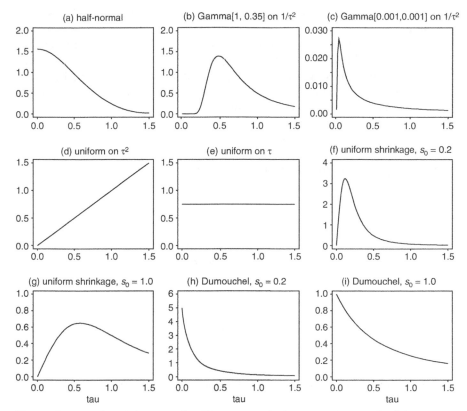

Figure 5.10 Alternative prior distributions on the between-unit standard deviation τ: see the text for discussion of each possible choice. (a) supports equality between units ($\tau = 0$) and discounts substantial heterogeneity ($\tau = 1$); (b) is based on an empirical summary of past meta-analyses and forces heterogeneity; (c) is an 'almost' improper prior that has been widely used but gives strong preference for small τ, (f) to (i) depend on the amount of evidence in the data, with $s_0 = 1$ representing weak evidence, and $s_0 = 0.2$ strong evidence.

(b) *Uniform on τ^2.* The uniform prior

$$p(\tau^2) \propto \text{constant}$$

is recommended by Gelman *et al.* (1995) and can be restricted to a suitable range to make it a proper distribution. Figure 5.10(d) shows its preference for high values of τ, which does not appear attractive.

(c) *Uniform on τ.* The uniform prior

$$p(\tau) \propto \text{constant}$$

is a natural contender and is shown in Figure 5.10(e). Nevertheless, it would be inappropriate to term this 'non-informative', as it is a fairly strong statement to declare that small values of τ are as likely as large values.

(d) *Uniform shrinkage priors.* Following Section 3.17, we assume an approximate normal likelihood with $y_k \sim N[\theta_k, s_k^2]$. A number of authors (Christiansen and Morris, 1997b; Natarajan and Kass, 2000; Daniels, 1999; Spiegelhalter, 2001) have investigated a prior on τ^2 that is equivalent to a uniform prior on the 'average' shrinkage

$$B_0 = s_0^2/(s_0^2 + \tau^2)$$

where s_0^2 is the harmonic mean of the s_k^2, *i.e.*

$$\frac{1}{s_0^2} = \frac{1}{K}\Sigma_k \frac{1}{s_k^2}.$$

Placing a uniform distribution on B_0 is equivalent to $1 - B_0 = \tau^2/(s_0^2 + \tau^2)$ having a uniform distribution. This leads to

$$p(\tau^2) = \frac{s_0^2}{(s_0^2 + \tau^2)^2},$$

$$p(\tau) = \frac{2\tau s_0^2}{(s_0^2 + \tau^2)^2}.$$

The uniform shrinkage prior distributions have the following properties:

	τ^2	τ
Mode	0	$s_0/3 = 0.57s_0$
First quartile	$s_0^2/\sqrt{3}$	$s_0/\sqrt{3} = 0.57s_0$
Median	s_0^2	s_0
Mean	–	$\pi s_0/2 = 1.57s_0$
Third quartile	$3s_0^2$	$\sqrt{3}s_0 = 1.73s_0$
Variance	–	–

The prior on τ^2 has an asymptote at 0, but the implied prior on τ returns to 0 at the origin.

Suppose $s_k^2 = \sigma_k^2/n_k$, so that

$$y_k \sim N[\theta_k, \ \sigma_k^2/n_k].$$

Three situations can be distinguished:

(i) $\sigma_k^2 = \sigma^2$, which is assumed known, such as the frequent adoption of $\sigma^2 = 4$. Then $s_0^2 = \sigma^2/\bar{n}$.

(ii) $\sigma_k^2 = \sigma^2$, which is unknown. σ^2 could then be given a standard Jeffreys prior $p(\sigma^2) \propto \sigma^{-2}$ – this induces an appropriate dependency between τ^2 and σ^2.

(iii) Each σ_k^2 is unknown. The σ_k^2 could then be assumed either exchangeable or independent. Within-unit empirical estimates $\hat{\sigma}_k^2$ can be used to estimate s_0^{-2} by

$$\frac{1}{s_0^2} = \frac{1}{K} \Sigma_k \frac{n_k}{\hat{\sigma}_k^2}.$$

Essentially, fixed effects are fitted first and then the average precision is used as an estimate of s_0^{-2}. This approach is illustrated in Examples 6.10 and 8.1.

In studies based on events we might equate s_0^2 to $4/n_0$, where n_0 represents the mean number of events in each study. Hence $s_0 = 0.2$ corresponds to large studies with an average of 100 events each, while $s_0 = 1.0$ corresponds to very small studies with an average of 4 events each. These priors are shown in Figures 5.10(f) and 5.10(g), showing that large studies lead to strong prior weight on low values of τ and hence an expectation of the studies showing 'similar' results.

(e) *DuMouchel priors.* DuMouchel (DuMouchel and Normand, 2000) has suggested a similar form to the uniform shrinkage prior but assuming a uniform prior for $s_0/(s_0 + \tau)$, which implies

$$p(\tau) = \frac{s_0}{(s_0 + \tau)^2},$$

$$p(\tau^2) = \frac{s_0}{2\tau(s_0 + \tau)^2}.$$

The distributions have the following properties:

	τ^2	τ
Mode	0	0
First quartile	$s_0^2/9$	$s_0/3$
Median	s_0^2	s_0
Mean	–	–
Third quartile	$9s_0^2$	$3s_0$
Variance	–	–

Note that the quartiles are at $B_0 = 0.1,\ 0.5,\ 0.9$, showing the DuMouchel prior gives preference to either strong or weak shrinkage. Figures 5.10(h) and 5.10(i) show the DuMouchel priors for $s_0 = 0.2$ and $s_0 = 1.0$, revealing the preference of these priors for both low and high values of τ.

In general our preference will be to use a uniform prior on τ as a baseline when there is reasonable information from the data. When prior information is strong or important a suitably informative prior can be chosen: the half-normal appears particularly attractive.

These points serve to underline the importance of carefully choosing and justifying the prior distributions used within a hierarchical setting, and subjecting those used to the type of sensitivity analysis adopted in Examples 6.10, 7.2, 8.1, 8.3 and 8.5.

5.8 EMPIRICAL CRITICISM OF PRIORS

The ability of subjective prior distributions to predict the true benefits of interventions is clearly of great interest, and Box (1980) suggested a methodology for comparing priors with subsequent data. The prior is used to derive a predictive distribution for future observations, and thus to calculate the chance of a result with lower predictive ordinate than that actually observed: when the predictive distribution is symmetric and unimodal, this is analogous to a traditional two-sided P-value in measuring the predictive probability of getting a result at least as extreme as that observed. With normal assumptions we can use (3.23) but substituting m for n, to give a pre-trial predictive distribution

$$ Y_m \sim \mathrm{N}\left[\mu,\ \sigma^2\left(\frac{1}{n_0}+\frac{1}{m}\right)\right]. \tag{5.5} $$

Given observed y_m, the predictive probability of observing a Y_m less than that observed is

$$ P(Y_m < y_m) = \Phi\left(\frac{y_m - \mu}{\sigma\sqrt{\dfrac{1}{n_0}+\dfrac{1}{m}}}\right), \tag{5.6} $$

and hence Box's generalised significance test is given by

$$ 2\min\left[P(Y_m < y_m),\ 1 - P(Y_m < y_m)\right]. $$

Another way of obtaining (5.6) is as the tail area associated with a standardised test statistic contrasting the prior and the likelihood, *i.e.*

$$ z_m = \frac{y_m - \mu}{\sigma\sqrt{\dfrac{1}{n_0}+\dfrac{1}{m}}}, $$

showing that Box's statistic explicitly acts as a measure of *conflict* between prior and data.

Example 5.5 *GREAT (continued): Criticism of the prior*

In Example 3.6, $\mu = -0.26$, $n_0 = 236.7$, $m = 30.5$, $\sigma = 2$ and hence the predictive distribution for the observed log(OR) has mean -0.26 and standard deviation 0.39. This is shown in Figure 5.11 with the observed OR $= 0.48$ ($y_m = \log(\text{OR}) = -0.74$) marked. Box's measure is twice the shaded area, which is $2\Phi((-0.74 + 0.26)/0.39) = 0.21$. We may also obtain this result as the standardised test statistic between prior and likelihood $z = -1.25$, with a two-sided *P*-value of 0.21. Thus there is no strong evidence for conflict between prior and data in the GREAT example.

There have been a number of prospective elicitation exercises for clinical trials, and many of these trials have now reported their results. Table 5.3 shows a selection of results, including the intervals for the prior distributions for treatment effects, the evidence from the likelihood, and Box's *P*-value summarising the conflict between the prior and the likelihood. The references for the prior assessments and the data are provided at the end of the section.

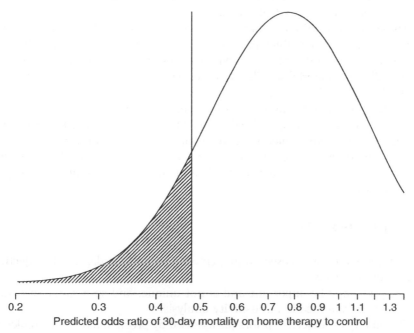

| 0.2 | 0.3 | 0.4 | 0.5 | 0.6 | 0.7 | 0.8 | 0.9 | 1 | 1.1 | 1.3 |

Predicted odds ratio of 30-day mortality on home therapy to control

Figure 5.11 Predictive distribution for observed OR in the GREAT trial with observed OR $= 0.48$ (log(OR) $= -0.74$) marked. Box's measure of conflict between prior and data is twice the shaded area $= 0.21$.

Table 5.3   A comparison of some elicited subjective prior distributions and the consequent results of the clinical trials. In each case a pooled prior was provided, assumed normal on a log(hazard ratio) scale – Box's P-value is calculated on this scale. This is transformed to a hazard ratio (HR) scale where HR < 1 corresponds to benefit of the new treatment: median and 95% intervals are given (note the gastric cancer results are reported with the inverse hazard ratio in Example 6.4).

Study	Prior		Likelihood		Z	P
	HR	95% interval	HR	95% interval		
CHART (Lung)[1]	0.76	(0.48, 1.19)	0.76	(0.63, 0.90)	0.00	1.00
CHART (HN)[1]	0.72	(0.44, 1.20)	0.95	(0.79, 1.14)	1.02	0.31
Thiotepa X1[2]	0.61	(0.37, 1.01)	1.11	(0.78, 1.59)	1.91	0.06
Osteosarcoma[3]	0.90	(0.55, 1.50)	1.07	(0.79, 1.45)	0.58	0.56
Gastric cancer[4]	0.88	(0.61, 1.28)	1.10	(0.87, 1.39)	1.00	0.32

Sources: [1]Example 6.6. [2]Spiegelhalter and Freedman (1986) and Richards *et al.* (1994). [3]Spiegelhalter *et al.* (1993) and Souhami *et al.* (1994). [4]Example 6.4.

Table 5.3 shows the generally poor experience obtained from prior elicitation. The clinicians are universally optimistic about the new treatments (median of prior hazard ratios less than 1), whereas only two of the trials – the CHART trials – eventually showed any evidence of benefit from the new treatment (likelihood hazard ratio less than 1), and only the CHART lung trial showed 'significant' benefit. The thiotepa trial shows particularly high conflict between data and prior, with the clinicians expecting a substantial benefit from thiotepa which failed to materialise. This also reflects the experience of Carlin *et al.* (1993) in their elicitation exercise.

Far from invalidating the Bayesian approach, such a conflict between prior and data only serves to emphasise the importance of pre-trial elicitation of belief; having these opinions explicitly recorded will help a data monitoring committee to focus on the difference between anticipated and actual results. Of course, the precise action to be taken in the face of considerable conflict will depend on the circumstances.

5.9   KEY POINTS

1. The use of a prior is based on judgement and hence a degree of subjectivity cannot be avoided.
2. The prior may be important and is not unique, and so a range of options should be examined in a sensitivity analysis.
3. The quality of subjective priors (as assessed by predictions) show predictable biases in terms of enthusiasm.
4. For a prior to be taken seriously by an external audience, its basis must be explicitly given. A variety of models exist for using historical data as a basis for prior distributions.

5. Archetypal priors, expressing both scepticism and enthusiasm, may be useful for identifying a reasonable range of prior opinion.
6. Great care is required in using default priors intended to be minimally informative.
7. Exchangeability assumptions lead to hierarchical models that are valuable in many situations, but such judgements should not be made casually.
8. Sensitivity analysis plays a crucial role in assessing the impact of particular prior distributions, whether elicited, derived from evidence, or reference, on the conclusions of an analysis.

EXERCISES

5.1. Consider tossing a drawing-pin (thumbtack) onto a flat surface.
    (a) Assess *your* beliefs about the true proportion of times that it will fall point-up, in terms of a best estimate, and low and high assessments.
    (b) Derive a beta prior distribution for this proportion based on these beliefs.
    (c) Use the conjugate beta-binomial model of Section 3.6.2 to update these beliefs after 12 tosses using the *same* hand.
5.2. Prior to the publication of the UK Medical Research Council RCT evaluating the use of high-energy neutrons for treatment of patients with tumours of the pelvic region (bladder, cervix, prostate and rectum) in 1991 a number of RCTs evaluating low-energy neutrons had been reported (Errington *et al.*, 1991). The results of these RCTs are summarised in Table 5.4. (a) Assuming balanced trials, approximate the log(hazard ratio) and its variance for each of these studies. (b) Use the 'method of moments' (3.37) to estimate the between-study variance τ^2. Use this historical evidence to establish a prior distribution for the MRC trial, assuming (c) the new trial is estimating the

Table 5.4 Summary of RCT evidence in terms of survival at 12 months for low-energy neutron therapy compared to conventional radiotherapy for tumours of the pelvic region.

Study	Year of publication	Site	Neutrons		
			Deaths(O)	Expected(E)	$V[O-E]$
Batterman	1982	Bladder and Rectum	34	32.6	5.3
Pointon	1985	Bladder	16	13.7	5.1
Duncan	1987	Bladder	26	20.1	6.7
Duncan	1987	Rectum (inoperable)	17	12.8	2.1
Duncan	1987	Rectum (recurrent)	10	7.3	2.0
Duncan	1987	Bladder	4	4.2	0.6

mean treatment effect of the previous trials, and (d) the new trial is exchangeable with the previous trials. The fact that the previous trials were low-energy, and the new trial high-energy, might lead one to doubt the exchangeability model.

(e) What model for systematic bias might be reasonable?

5.3. In Exercise 5.2, on average the oncologists claimed that they required the survival rate for neutron therapy to be 61.5%, relative to a 1-year survival rate of 50% in the control group, before considering it for routine treatment. The range of equivalence was therefore taken to be from 50% to 61.5%. For each of the situations modelled, obtain the prior probabilities of no benefit of neutrons relative to conventional therapy, the range of equivalence, and clinical benefit in favour of neutron therapy.

5.4. In addition to the meta-analysis in Exercise 5.2, the beliefs and clinical demands of ten oncologists were elicited before the final analysis of the high-energy trial data. Table 5.5 summarises the elicited prior distributions for all ten oncologists for the 1-year survival rate on neutron therapy compared to a 50% survival rate with conventional therapy.

(a) Calculate an average histogram.

(b) Transform this to a histogram on the log(hazard ratio) scale using the techniques in Example 5.1.

(c) Fit a normal distribution to this distribution by matching the mean and variance or by some other method.

(d) Given the disagreement between the oncologists, do you think it reasonable to create such a pooled distribution?

5.5. Prior to the publication of the HAI RCT considered in Exercise 2.7, results from five previous RCTs had been published, and these are summarised in terms of overall survival in Table 5.6. (a) For each trial, estimate the

Table 5.5 Elicited prior beliefs in terms of percentage survival at 12 months for high-energy neutron therapy compared to a 50% survival rate for conventional radiotherapy for tumours of the pelvic region.

ID	Neutron 1-year survival rate (%)																	
	15−	20−	25−	30−	35−	40−	45−	50−	55−	60−	65−	70−	75−	80−	85−	90−	95−	Total
1	0	0	0	10	20	25	15	15	10	5	0	0	0	0	0	0	0	100
2	0	0	5	10	20	35	30	0	0	0	0	0	0	0	0	0	0	100
3	0	0	0	0	0	20	60	20	0	0	0	0	0	0	0	0	0	100
4	0	0	0	0	0	30	30	30	10	0	0	0	0	0	0	0	0	100
5	0	0	5	10	25	20	15	10	10	5	0	0	0	0	0	0	0	100
6	0	0	0	0	0	0	5	0	0	0	15	20	20	15	10	10	5	100
7	5	5	10	25	25	15	5	2.5	2.5	2.5	2.5	0	0	0	0	0	0	100
8	0	0	0	5	5	10	25	25	15	5	2.5	2.5	2.5	2.5	0	0	0	100
9	0	0	0	5	5	10	25	25	15	5	2.5	2.5	2.5	2.5	0	0	0	100
10	0	0	0	15	15	25	20	20	5	0	0	0	0	0	0	0	0	100

log(hazard ratio) and the effective number of events assuming $\sigma = 2$. Obtain a prior distribution for the log(hazard ratio) for overall survival of HAI compared to control patients, assuming (b) a common effect in all trials, (c) that the past trials are exchangeable with the current trial.

5.6. Sutton *et al.* (2000, p. 261) consider 17 single-arm studies of either radiotherapy alone (RTx) following surgery for childhood medulloblastoma, or radiotherapy together with adjuvant chemotherapy (RTx + Chm) following surgery. Table 5.7 displays the 5-year survival rates together with standard errors for all 17 studies.

Table 5.6 Summary of RCT evidence in terms of overall survival, prior to 1994, for HAI compared to control for the treatment of non-resectable liver metastases associated with primary colorectal cancer.

Study	Year publication	HAI		Control		$O-E$	$V[O-E]$
		Deaths	Total	Deaths	Total		
MSKCC	1987	43	45	48	48	−5.8	21.9
NCCTG	1990	39	39	35	35	−1.0	17.9
NCI	1987	25	32	26	32	−2.7	12.5
City of Hope	1986	9	9	6	6	−2.3	3.3
France	1992	72	81	78	82	−14.2	36.4

Table 5.7 Five-year survival rates and standard errors for single-arm studies considering either radiotherapy alone (RTx) or radiotherapy together with adjuvant chemotherapy (RTx + Chm) following surgery for childhood medulloblastoma.

Study	RTx + Chm		RTx	
	S_5	$SE(S_5)$	S_5	$SE(S_5)$
1	0.83	0.030	−	−
2	0.82	0.120	−	−
3	0.96	0.039	−	−
4	0.82	0.384	−	−
5	0.55	0.188	−	−
6	0.64	0.170	−	−
7	0.26	0.196	−	−
8	0.60	0.097	−	−
9	0.36	0.170	−	−
10	0.93	0.120	−	−
11	−	−	0.71	0.184
12	−	−	0.48	0.223
13	−	−	0.41	0.087
14	−	−	0.32	0.057
15	−	−	0.34	0.080
16	−	−	0.71	0.068
17	−	−	0.33	0.071

   (a) Looking at the data, do you think a pooled effect is a reasonable assumption?
   (b) Estimate the between-study variance for each treatment using (3.37).
   (c) Assuming a normal random-effects model, estimate a prior distribution for the 5-year survival in a new study, assuming exchangeability with the previous studies.
   (d) Combine these two prior distributions into a prior for the difference in the 5-year survival rate, *i.e.* RTx + Chm − RTx, in a proposed clinical trial.
   (e) Is normality a reasonable assumption for the random-effects distribution?

5.7. The trial discussed in Exercise 5.2 ended by yielding an estimated hazard ratio of 1.52 (95% CI from 0.91 to 2.50), *i.e.* in favour of the control group (Errington *et al.*, 1991).
   (a) For the data-based prior using all six previous studies, assess the conflict of these prior distributions, using the methods of Section 5.8.
   (b) Repeat this for oncologists 6 and 7.

5.8. Verify for a normal model in Section 5.4, when there is a single historical study, the assumptions under which exchangeability, bias and discounting can lead to the same prior distribution. Does this hold for multiple studies?

5.9. Plot three half-normal prior distributions for a model parameter τ which have the properties that:
   (a) the mean of τ is 1.5;
   (b) the median is 3; and
   (c) the probability of τ being greater than 1 is 5%.

5.10. For the magnesium meta-analysis in Example 3.13 calculate and plot DuMouchel and uniform shrinkage prior distributions for the random-effects standard deviation τ.

6

Randomised Controlled Trials

A Bayesian: one who asks you what you think before a clinical trial in order to tell you what you think afterwards. (Senn, 1997b)

6.1 INTRODUCTION

Randomised controlled trials are traditionally considered the 'gold standard' for evaluation of health-care interventions, and have provided fertile territory for arguments between alternative statistical philosophies. In this chapter we consider a number of specific issues in which a distinct Bayesian approach is identifiable: these include the role of decision theory, ethics of randomisation, use of historical controls, selection of sample size, monitoring sequential studies, subset analysis, alternative designs and so on. Some of the strongest arguments for the Bayesian approach have been made in this context, with notable examples being Cornfield (1976), Berry (1993) and Kadane (1995). Each of these authors has emphasised the internal consistency of the Bayesian approach, and welcomed the need for explicit prior distributions and loss functions as producing scientific openness and honesty: see Section 6.13 for additional references by these and other authors.

The issues in this chapter are largely common to trials both in the public sector and in the pharmaceutical industry. For industry-sponsored trials we shall use the standard language of drug development: phase I studies deal with identifying a safe dose, usually on healthy volunteers; phase II studies are concerned with finding an effective dose; phase III studies are intended to prove treatment benefit over an appropriate control; and phase IV studies monitor the use and possible side-effects of a drug in routine use. This structure is necessarily rather simplistic, and there are increasing moves toward hybrid studies in order to speed up the drug development process. Parallel phases of development can be given for complex

Bayesian Approaches to Clinical Trials and Health-Care Evaluation D. J. Spiegelhalter, K. R. Abrams and J. P. Myles
© 2004 John Wiley & Sons, Ltd ISBN: 0-471-49975-7

public health interventions (Campbell *et al.*, 2000): in phase I an intervention is developed possibly through a theoretical model; in phase II explanatory trials in tightly controlled situations seek to demonstrate the potential efficacy of the intervention; in phase III pragmatic trials evaluate its costs and effectiveness in practice; and in phase IV the intervention is rolled out into routine use.

We shall begin by considering the basic issue of whether a trial is for inference or decision (Section 6.2), and then investigate the role of null hypotheses and their relation to the demands set of a new intervention (Section 6.3). The ethics of randomisation are then viewed from a Bayesian perspective (Section 6.4). A substantial section explores a number of ways in which prior opinion can be incorporated into sample-size calculations (Section 6.5), followed by a full discussion of the many ways to tackle the important issue of trial monitoring (Section 6.6), and the possible use of sceptical priors in deciding whether a confirmatory trial is necessary (Section 6.7). Apart from repeated looks at the data, 'multiplicity' features in many aspects of trial design and analysis, and we briefly discuss multiple subsets, outcomes, centres and trial arms (Section 6.8). The use of historical control groups fits naturally into a Bayesian perspective and is treated in some detail (Section 6.9); different trial designs are then examined, for example data-dependent allocation (Section 6.10) and multiple N-of-1 studies (Section 6.11). We only briefly consider phase I and II studies (Section 6.12), and discussion about the regulatory context is left until we consider policy decisions (Chapter 9).

6.2   USE OF A LOSS FUNCTION: IS A CLINICAL TRIAL FOR INFERENCE OR DECISION?

There has been a heated dispute about whether a clinical trial should be considered as a *decision* problem, with an accompanying loss function, or as an *inference* problem in which no explicit loss function is developed and conclusions are based solely on the posterior distributions of quantities of interest. This has been a point of clear distinction between different schools of Bayesianism (Section 3.20). Here we briefly review the arguments.

1. *A clinical trial should be a decision.* Lindley (1994) categorically states that 'Clinical trials are not there for inference but to make decisions', while Berry (1994) states that 'deciding whether to stop a trial requires considering why we are running it in the first place, and this means assessing utilities'. Healy (1978) considers that 'the main objective of almost all trials on human subjects is (or should be) a decision concerning the treatment of patients in the future'. The potential role for explicit statement of a loss function is a running theme throughout discussions on sample size (Section 6.5), sequential analysis (Section 6.6.4), adaptive allocation (Section 6.10) and payback from research programmes (Section 9.10), and many would argue that the

eventual decision is inseparable from the design and analysis of a study.

From an economic perspective, it is claimed that a utility approach to clinical trial design and analysis is necessary in order to prevent conclusions based on inferential methods leading to health or monetary losses. This perspective derives from the observation made in Section 3.14 that only the expected utility of a decision is relevant, and expressions of uncertainty are, theoretically, of no concern except when deciding whether to collect further evidence. This echoes the original work on pragmatic clinical trials by Schwartz *et al.* (1980), in which it was argued that *P*-values and interval estimates are irrelevant to trials that guide decisions. The role for decision theory in health policy and regulation will be covered in Section 9.11.

The explicit use of utility functions within the design and monitoring of clinical trials is controversial but has been explored in a number of contexts: for example, Berry and Stangl (1996a) discuss the problems of whether to stop a phase II trial based on estimating the number of women in the trial and who will respond in the future; whether to continue a vaccine trial by estimating the number of children who will contract the disease; and the use of adaptive allocation in a phase III trial such that at each point the treatment which maximises the expected number of responders is chosen.

2. *A clinical trial provides an inference.* Armitage (1985), Breslow (1990), DeMets and Lan (1994), Simon (1977) and Orourke (1996) all describe how it is unrealistic to place clinical trials within a decision-theoretic context, primarily because the impact of stopping a trial and reporting the results cannot be predicted with any confidence: Peto (1985), in the discussion of Bather (1985), states that 'Bather, however, merely assumes... "it is implicit that the preferred treatment will then be used for all remaining patients" and gives the problem no further attention! This is utterly unrealistic, and leads to potentially misleading mathematical conclusions'. Peto goes on to argue that a serious decision-theoretic formulation would have to model the subsequent dissemination of a treatment.

3. *It depends on the context.* Whitehead (1997b, p. 208) points out that the theory of optimal decision-making only exists for a single decision-maker, and that no optimal solution exists when making a decision on behalf of multiple parties with different beliefs and utilities. He therefore argues that internal company decisions at phase I and phase II of drug development may be modelled as decision problems, but that phase III trials cannot (Whitehead, 1993).

Our personal view is that the context of evaluation often means that the investigators who design and carry out a study are generally not the same body who make decisions on the basis of the evidence (Section 3.1), and so, taking a pragmatic rather than ideological perspective, our general separation of inference and decision appears reasonable.

6.3 SPECIFICATION OF NULL HYPOTHESES

Attention in a trial usually focuses on the null hypothesis of treatment equivalence expressed by $\theta = 0$, but realistically this is often not the only hypothesis of interest. Increased costs, toxicity and so on may mean that a certain improvement would be necessary before the new treatment could be considered clinically superior, and we shall denote this value θ_S. Similarly, the new treatment might not actually be considered clinically inferior unless the true benefit were less than some threshold denoted θ_I. The interval between θ_I and θ_S has been termed the 'range of equivalence' (Freedman *et al.*, 1984); often θ_I is taken to be 0.

This is not a specifically Bayesian idea (Armitage, 1989) and can be considered as representing an interval null hypothesis. Figure 6.1 shows the

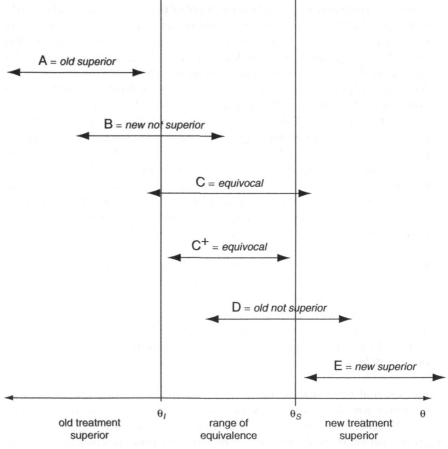

Figure 6.1 Possible situations at any point in a trial's progress, derived from superimposing an interval estimate (say, 95%) on the range of equivalence.

possible situations one could be in at any stage of a trial when calculating a 95% interval for a treatment benefit.

A: We are confident that the old treatment is clinically superior.
B: The new treatment is not superior, but the treatments could be clinically equivalent.
C: We are substantially uncertain as to the two treatments – this is essentially a position of 'equipoise'.
C^+: We are confident the two treatments are clinically equivalent – as applied to equivalence studies (Section 6.11).
D: The old treatment is not superior, but the treatments could be clinically equivalent.
E: We are confident that the new treatment is clinically superior.

It could be argued that if one really wants to convince people of the clinical superiority of a treatment, then one should aim for conclusion E in design and monitoring, even though this demands increased sample sizes and requires a highly significant (in the traditional sense) result.

Example 6.1   *CHART (continued): Clinical demands for new therapies*

References: Parmar *et al.* (1994, 2001) and Spiegelhalter *et al.* (1994). See Example 5.1 for details of the trials and the elicitation process.

Loss function or demands: No formal loss function was elicited, but a pre-trial survey was carried out of 11 clinicians participating in the trials. The clinicians were given the following instructions (Parmar *et al.*, 1994):

Suppose you had been told on good authority the exact absolute improvement [in 2-year survival rates] you would obtain by treating patients with the CHART regimen. If this was exactly zero improvement you would presumably use your standard radical radiotherapy in the future. If there was an absolute improvement of 20% you would presumably use CHART. Somewhere in between these figures there is likely to be a difference where you would change from standard therapy to CHART. There may be a range of differences where the decision would not be clearcut, i.e. a range where you feel the two regimens are approximately equivalent. Please mark your change-over point or the range on the scale of treatment differences shown below.

The upper and lower values for the ranges were averaged and the following results were obtained.

Lung trial. The participants would be willing to use CHART routinely if it conferred at least 13.5% improvement in 2-year survival (from a baseline of 15%), and unwilling if less than 11% improvement. Thus the range of equivalence is from 11% to 13.5%: from (2.33) this is equivalent to hazard ratios (HR) from 0.66 to 0.71, or log(HR) from −0.41 to −0.34.

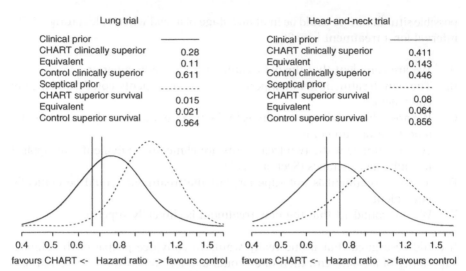

Figure 6.2 Clinical and sceptical priors superimposed on an assessed average clinical range of equivalence. Probabilities of lying below, within and above the range of equivalence are given both for clinical and sceptical priors. The juxtaposition of the clinical priors and ranges of equivalence suggests a reasonable basis for randomisation.

Head-and-neck trial. The participants would be willing to use CHART routinely if it conferred a 13% improvement in 2-year recurrence-free rate (from a baseline of 45%), and unwilling if less than 10% improvement. Thus the range of equivalence is from 10% to 13%, equivalent to HR from 0.68 to 0.75, or log(HR) from −0.38 to −0.29. The average ranges of equivalence are shown in Figure 6.2, with the clinical and sceptical priors derived previously. The average range of equivalence is reasonably central to the clinical prior, suggesting, on average, a reasonable basis for randomisation.

One advantage of the Bayesian approach is that the posterior distribution can be juxtaposed to the clinical demands being made in order to graphically display the current probabilities concerning the status of treatments. There is also no reason why the 'goalposts' shown in Figure 6.1 should not change as a study progresses and more is learnt about, for example, the side-effects of treatments. However, in order to prevent subjective bias, it may be better for those responsible for specifying the 'range of equivalence' to be blind to the data. Elicitation of such intervals can be carried out at the same time as elicitation of prior beliefs (Section 5.2) and uses very similar techniques: see Example 6.1. The crucial aspect is that those whose opinions are being elicited must be very clear in their distinction between *demands*, as expressed in their range of equivalence, and their *expectation or beliefs*, as represented by the prior distribution. Two factors increase the potential for confusion: demands and beliefs are often quantitatively similar (indeed, we argue below that this is the ethical basis for randomisation), and the loose usage of words such as 'the difference hoped for', which carries connotations both of desire and expectation. It follows that such terms must be strictly avoided!

6.4   ETHICS AND RANDOMISATION: A BRIEF REVIEW

6.4.1   Is randomisation necessary?

Randomisation has two traditional justifications: it ensures treatment groups are directly comparable (up to the play of chance), and it provides a fundamental basis for the probability distributions underlying conventional statistical procedures. Since Bayesian probability models are derived from subjective judgement, and hence do not require any underlying physical justification for a randomisation mechanism, the latter requirement is irrelevant. This has led some to question the need for randomisation at all, provided alternative methods of balancing groups can be established. For example, Urbach (1993) argues that a 'Bayesian analysis of clinical trials affords a valid, intuitively plausible rationale for selective controls, and marks out a more limited role for randomisation than it is generally accorded'. It has even been claimed that 'Randomised trials are inherently unethical' (Berry, 1989a). Papineau (1994) refutes Urbach's position and claims that, despite it not being essential for statistical inference, experimental randomisation forms a vital role in drawing causal conclusions (Rubin, 1978). The relationship between randomisation and causal inferences is beyond the scope of this book, but in general the need for sound experimental design appears to dominate philosophical statistical issues (Hutton, 1996). In fact, Berry and Kadane (1997) suggest that if there are several parties who make different decisions and observe different data, randomisation may be a strictly optimal procedure since it enables each observer to draw their own appropriate conclusions.

The extent to which careful analysis of high-quality databases can complement or even replace randomised trials is a delicate issue: for example, Howson and Urbach (1989) and Hlatky (1991) argue in favour of databases, while Byar (1980) puts an opposing view. Although a full discussion is outside the scope of this book, we nevertheless point out that Bayesian methods provide a natural basis for synthesising data from randomised and non-randomised studies: see the discussion on the use of historical data (Section 3.16), historical controls (Section 6.9) and cross-design synthesis (Section 8.4).

6.4.2   When is it ethical to randomise?

If we agree that randomisation is useful, then the issue arises of when it is ethical to randomise. This is closely associated with the process of deciding when to stop a trial (Section 6.6) and is often represented as a balance between *individual* and *collective* ethics (Pocock, 1992; Palmer and Rosenberger, 1999): individual ethics would suggest that it is inappropriate to randomise a patient to a treatment near the end of a trial in which one could be reasonably confident as to another treatment's superiority, while collective ethics could argue that such a benefit will only be available for future patients if the current trial runs long enough for the findings to be convincing to a wide range of clinical opinion. See Edwards *et al.* (1998) for a full review of issues concerning the ethics of randomisation in clinical trials.

Freedman (1987) introduced the idea of *professional equipoise*, in which disagreement among the medical profession makes randomisation ethical. The trial design of Kadane (1996) is an expression of this principle, in that only a treatment that at least one clinician thought optimal could be given to a patient (although unfortunately a programming error meant that some patients were allocated to treatments that *all* clinicians felt were sub-optimal). Perhaps a more appealing approach is the 'uncertainty principle' which is often argued as a basis for ethical randomisation (Byar *et al.*, 1990): this may be thought of as 'personal equipoise' in which the clinician was uncertain as to the best treatment for the patient in front of them. However, a quantified degree of uncertainty is not specified. Senn (2002) argues that it is reasonable for a society to restrict new interventions to trials, and in those trials it is ethical to randomise even when one believes in the superiority of the new treatment.

The Bayesian approach can be seen as formalising the uncertainty principle by explicitly representing, in theory, the judgement of an individual clinician that a treatment may be beneficial – this could be provided by superimposing the clinician's posterior distribution on the range of equivalence (Section 6.3) relevant to a particular patient (Spiegelhalter *et al.*, 1994). It has been argued that a Bayesian model naturally formalises the individual ethical position (Lilford and Jackson, 1995; Palmer, 1993), in that it explicitly confronts the personal belief in the clinical superiority of one treatment. Berry (1993), however, has suggested that if patients were honestly presented with numerical values for their clinician's belief in the superiority of a treatment, then few might agree to be randomised. One option might be to randomise but with a varying probability that is dynamically weighted towards the currently favoured treatment (Section 6.10).

Chaloner and Rhame (2001) consider the roles of professional and individual equipoise, and suggest scenarios which indicate different bases for ethical randomisation. Fifty-eight opinions elicited before a trial showed a wide range of responses, and the acknowledged variability in clinical opinion suggests that a suitable aim in conducting a trial is to bring disparate opinions into agreement: Chaloner and Rhame (2001) quote Byar as saying 'We may reasonably ask, if we do a study that convinces us but convinces no one else and is then ignored or requires confirmation by yet another study, whether we have really acted in the most ethical fashion in the long run'. Pocock and White (1999) consider the situation in which one has a 'significant' effect in a trial, when further random-isation is 'unethical, but only if the statistically significant difference is genuine (in many cases it is not) and if the new treatment would indeed be given to future patients (which is by no means inevitable)'. We largely agree with the advice of Kass and Greenhouse (1989), who claim that 'the purpose of a trial is to collect data that bring to conclusive consensus at termination opinions that had been diverse and indecisive at the outset' and go on to state that 'randomisation is ethically justifiable when a cautious reasonable sceptic would be unwilling to state a preference in favour of either the treatment or the control'. This approach leads naturally to the development of sceptical prior distributions (Section 5.5.2) and their use in monitoring sequential trials (Section 6.6.2).

6.5   SAMPLE SIZE OF NON-SEQUENTIAL TRIALS

In this section we consider the Bayesian contribution to selecting the sample size of a clinical trial which will not be subject to interim monitoring: there is particular emphasis on 'hybrid' methods in which prior information is formally used but the final analysis is carried out in a classical framework. In some contexts this may be quite appropriate, as there may be substantial prior infor-mation that cannot be included in the final report for, say, regulatory purposes.

This section does contain a number of rather complex expressions for quan-tities of interest, but the content appears too important for this to be a 'starred' section. On a technical note, the formulae we present follow the traditional formulation in which interest focuses on a parameter θ and $\theta > 0$ indicates benefit of the experimental treatment. We recognise that in many of our examples $\theta < 0$ has represented such benefit, and furthermore in other cases we might be using thresholds other than 0. Care must therefore be taken when using the formulae in this chapter – it may be best to first transform the particular problem being analysed into the standard formulation adopted here. Details of these transformations are given in Section 6.5.4.

It could be argued that elicitation of prior beliefs and demands from a broad community of stakeholders is necessary not only in order to undertake a specifically Bayesian approach to design and analysis, but also more generally as part of good research practice. A potential consequence of ignoring this source of judgement is that trials may be designed on the basis of over-enthusiastic beliefs and demands, and hence fail to convince others and modify health-care policy or practice.

6.5.1   Alternative approaches to sample-size assessment

In Section 4.1 we described a taxonomy of six broad statistical approaches to the evaluation of health-care interventions. Here we focus on how the four main viewpoints (ignoring the Bayesian hypothesis-testing and classical decision-theory approaches) deal with selecting the sample size of a fixed-size experiment: the design and monitoring of sequential studies will be covered in Section 6.6. A hybrid philosophy is also included.

Fisherian. In principle there is no need for preplanned sample sizes, but a choice may be made by selecting a particular precision of measurement and informally trading that off against the cost of experimentation.

Neyman–Pearson. The first stage is to set up a null hypothesis (Section 6.3), and then specify an alternative hypothesis H_A: $\theta = \theta_A$ that the trial is being designed to detect. A variety of opinions have been expressed about the interpretation of θ_A (Spiegelhalter *et al.*, 1994), including a 'minimum clinically significant difference', a 'worthwhile difference' and a difference 'thought likely to occur'. These ideas tend to conflate the demands made of the new treatment and the expectations of its benefit (Section 6.3), and this combined role of the alternative is reflected in its common definition as a difference that is 'both realistic and important' (within a Bayesian framework these properties are clearly separated). The sample size is then selected to have reasonable power to detect this alternative hypothesis. Power is generally set to 80% or 90%: formula (2.38) can be used to derive the necessary sample size in simple circumstances. In practice the choice of alternative may be influenced by available resources.

Hybrid classical and Bayesian. Considerable attention has been paid to a hybrid approach in which it is assumed that a traditional analysis will take place at the end of the trial, and the prior distribution is used solely for the design.

It may be helpful to consider the joint probability distribution of hypotheses and outcomes displayed in Table 6.1. In a traditional framework these are point hypotheses and the study is designed around the Type I error $\alpha = p((D_1|H_0)$, and the power $1 - \beta = p(D_1|H_1)$. However, if we are prepared to acknowledge prior

Table 6.1 Joint probability distribution of hypotheses and outcomes of a hypothesis test.

		Truth		
		H_0	H_1	
Outcome	D_0 : do not reject H_0	$p(D_0, H_0) =$ P(correct negative)	$p(D_0, H_1) =$ P(false negative)	$p(D_0)$
	D_1 : reject H_0	$p(D_1, H_0) =$ P(false positive)	$p(D_1, H_1) =$ P(correct positive)	$p(D_1)$
		$p(H_0)$	$p(H_1)$	1

probabilities for the hypotheses, then it would appear reasonable to focus also on the probability of rejecting H_0 *and* this being the correct decision, *i.e.* the joint probability $p(D_1,H_1)$. Since $p(D_1,H_1) = p(D_1|H_1)\,p(H_1) = (1 - \beta)\,p(H_1)$, this simply means adjusting the power by the initial probability of H_1: the problem with using only the conditional power $p(D_1|H_1)$ is that no account is taken of the plausibility of the alternative and hence there is a temptation to delude oneself into designing trials to detect implausible hypotheses.

The unconditional probability of getting a 'positive' conclusion can be expressed as

$$p(D_1) = p(D_1, H_0) + p(D_1, H_1),$$

and the first term, which is the probability $p(D_1, H_0) = p(D_1|H_0)\,p(H_0)$ of a false positive result, will generally be very small provided that $\alpha = p(D_1|H_0)$ is small and the prior opinion is substantially supportive of H_1 (as will often be the case preceding a trial). Thus

$$p(D_1) \approx p(D_1|H_1)\,p(H_1); \tag{6.1}$$

and so the 'prior-adjusted power' $(1 - \beta)\,p(H_1)$ will often also be close to the unconditional probability of the trial getting a 'significant' result.

Things get a little more complicated in the more general case when the hypotheses are composite, for example H_0: $\theta < 0$ and H_A: $\theta > 0$. Here the classical power is given by a curve $p(D_1|\theta)$, and we wish to make use of a continuous prior distribution $p(\theta)$.

A number of means of incorporating the prior are possible.

1. One can plot the conditional power curve and superimpose the prior distribution as an informal guide to the relative plausibility of alternative hypotheses. This might prevent a study being designed around an alternative that was clearly grossly optimistic.
2. The prior mean μ might simply be taken as a point alternative hypothesis θ_A, representing a 'plausible and worthwhile difference', although this does not acknowledge the current uncertainty about θ expressed by the prior.
3. The whole classical power curve $p(D_1|\theta)$ can be averaged with respect to the prior distribution to obtain an 'expected' or 'average' classical power $p(D_1) = \int p(D_1|\theta)\,p(\theta)\,d\theta$. This will give the unconditional probability of rejecting H_0. From the discussion above, we might expect this to be a reasonable approximation to the prior-adjusted power $p(D_1,H_1)$ if $p(\theta)$ does not give substantial probability to values of $\theta < 0$.
4. The classical power curve can be averaged with respect to the prior distribution $p(\theta|H_1) = p(\theta|\theta > 0)$, *i.e.* conditional on H_1 being true (since $p(\theta|\theta > 0) = p(\theta,\ \theta > 0)/p(\theta > 0)$, this can be obtained by restricting the prior to $\theta > 0$ and renormalising it to have total probability 1). Brown *et al.*

(1987) recommend this technique as predicting the chance of *correctly* detecting a positive improvement, rather than the overall chance $p(D_1)$ of getting a positive result regardless of the truth. But this method suffers from the same difficulty as the original classical power calculation, in that no account is taken of the plausibility of H_1.

5. The predictive distribution over the possible powers could be displayed as an aid to deciding appropriate sample sizes.

We shall illustrate these options in the following sections, using normal likelihoods and priors.

Prior distributions might be from any of the sources described in Chapter 5, for example subjective assessments (Ten Centre Study Group, 1987), a single previous study (Brown *et al.*, 1987), or a meta-analysis of previous results (DerSimonian, 1996): Example 6.4 illustrates the use of subjective opinion. Most of the applications have assumed a conventional analysis, although Bryant and Day (2000) suggest that a suitable Bayesian perspective is for a trial to be large enough to enable a sceptic and an enthusiast to be brought into consensus.

Finally, it is natural to express a cautionary note on projecting from previous studies (Korn, 1990), and possible techniques for discounting past studies are very relevant (Section 5.4).

Proper Bayesian. As in the Fisherian approach, there is in principle no need for preplanned sample sizes (Lilford *et al.*, 1995). Alternatively, it is natural to focus on the eventual precision of the posterior distribution of the treatment effect: for normal assumptions this is straightforward to calculate. There is an extensive literature on non-power-based Bayesian sample-size calculations (Joseph *et al.*, 1997).

When working within a hypothesis-testing framework, all the above discussion on hybrid classical and Bayesian methods holds, except that the final conclusion of whether the result is 'significant' or not will be based on a posterior distribution rather than a classical analysis. One is still faced with a variety of means of incorporating the prior distribution, although since the conclusions are going to include that prior it seems natural to use its full form and calculate expected power. The necessary formulae for normal likelihoods and priors are provided in Section 6.5.3.

Lee and Zelen (2000) propose a method based on obtaining a high posterior probability of an effective treatment after a 'significant' result, using the analysis described in Section 3.10, *i.e.* by trying to fix $p(H_1|D_1)$. This has been criticised by Simon (2000) and Bryant and Day (2000) as ignoring the actual data observed and hence violating the likelihood principle.

Decision-theoretic Bayesian. If we are willing to express a utility function for the cost of experimentation and the potential benefit of the treatment, then sample sizes can be chosen to maximise the expected utility. Lindley (1997) and discussants argue strongly for this position. Detsky (1985) conducted an early attempt to model the impact of a trial in terms of future lives saved, which required modelling beliefs about the future number to be treated and the true benefit of the treatment, while Claxton *et al.* (2000) and Gittins and Pezeshk (2000), for example, show how sample sizes could be explicitly determined by a trade-off between the cost of the trial and the expected future benefit: for further references, see Section 6.13. This approach also attempts to answer the question 'what is the expected net benefit from carrying out the trial?' (Section 9.10). An intermediate 'information-theoretic' position is taken by Lindley (1997) who does not attempt to model the future benefit of a trial, and instead trades off the information in the posterior distribution against the cost of sampling.

6.5.2   'Classical power': hybrid classical–Bayesian methods assuming normality

We now assume we have a prior distribution to use in our study design, but that the conclusions of the study will be entirely classical and will not make use of the prior, perhaps because of submission to a regulatory authority. Suppose we have a normal prior $\theta \sim N[\mu, \sigma^2/n_0]$ and our future data Y_n have distribution $Y_n \sim N[\theta, \sigma^2/n]$, and we wish to calculate the predictive probability of obtaining a classically 'significant' result when testing the null hypothesis $\theta < 0$. Under a classical analysis (Section 2.5), H_0 will be rejected when the parameter estimate Y_n obeys

$$Y_n > -\frac{1}{\sqrt{n}} z_\epsilon \sigma; \tag{6.2}$$

this event, denoted S_ϵ^C, will occur with probability

$$P(S_\epsilon^C | \theta) = \Phi\left[\frac{\theta \sqrt{n}}{\sigma} + z_\epsilon\right], \tag{6.3}$$

which is the classical power curve previously given in (2.37).

We can plot (6.3) superimposed on the prior $p(\theta)$, which can reveal the relative plausibility of the potential alternative hypotheses and suggest whether the trial is based on over-optimistic assumptions (see Example 6.2). If we wish to calculate the overall unconditional probability of a 'significant' result S_ϵ^C we can integrate (6.3) with respect to the prior. However, it is analytically more straightforward to use the the predictive distribution (3.23)

$$Y_n \sim N\left[\mu, \sigma^2\left(\frac{1}{n_0} + \frac{1}{n}\right)\right]$$

to directly evaluate the chance of the critical event (6.2) occurring, which can be shown to be

$$P(S_\epsilon^C) = \Phi\left[\sqrt{\frac{n_0}{n_0 + n}}\left(\frac{\mu\sqrt{n}}{\sigma} + z_\epsilon\right)\right]. \tag{6.4}$$

The relationship to the power curve (6.3) is clear. As $n_0 \rightarrow \infty$, the prior tends to a lump on μ and $P(S_\epsilon^C)$ tends to the classical power evaluated at the prior mean μ. However, finite n_0 will mean that the expected power is less than the classical power evaluated at the prior mean μ, provided the classical power is greater than 50%. This may be a more realistic assessment of the chance that the trial will yield a positive conclusion.

We note that Table 6.1 can be extended to allow 'equivocal' decisions, and that the necessary probabilities can be calculated using tail areas of the bivariate normal distribution (Spiegelhalter and Freedman, 1986).

Example 6.2   *Bayesian power: Choosing the sample size for a trial*

We revisit Example 2.6, in which a trial for a new cancer treatment is designed to have 80% power to detect a log(hazard ratio) $\theta_A = 0.56$, requiring 100 events when assuming a two-sided α of 0.05. Consider an archetypal enthusiastic prior (Section 5.5.3) centred on the alternative hypothesis and with 5% prior probability that $\theta < 0$. Hence $\theta \sim N[\mu, \sigma^2/n_0]$ where $\mu = 0.56$, $\sigma = 2$ and $\mu - 1.645\sigma/\sqrt{n_0} = 0$, so that $n_0 = 1.645^2\sigma^2/\mu^2 = 34.5$. The classical power curve and the prior are shown on Figure 6.3: the power at the prior mean is 80% as designed, the expected power (6.4) averaging over the entire prior distribution is 0.66, showing the decline from the conditional value of 0.80. If we took the approach recommended by Brown *et al.* (1987) we would average the power curve with respect to the conditional prior $p(\theta|H_1) = p(\theta|\theta > 0)$; this is not straightforward to calculate and is perhaps easiest to evaluate using Monte Carlo methods (Section 3.19.1), from which we find, using the notation of Table 6.1, that $p(D_1|H_1) = 0.70$. Such a value might have been predicted, since we know that $p(H_1) = 0.95$, $p(D_1) = 0.66$, and from (6.1) that $p(D_1) \approx p(D_1|H_1)p(H_1)$.

6.5.3   'Bayesian power'

Suppose we have the same normal prior and likelihood as in Section 6.5.2 but now wish to carry out a fully Bayesian analysis in which the prior will be

(a) Classical (solid) and Bayesian (dashed) power curves

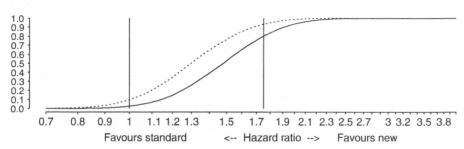

Favours standard   <-- Hazard ratio -->   Favours new

(b) Enthusiastic prior

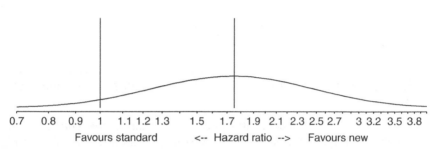

Favours standard   <-- Hazard ratio -->   Favours new

Figure 6.3   Power curves (a) for testing $H_1 : \theta > 0$, designed to have classical power of 80% at $\theta_A = 0.56$ (HR = 1.75). The Bayesian power curve in (a) assumes that the enthusiastic prior shown in (b) is to be included in the analysis.

incorporated. We wish to calculate the predictive probability of obtaining a 'significant' Bayesian result when testing the null hypothesis $\theta < 0$ against an alternative $\theta > 0$, and we shall denote such 'Bayesian significance' as $S_\epsilon^B \equiv P(\theta < 0 | \text{data}) < \epsilon$.

Assuming a future parameter estimate Y_n, we will obtain the posterior distribution

$$\theta | Y_n \sim \mathrm{N}\left[\frac{n_0 \mu + n Y_n}{n_0 + n}, \frac{\sigma^2}{n_0 + n}\right],$$

and so S_ϵ^B will occur when the parameter estimate Y_n obeys

$$Y_n > \frac{-\sqrt{n_0 + n} \, z_\epsilon \, \sigma - n_0 \mu}{n}. \qquad (6.5)$$

For a particular true value of θ, $Y_n \sim \mathrm{N}[\theta, \sigma^2/n]$, and hence it can be easily shown that this event will occur with probability

$$P(S_\epsilon^B | \theta) = \Phi\left[\frac{\theta\sqrt{n}}{\sigma} + \frac{\mu n_0}{\sigma\sqrt{n}} + \sqrt{\frac{n_0 + n}{n}} z_\epsilon\right]. \qquad (6.6)$$

With vague prior opinion, $n_0 \to 0$ and we are left with the standard classical power curve given in (2.37).

Just as in Section 6.5.2, we can plot (6.6) superimposed on the prior $p(\theta)$. To calculate the overall unconditional probability of a 'significant' result S_ϵ^B it is again analytically more straightforward to use the the predictive distribution of Y_n to evaluate the chance of the critical event (6.5) occurring:

$$\begin{aligned} P(S_\epsilon^B) &= P\left(Y_n > \frac{-\sqrt{n_0 + n}z_\epsilon\sigma - n_0\mu}{n}\right) \\ &= \Phi\left[\frac{\mu\sqrt{n_0 + n}\sqrt{n_0}}{\sigma\sqrt{n}} + \sqrt{\frac{n_0}{n}}z_\epsilon\right]. \end{aligned} \qquad (6.7)$$

Example 6.3   *Bayesian power (continued): Choosing the sample size for a trial*

If we are willing to include the prior distribution in the analysis then we obtain the Bayesian power curve (6.6) shown as a dashed line in Figure 6.3(a), which is substantially higher than the classical power curve due to the prior giving a 'head start'. The power at the alternative hypothesis $\theta_A = 0.56$ is 0.93, while the chance of a false rejection of $\theta = 0$ has risen from 0.025 to 0.10 – this inflated chance of a Type I error illustrates the danger of getting the prior 'wrong'. The expected Bayesian power (6.7), averaged with respect to the prior distribution in Figure 6.3(b), is 0.78.

6.5.4   Adjusting formulae for different hypotheses

All the formulae provided so far have assumed that $\theta > 0$ indicates superior performance of the innovative treatment and therefore is the alternative hypothesis of interest – this has simplified the exposition but clearly will not hold in all situations. One option is to redefine the outcome measures and parameters so that θ has the required properties. Alternatively, one can transform the formulae provided, and we now consider the necessary transformations when different hypotheses are being considered.

- **Non-zero threshold.** Suppose the null hypothesis is $H_0: \theta < \theta_0$ and the alternative $H_1: \theta > \theta_0$. Each of the previous formulae can be transformed by subtracting θ_0 from the prior mean μ, the observed statistic y_m and, in conditional power calculations, the parameter θ. For example, suppose in

Example 6.2 that the threshold of interest was changed to $\theta = 0.2$, *i.e.* the posterior interval would need to lie wholly above a log(hazard ratio) of $0.2\,(\mathrm{HR} = 1.22)$ before H_0 is rejected. The conditional power at the alternative hypothesis $\theta_A = 0.56$ is now only 0.56, obtained from transforming (6.6), while the expected power is found from (6.7) to be 0.53.

- **Reversal of hypotheses.** As we have seen in most of our examples, it is common to express benefit from the new intervention as a reduction in risk, and hence on a logarithmic scale to set $H_1 : \theta < 0$. Thus a 'significant' result will be obtained if a final interval lies wholly below 0. If, for example, we were adopting a fully Bayesian approach this would be equivalent to the event $P(\theta > 0|\mathrm{data}) < \epsilon$, which we shall denote S_ϵ^{B-}. Now

$$S_\epsilon^{B-} \equiv [P(\theta > 0|\mathrm{data}) < \epsilon] \equiv [P(\theta < 0|\mathrm{data}) > 1 - \epsilon]$$

and hence, for example,

$$P(S_\epsilon^{B-}) = 1 - P(S_{1-\epsilon}^{B}).$$

Therefore the formulae provided can be transformed by substituting $1 - \epsilon$ for ϵ, and subtracting the result from 1.

For example, suppose in Example 6.2 that the threshold of interest was changed to $\theta = 0.69$, $\mathrm{HR} = 2$, and furthermore we were interested in the expected power to reject the null hypothesis $H_0 : \theta > \theta_0$, *i.e.* we are interested in values of θ with an odds ratio less than 2. Using both transformations on (6.7) leads to

$$P(S_\epsilon^{B-}) = 1 - P(S_{1-\epsilon}^{B}) = 1 - \Phi\left[\frac{(\mu - \theta_0)\sqrt{n_0 + n}\sqrt{n_0}}{\sigma\sqrt{n}} + \sqrt{\frac{n_0}{n}}z_{1-\epsilon}\right]. \qquad (6.8)$$

Then from (6.8) we find the expected power is 0.24: such a low value might be anticipated from the substantial prior support for H_0.

Example 6.4   *Gastric: Sample size for a trial of surgery for gastric cancer*

Reference: Fayers *et al.* (2000).

Intervention: Radical (D2) compared to conventional (D1) surgery for gastric cancer.

Aim of study: Evidence from Japan suggested that more radical surgery was a possible explanation for the better survival rates of patients with gastric cancer, and the UK Medical Research Council initiated a randomised trial to compare survival following radical and conventional surgery.

Study design: Two-group parallel RCT.

Outcome measure: Hazard ratio of death (HR > 1 favours radical treatment).

Planned sample size: The trial was designed under the assumption that the minimum clinically significant difference was a 13.5% improvement in 5-year survival from 20% to 33.5% in patients undergoing conventional surgery – this value for the alternative hypothesis was based on the opinion of the trial team. This is equivalent to a hazard ratio of $\log{(0.20)}/\log{(0.335)} = 1.47$ (Section 2.4.2), or $\log{(HR)} = 0.39$. For the trial to be able to detect a 13.5% difference at the 5% significance level with 90% power, the necessary number of events (*i.e.* deaths) is $n = \sigma^2(1.96 + 1.28)^2/0.39^2 = 276$, when taking $\sigma = 2$ (Section 2.4.2 and (2.38)). The trial was designed to have 200 patients per arm which was predicted to yield this number of events.

Statistical model: For planning purposes, the normal approximation of Section 2.4.2 was adopted, while for analysis a full Cox regression was used to obtain a likelihood for log(HR).

Prospective analysis?: Yes.

Prior distribution: In addition to the three surgical members of the trial steering committee, a further 23 surgeons had their beliefs regarding the likely benefit/harm of radical compared to conventional surgery elicited, both at the start of the trial and later when the trial had stopped but had not yet been published. Fayers *et al.* (2000) shows each individual's prior distribution on a scale representing improvement in 5-year survival, elicited using a similar questionnaire to that of Parmar *et al.* (1994); see Example 5.1. The average distribution had a prior mean of 9.4% improvement over their average assessed control 5-year survival of 21%, although skewness in the distributions gives rise to a median of around 4%. Assuming a baseline survival of 21%, the distribution for an improvement p can be transformed to a log(HR) scale by $\log{(HR)} = \log{(\log{(0.21)}/\log{(0.21 + p)})}$ as in Example 5.1: fitting a normal distribution to the transformed histogram yields a prior with mean $\mu = 0.12$ and standard deviation $\sigma/\sqrt{n_0} = 0.19$, and so $n_0 = 4/0.19^2 = 111$. This corresponds to a hazard ratio of 1.13 (95% interval from 0.78 to 1.64). This distribution is shown in Figure 6.4(a), revealing that the probability of exceeding the alternative hypothesis of $HR = 1.47$ is 8%. Hence, the overall prior beliefs for the surgeons reveal the trial has been designed around a rather optimistic target.

Figure 6.4(b) shows the power curve (6.3) for the trial based on an expected $n = 276$ events, with 90% power at the alternative hypothesis of 1.47. Juxtaposing with Figure 6.4(a) shows that the surgeons' belief is concentrated in an area of rather low power. Indeed, (6.4) shows that the expected power is only 30%, which rises marginally to 31% if a Bayesian final analysis is undertaken (6.7). Even if the surgeons were considerably more optimistic, and their prior mean was set to the alternative hypothesis of HR = 1.47, then the expected power would rise to only 45%.

Loss function or demands: No, but as well as eliciting the beliefs of the surgeons, the authors elicited their demands for radical surgery: around a 10% improvement was judged to be necessary before wishing to routinely implement the more radical surgery, which is more extensive and has extra risk of complications and resource usage.

Computation/software: Conjugate normal model.

Evidence from study: The trial recruited the full 200 patients on each arm, and eventually 281 events were observed (137 under D1, 144 under D2), with a result slightly in favour of the conventional surgery. The observed hazard ratio, based on a Cox regression, was 0.91 (95% CI from 0.72 to 1.15), equivalent to a log(HR) of -0.09 (standard error 0.11, equivalent to an effective number of events of $m = \sigma^2/0.11^2 = 278$, almost exactly the same as the actual number of events observed). The 5-year survival rate in those patients undergoing conventional surgery was 30%, considerably higher than the 20% expected before the trial started. This likelihood is displayed in Figure 6.4(c).

Bayesian interpretation: Figure 6.4(d) displays the predictive distribution for the observed hazard ratio, derived using the methods described in Section 3.13. The probability of observing a result as extreme as that observed is 0.32, twice the shaded area shown in Figure 6.4(d). From Section 5.8 this is Box's measure of conflict between prior and likelihood, and is not particularly extreme even though the prior expectation of a benefit from D2 conflicted with the observed hazard ratio.

Comments: Fayers *et al.* (2000) carried out a second elicitation exercise when the trial was complete but before the results were announced, and found there was still considerable optimism among the clinical collaborators. They conclude that although opinions change over time, those involved in a clinical trial tend to be optimistic and if their prior expectations are used as a naive basis for sample-size calculations, the trial could result in too small a sample size. Nevertheless, in this example the alternative hypothesis was judged to be optimistic even by the participants. A more realistic assessment of the trial's chances of success might be made by taking into account their full uncertainty.

It is also important to monitor such a trial so that it does not continue unnecessarily – in this example the trial might have been stopped and

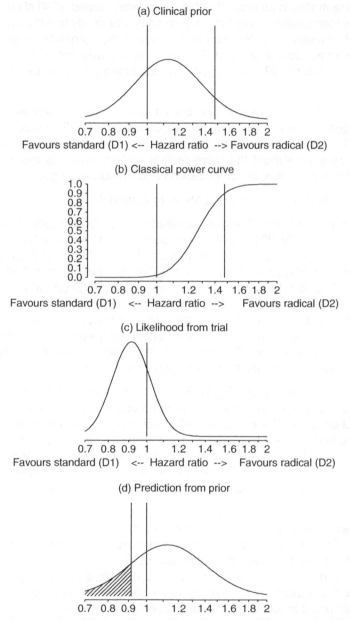

Figure 6.4 The prior assessment (a) for D2 trial in gastric cancer surgery shows some expectation of benefit, but the alternative hypothesis of 1.47 around which the trial has been designed is clearly very optimistic (b). The eventual trial result (c) showed no clear evidence for benefit. The predictive distribution derived from the prior (d) shows that the observed result (HR = 0.91) was not particularly surprising, given the prior opinion as expressed by (a).

rejected an 'important difference' some time before the eventual conclusion. However, as we shall see in Section 6.6.2, it may be more appropriate to monitor using the clinical prior, in order to ensure that the
negative finding is convincing even to enthusiasts.

6.5.5   Predictive distribution of power and necessary sample size

Consider the classical power formula given in (2.37). If we express uncertainty
over the parameters as a prior distribution, then the power can be considered as
an unknown quantity with a distribution induced by this prior. This *predictive*
distribution over the power can best be obtained by simulation methodology:
essentially the unknown parameters are simulated from their prior distribution,
plugged into the formula for the power, and the result recorded. After many
iterations of this procedure a distribution over possible powers is obtained. This
is essentially a *Monte Carlo* procedure (Section 3.19.1) and is illustrated in
Example 6.5.

Example 6.5   *Uncertainty: Predictive distribution of power*

Assume that a randomised trial is planned with n patients in each of two
arms, using a response with standard deviation $\sigma = 1$; hence, the variance
of a contrast between two patients is $2\sigma^2$. The trial is aimed to have Type I
error (two-sided α) of 5%, and 80% power to detect a true difference of
$\theta = 0.5$ in mean response between the groups.

From (2.38) the necessary sample size per group is

$$n = \frac{2\sigma^2}{\theta^2}(z_{0.8} - z_{0.025})^2$$

where $z_{0.8} = 0.84$, $z_{0.025} = -1.96$; note that this differs slightly from (2.38)
as here σ is the standard deviation of a single response.

The necessary sample size is $n = 63$. Suppose, however, that we wish to
express uncertainty concerning both θ and σ. For θ we assess a prior mean
of 0.5 and prior standard deviation of 0.1, while for σ we assume a prior mean
of 1 and standard deviation of 0.3. θ and σ are assumed to be independent
and normally distributed (subject to the constraint of σ being positive).

Using Monte Carlo methods we simulate values of θ and σ from their prior
distributions, substitute them in the sample-size formula above, and so
obtain a predictive distribution over n. This distribution has the properties
shown in Table 6.2 and is plotted in Figure 6.5 – it is clear that there is huge
uncertainty as to the appropriate sample size.

Table 6.2 Properties of predictive distributions of necessary sample size n for fixed power of 80%, and power for fixed sample size $n = 63$.

	Median	95% interval
n	62.5	9.3 to 247.2
Power (%)	80	29 to 100

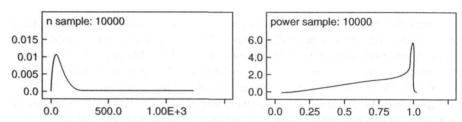

Figure 6.5 Predictive distributions from WinBUGS for necessary sample size n to achieve 80% power, and power for $n = 63$ patients per group.

For fixed n, the power is

$$\text{power} = \Phi\left(\sqrt{\frac{n\theta^2}{2\sigma^2}} + z_{0.025}\right).$$

If we decide to use 63 patients per group, we can simulate potential values for the power using the same methodology. The results are again presented in Table 6.2 and plotted in Figure 6.5, and show that although the median power is 80%, a trial of 63 patients per group could be seriously underpowered. We can calculate other quantities that could give insight into the planned sample size: for example, that there is a 37% chance that the power is less than 70%.

6.6   MONITORING OF SEQUENTIAL TRIALS

6.6.1   Introduction

Whether or not to stop a trial early is a complex ethical, financial, organisational and scientific issue, in which statistical analysis plays a considerable role. Section 4.3 has already demonstrated that sequential analysis might be considered the 'front line' between Bayesian and frequentist approaches, and the monitoring of sequential trials has been said to reach 'to the very foundations of the two paradigms' (Etzioni and Kadane, 1995).

Recommendations concerning early stopping or changes in the conduct of trials increasingly rest in the hands of independent committees known as data and safety monitoring boards or data monitoring committees (DMC). We shall adopt the latter term. In Section 6.6.6 we shall discuss the relevance of the Bayesian perspective to the deliberations of a DMC, where we shall emphasise the ability to incorporate external evidence and formally account for the desire to bring the trial to a conclusive result.

Four main statistical approaches can be identified, again corresponding to the four main entries in Table 4.1:

- *Fisherian.* This is perhaps best exemplified in trials influenced by the Clinical Trial and Services Unit in Oxford, in which protocols generally state (Collins *et al.*, 1995) that the DMC should only alert the steering committee to stop the trial on efficacy grounds if there is '*both* (a) "proof beyond reasonable doubt" that for all, or for some, types of patient one particular treatment is clearly indicated . . . *and* (b) evidence that might reasonably be expected to influence the patient management of many clinicians who are already aware of the results of other main studies'. There is no formal expression of what evidence is required to establish 'proof beyond reasonable doubt' (although $2P < 0.001$ is mentioned as a possible criterion). We also note the explicit, though again informal, appeal to the idea that the results should be convincing to a broad spectrum of opinion, and its close relation to the quote by Kass and Greenhouse (1989) on the need for trials to bring 'conclusive consensus' (Section 6.4.2).

- *Neyman–Pearson.* This classical method attempts to retain a fixed Type I error through prespecified stopping boundaries or guidelines which may be used at prespecified analysis times ('group-sequential methods') or with continuous monitoring. Group-sequential methods boundaries include those of O'Brien and Fleming, which are very conservative at early interim analyses, and Pocock, which have constant nominal 'significance', while continuous methods include alpha-spending functions and triangular boundaries. See Whitehead (1997a) for a detailed review. DeMets (1984) states that 'while they are not stopping rules, such methods can be useful in the decision-making process', although regulatory authorities require good reasons for not adhering to such boundaries (International Conference on Harmonisation E9 Expert Working Group, 1999).

  Objections to this approach from both Fisherian and Bayesian perspectives have already been covered in Section 4.3. In addition, there is no agreed method of estimation following a sequential trial (Freedman, 1996), although frequentist sequential rules are 'prone to exaggerate magnitude of treatment effect' (Pocock and Hughes, 1989) since they would tend to stop when on a random high; Pocock and White (1999) term the tendency for early extreme results to become less impressive as 'regression to the truth'. Armitage (1991a) agrees that adjusted *P*-values are 'too tenuous to be quoted in an authoritative analysis of the data', but still considers frequency properties of stopping rules may be useful guides for 'mental adjustment'.

In practice, a DMC will need to take into account multiple sources of evidence when making its judgement and, if working within the traditional Neyman–Pearson paradigm, classical sequential analysis may be a useful warning against over-interpretation of naive *P*-values. Freidlin *et al.* (1999) provide a useful analysis, pointing out that the role of a trial is to change practice and warning of over-strict adherence to formal stopping procedures.

- *Proper Bayesian.* Probabilities derived from a posterior distribution may be used for monitoring, without formally prespecifying a stopping criterion or even prespecifying a sample size (Berry, 1993). It is natural to use the posterior probabilities of hypotheses of interest as a basis for monitoring (Section 6.6.2), although this may be supplemented by making predictions of the possible consequences of continuing (Section 6.6.3). As for trials with fixed sample size, a hybrid strategy is possible in which prior distributions may be used at the design stage but assuming a Neyman–Pearson analysis (McPherson, 1982). However, if external evidence becomes available during a clinical trial it can be argued that this should be incorporated into a prior distribution.

  There is no direct implication of the Bayesian approach on trial size. Matthews (1995) and Edwards *et al.* (1997) have suggested that small, open trials fit well into a Bayesian perspective in which all evidence contributes and there is no demand for high power to reject hypotheses. Alternatively, monitoring with a sceptical prior may demand larger than standard sample sizes in order to convince an archetypal sceptic about treatment superiority.

- *Decision-theoretic Bayesian.* This assumes we are willing to explicitly assess the losses associated with consequences of stopping or continuing the study, and therefore the trial requires a full specification of the 'patient horizon', the allocation rule and so on. This approach also quantifies the expected benefit of the trial and therefore helps decide whether to conduct the trial at all – see Sections 6.6.4 and 9.10.

6.6.2   Monitoring using the posterior distribution

Following the 'proper Bayesian' approach, it is natural to consider terminating a trial when one is confident that one treatment is better than the other, and this may be formalised by assessing the posterior probability that the treatment benefit θ lies above or below some boundary, such as the ends of the range of equivalence described in Figure 6.1. For example, when comparing two treatments in which θ represents success rates, we might consider stopping in favour of the new treatment and concluding $\theta > 0$ when the posterior probability that $\theta < 0$ is less than some threshold ϵ (we note we are not using α to denote our tail area in order to avoid confusion with expressions for Type I error). In

Section 6.5.3 we denoted this event S_ϵ^B, and for normal prior and likelihood this will occur if the parameter estimate y_m obeys

$$y_m > \frac{-\sqrt{n_0 + m}\, z_\epsilon\, \sigma - n_0\mu}{m};$$ (6.9)

this is equivalent to (6.5) but seen as a retrospective assessment of observed data y_m rather than a prospective view of future data Y_n. Applications of this procedure have been reported in a wide variety of trials (Section 6.13).

   We have already discussed how a well-designed trial should contain sufficient evidence to bring both a sceptic and an enthusiast to broadly the same conclusions (Section 6.4.2) as to whether the treatment is effective or not. This idea may be formalised in the following way, using the concept of sceptical and enthusiastic priors (Section 5.5).

- First, stopping with a 'positive' result (*i.e.* in favour of the new treatment) might be considered if a posterior based on a *sceptical* prior suggested a high probability of treatment benefit.

- Second, stopping with a 'negative' result (*i.e.* that is equivocal or in favour of the standard treatment) may be based on whether the results were sufficiently disappointing to make a posterior based on an *enthusiastic* prior rule out a treatment benefit.

In other words, we should stop if we have convinced a reasonable adversary that they are wrong. Fayers *et al.* (1997) provide a tutorial on such an approach, and Example 6.6 describes its application by a DMC for two cancer trials. In addition, Example 6.7 considers a trial in which the data overwhelmed an optimistic prior centred on a 40% risk reduction, and hence justified assuming a negative result and early stopping with a conclusion of no treatment benefit.

   It is worth considering in more detail the use of a sceptical prior as a basis for monitoring, particularly as it encourages an explicit comparison with classical sequential methods. Suppose we assume a sceptical prior for a treatment difference

$$\theta \sim N\left[0, \frac{\sigma^2}{n_0}\right],$$

and we would consider stopping the trial when the event S_ϵ^B occurs, *i.e.* $P(\theta < 0|\text{data}) < \epsilon$, or equivalently when a symmetric $100(1 - 2\epsilon)\%$ interval lies wholly above 0. From (6.9) this will occur when

$$y_n > \frac{-\sqrt{n_0 + m}\, z_\epsilon\, \sigma}{m}.$$ (6.10)

Let $z_m = y_m\sigma/\sqrt{m}$ be the standardised classical test statistic. Then (6.10) can be rearranged as

$$z_m > -z_\epsilon \sqrt{1 + \frac{n_0}{m}}. \tag{6.11}$$

The term $\sqrt{1 + n_0/m}$ is a multiplier of the 'naive' critical value $-z_\epsilon$, and demonstrates how the sceptical prior opinion introduces conservatism through increasing the critical value.

Suppose $2\epsilon = 0.05$ and hence $-z_\epsilon = 1.96$, and the maximum intended sample size of the trial is n. In Section 5.5.2 we argued that a reasonable 'handicap' might be $n_0/n = 0.26$, based on a trial with 90% power to detect an 'optimistic' difference. Substituting into (6.11), we stop and reject H_0 when

$$z_m > 1.96\sqrt{1 + 0.26\frac{n}{m}}. \tag{6.12}$$

The boundary is a function solely of the proportion m/n of the trial that has been completed, and is shown in Figure 6.6. Assuming a sceptical prior thus

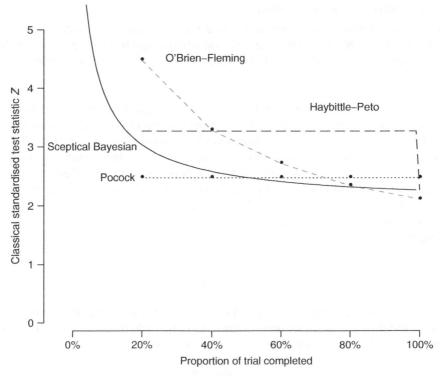

Figure 6.6 Monitoring boundaries for a sceptical prior opinion with $2\epsilon = 0.05$ and handicap 0.26. This is compared to Pocock and O'Brien–Fleming boundaries assuming five equally spaced analyses, and the Haybittle–Peto boundary in which a difference of three standard errors is sought at all interim analyses, and then an unadjusted P-value adopted at the end of trial.

provides a handicap to early stopping: explicit comparison with boundaries obtained by classical sequential methods is made in Figure 6.6 and the qualitative similarity is clear, while a quantitative investigation is made in Section 6.6.5. Other comparisons with frequentist procedures have been carried out by Freedman and Spiegelhalter (1989), DerSimonian (1996) and Freedman *et al.* (1994).

It is also possible to use 'robust priors' (Section 5.6) in which the set of prior distributions leading to a specific conclusion are identified at each interim analysis (Greenhouse and Wasserman, 1995; Carlin and Sargent, 1996). In addition, posterior probabilities of two responses can be monitored jointly and stopping considered when an event of interest, such as either outcome occurring (Etzioni and Pepe, 1994), exceeds a certain threshold. This monitoring scheme has also been proposed for single arm studies and for phase I and II trials (Section 6.12).

Although monitoring using posterior distributions appears intuitive, criticisms of this procedure include its lack of explicit loss function (Section 6.6.4), its sampling properties, and its dependence on the prior (Section 6.6.5).

Example 6.6 *CHART (continued): Monitoring trials using sceptical and enthusiastic priors*

Reference: Parmar *et al.* (1994, 2001) and Spiegelhalter *et al.* (1994). This example has previously been considered in Examples 5.1, 5.3 and 6.1.

Evidence from study: For the lung cancer trial, the data reported at each of the annual meetings of the independent DMC is shown in Table 6.3: the final row is that of the published analysis. Recruitment stopped in early 1995 after 563 patients had entered the trial. It is clear that the extremely beneficial early results were not retained as the data accumulated, although a clinically important and statistically significant difference was eventually found. Perhaps notable is that the DMC recommended continuation of the trial even when the two-sided *P*-value was 0.001, *i.e.* when the data had crossed the Haybittle–Peto boundary.

Table 6.3 Summary data reported at each meeting of the CHART lung trial DMC. Under a proportional hazards assumption with hazard ratio HR, the 2-year survival improvement, s, over a baseline of 15%, obeys HR $= \log (0.15 + s) / \log (0.15)$, which can be rearranged to $s = 0.15^{HR} - 0.15$.

Date	No. patients	No. deaths	Hazard ratio Estimate	Hazard ratio (95% CI)	2-year % survival improvement Estimate	2-year % survival improvement (95% CI)	Two-sided *P*-value
1992	256	78	0.55	(0.35 to 0.86)	20	(5 to 36)	0.007
1993	380	192	0.63	(0.47 to 0.83)	15	(6 to 26)	0.001
1994	460	275	0.70	(0.55 to 0.90)	12	(4 to 20)	0.003
1995	563	379	0.75	(0.61 to 0.93)	9	(3 to 16)	0.004
1996	563	444	0.76	(0.63 to 0.90)	9	(3 to 15)	0.003

Table 6.4 Summary data reported at each meeting of the CHART head-and-neck trial DMC. Two-year survival improvements are based on a baseline of 45% disease-free survival.

Date	No. patients	No. events	Hazard ratio		2-year % survival improvement		Two-sided P-value
			Estimate	(95% CI)	Estimate	(95% CI)	
1992	531	188	0.91	(0.68, 1.21)	3	(−7, 11)	0.50
1993	674	293	0.92	(0.73, 1.16)	3	(−5, 11)	0.16
1994	791	387	0.89	(0.72, 1.09)	4	(−3, 11)	0.20
1995	918	464	0.92	(0.76, 1.11)	3	(−4, 10)	0.33
1996	918	485	0.95	(0.79, 1.14)	2	(−5, 8)	0.52

For the head-and-neck cancer trial, the data reported at each meeting of the independent DMC are shown in Table 6.4. There was no strong evidence of benefit shown at any point in the study.

Bayesian interpretation: For the lung trial, the DMC was presented with survival curves, and posterior distributions and tail areas arising from a reference prior (uniform on a log(HR) scale). In view of the positive findings, the posterior distribution resulting from the sceptical prior derived in Example 5.3 was presented, in order to check whether the evidence was sufficient to persuade a reasonable sceptic.

Figure 6.7 shows the sceptical prior distributions at the start of the lung cancer trial, and the likelihood (essentially the posterior under the reference prior) and posterior for the results available in subsequent years. Under the reference prior there is substantial reduction in the estimated effect as the extreme early results are attenuated, while the sceptical results are remarkably stable and the initial estimate in 1992 is essentially unchanged as the trial progresses. The detailed results under the sceptical prior are shown in Table 6.5. Before the trial the clinicians were demanding a 13.5% improvement before changing treatment: however, the inconvenience and toxicity were found to be substantially less than expected and so probabilities of improvement are shown for 0% and 7%, around half the initial demands. Such 'shifting of the goalposts' is entirely reasonable provided it is not based on the primary outcome results.

Table 6.5 Estimates presented to CHART DMC in successive years (apart from 1996, which are the final published data) for lung cancer trial, obtained under a sceptical prior distribution. Posterior probabilities are presented for 'no improvement from CHART' (analogous to one-sided P-values), and for 'practically significant improvement from CHART'.

Date	No deaths	Estimated hazard ratio (HR)	2-year % survival improvement (95% CI)		P (imp. < 0%) i.e. HR > 0	P (imp. > 7%) i.e. HR < 0.80
1992	78	0.79	7	(−1 to 17)	0.048	0.56
1993	175	0.73	10	(3 to 18)	0.006	0.73
1994	275	0.78	8	(2 to 15)	0.009	0.60
1995	379	0.80	7	(1 to 13)	0.010	0.48
1996	444	0.81	7	(2 to 12)	0.003	0.52

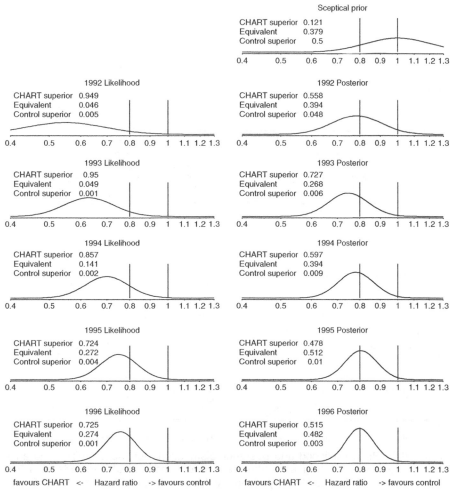

Figure 6.7 Prior, likelihood and posterior distributions for the CHART lung cancer trial assuming a sceptical prior. The likelihood becomes gradually less extreme, providing a very stable posterior estimate of the treatment effect when adopting a sceptical prior centred on a hazard ratio of 1. Demands are based on a 7% improvement from 15% to 22% 2-year survival, representing a hazard ratio of 0.80.

The sceptical posterior distribution is centred around these clinical demands, showing that these data should persuade even a sceptic that CHART both improves survival and, on balance, is the pragmatic treatment of choice.

Since the results for the head-and-neck trial were essentially negative, it is appropriate to monitor the trial assuming a enthusiastic prior in order to see if it is sufficiently convincing even to optimists. The results are shown in

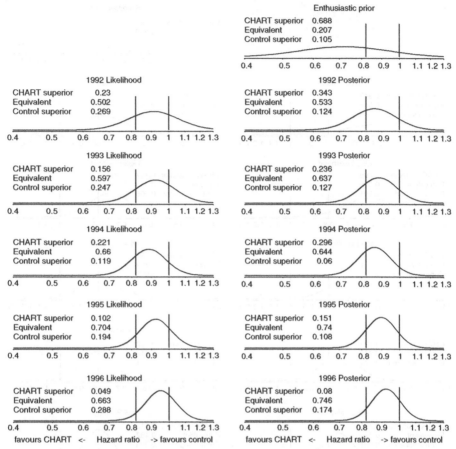

Figure 6.8   Prior, likelihood and posterior distributions for the CHART head-and-neck cancer trial assuming an enthusiastic prior, and clinical demands of a 7% improvement from 45% to 52% 2-year survival, equivalent to a hazard ratio of 0.82.

Figure 6.8, using the clinical prior derived in Example 5.1. The initial clinical demands were a 13% improvement in survival from 45% to 58%, but in parallel with the lung trial we have reduced this to a 7% improvement. The results remain equivocal, and should be sufficient to convince a reasonable enthusiast that, on the basis of the trial evidence, CHART is not of clinical benefit in head-and-neck cancer.

Sensitivity analysis: The three priors provide the sensitivity analysis.

Comments: There are two important features of the prospective Bayesian analysis of the CHART trial. First, while classical stopping rules may well have led the DMC to stop the lung trial earlier, perhaps in 1993 when the two-sided *P*-value was 0.001, this would have overestimated the benefit.

The DMC allowed the trial to continue, and consequently produced a strong result that should be convincing to a wide range of opinions. Second, after discovering that the secondary aspects of the new treatment were less unfavourable than expected, the DMC is allowed to 'shift the goalposts' and not remain with unnecessarily strong clinical demands.

6.6.3   Monitoring using predictions: 'interim power'

Investigators and funders are often concerned with the question – given the data so far, what is the chance of getting a 'significant' result? This is closely related to the concept of 'futility', and the traditional approach to this question is 'stochastic curtailment' (Halperin *et al.*, 1982) which calculates the conditional power of the study, given the data so far, for a range of alternative hypotheses: this might also be termed 'interim power'.

The following formulae assume we are interested in predicting whether future data will result in a posterior probability, or a one-sided *P*-value, for the null hypothesis H_0: $\theta < 0$, being less than ϵ, *i.e.* either the event S_ϵ^B or S_ϵ^C. One can make the appropriate adjustments for H_0: $\theta > 0$ and non-zero thresholds using the methods described in Section 6.5.4.

'Hybrid' predictions: using a prior and current data to predict a future classical analysis. It is straightforward to calculate predictive probabilities of eventual classical conclusions if we assume a normal likelihood. Suppose we have observed a parameter estimate y_m based on our current sample size m, and are considering a further n observations which will yield a parameter estimate Y_n. Then, since

$$\frac{my_m + nY_n}{m+n} \sim \mathrm{N}\left[\theta, \; \frac{\sigma^2}{m+n}\right],$$

after these observations we shall have a classically 'significant' result S_ϵ^C provided that

$$Y_n > \frac{-\sqrt{m+n}\, z_\epsilon\, \sigma - my_m}{n}. \tag{6.13}$$

Since $Y_n \sim \mathrm{N}[\theta, \; \sigma^2/n]$, the probability of this occurring, as a function of the observed data and unknown θ, is

$$P(S_\epsilon^C | y_m, \theta) = \Phi\left[\frac{\sqrt{n}\,\theta}{\sigma} + \frac{m\,y_m}{\sigma\sqrt{n}} + \sqrt{\frac{m+n}{n}}\, z_\epsilon\right]; \tag{6.14}$$

we note that this is exactly the form of the pre-trial Bayesian power curve (6.6) but replacing the 'imaginary' prior data with the observed real data. Equation (6.14) is known as the 'conditional power curve' and forms the basis for a stochastic curtailment procedure, in which this curve may be plotted and its value examined at the null, alternative and other values of θ.

It does not, however, seem reasonable to condition on a hypothesis that is no longer tenable (Spiegelhalter *et al.*, 1986; Dignam *et al.*, 1998). From a Bayesian perspective it is natural to average such conditional powers with respect to the current posterior distribution, just as the pre-trial power was averaged with respect to the prior to produce the average or expected power (Section 6.5). By again using the predictive distribution (3.24) of Y_n we can calculate the probability of S_ϵ^C to be

$$
p(S_\epsilon^C|y_m, \text{ prior}) = \Phi\left(\sqrt{\frac{n_0 n}{(n_0 + m)(n_0 + m + n)}} \, \frac{\sqrt{n_0}\mu}{\sigma} \right.
$$
$$
\left. + \sqrt{\frac{m(n_0 + m + n)}{n(n_0 + m)}} \, \frac{\sqrt{m}y_m}{\sigma} + \sqrt{\frac{(m + n)(n_0 + m)}{n(n_0 + m + n)}} \, z_\epsilon \right).
$$

(6.15)

We note that if $m = 0$ there are no current data and (6.15) can be shown to reduce to the pre-trial average classical power given by (6.4).

Bayesian predictions: using a prior and current data to predict a future Bayesian analysis. In a fully Bayesian analysis the posterior distribution will eventually be

$$
\theta|y_m, Y_n \sim \text{N}\left[\frac{n_0\mu + my_m + nY_n}{n_0 + m + n}, \frac{\sigma^2}{n_0 + m + n} \right].
$$

Having observed Y_n, we shall assume that we are interested in a 'significant' result S_ϵ^B which we have defined as the event $p(\theta < 0|y_m, Y_n) < \epsilon$, *i.e.* the tail area of the posterior is less than ϵ. This result will occur if

$$
Y_n > \frac{-\sqrt{n_0 + m + n} \, z_\epsilon \, \sigma - (n_0\mu + my_m)}{n}.
$$

(6.16)

Since $Y_n \sim \text{N}[\theta, \sigma^2/n]$, the probability of this event occurring, as a function of the observed data and unknown θ, is

$$
P(S_\epsilon^B|y_m, \theta) = \Phi\left[\frac{\sqrt{n}\theta}{\sigma} + \frac{my_m}{\sigma\sqrt{n}} + \frac{n_0\mu}{\sigma\sqrt{n}} + \sqrt{\frac{n_0 + m + n}{n}} \, z_\epsilon \right].
$$

(6.17)

Equation (6.17) can be thought of as a general form of all the other conditional power curves we have previously derived: if $n_0 = 0$ we have no prior input and we obtain the classical conditional power curve in (6.14); if $m = 0$ we obtain the Bayesian power curve in (6.6); while if $n_0 = 0$, $m = 0$ we obtain the standard power curve in (6.3).

Expression (3.24) gives the predictive distribution of Y_n, and from this we can calculate the unconditional probability of S_ϵ^B to be

$$p(S_\epsilon^B | y_m, \text{ prior}) = \Phi \left[\frac{\sqrt{n_0 + m + n}}{\sqrt{(n_0 + m)n}} \frac{(n_0 \mu + m y_m)}{\sigma} + \sqrt{\frac{n_0 + m}{n}} z_\epsilon \right]. \qquad (6.18)$$

Classical predictions: using only current data to predict a future classical analysis. If we wish to ignore prior opinion both in the prediction and in the reporting then we can set $n_0 = 0$ in either (6.15) or (6.18) and obtain a predictive probability of a significant result as

$$p(S_\epsilon^C | y_m) = \Phi \left[\frac{\sqrt{m + n}}{\sqrt{n}} \frac{\sqrt{m} y_m}{\sigma} + \sqrt{\frac{m}{n}} z_\epsilon \right]. \qquad (6.19)$$

This can be expressed solely in terms of the current standardised test statistic $z = \sqrt{m} y_m / \sigma$ and the fraction $f = m/(m + n)$ of the trial so far completed, to give the probability that the future tail area below 0 is less than ϵ as

$$p(S_\epsilon^C | y_m) = \Phi \left[\frac{z + \sqrt{f} z_\epsilon}{\sqrt{1 - f}} \right]. \qquad (6.20)$$

Values of this quantity are plotted in Figure 6.9, which reveals that predicted probabilities of success are often surprisingly low.

The technique has been used with results that currently show approximate equivalence between treatments to justify the 'futility' of continuing a trial (Ware *et al.*, 1985), and may be particularly useful for DMCs and funders when accrual or event rates are lower than expected (Korn and Simon, 1996; Abrams, 1998). Example 6.7 provides a practical illustration of its use by a DMC. The method does not, strictly speaking, require a Bayesian justification, since the predictions can be based on a 'pivotal quantity' that does not depend on the parameter (Armitage, 1989): the 'B-value' of Lan and Wittes (1988) enables calculation of the predictive probability of significance. Frei *et al.* (1987) and Hilsenbeck (1988) provide practical examples of stopping studies due to the futility of continuing; see Section 6.13 for further references.

In spite of the attraction of making such predictions at interim analyses, we follow Armitage (1991b) in warning against using this predictive procedure as any kind of formal stopping rule. It gives an undue weight to 'significance', and makes strong assumptions about the direct comparability of future data with

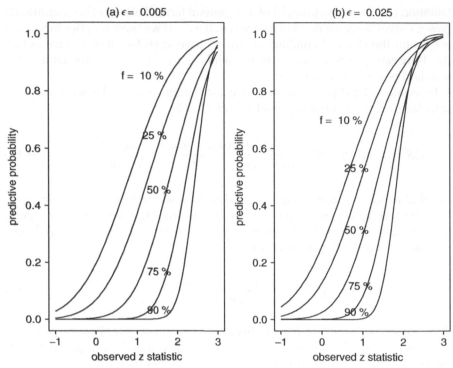

Figure 6.9   Predictive probability $\Phi[(z + \sqrt{f}z_\epsilon)/\sqrt{1-f}]$ of obtaining a classically significant result (two-sided $P = 0.01$ or 0.05, *i.e.* $\epsilon = 0.005$ or 0.025), given a fraction f of the study completed ($f = 10\%$, 25%, 50%, 75% and 90%) and current standardised test statistic z. For example, if one is half-way through a study ($f = 50\%$), and the treatment effect is currently one standard error away from 0 ($z = 1$), then based on this information alone there is only a 29% chance that the trial will eventually show a significant (two-sided $P = 0.05$) benefit of treatment.

those data already observed – for example, if future data involve extended follow-up there may be undue reliance on an assumption of proportional hazards.

Example 6.7   *B-14: Using predictions to monitor a trial*

Reference: Dignam *et al.* (1998).

Intervention: Long-term tamoxifen therapy for prevention of recurrence of breast cancer.

Aim of study: To estimate disease-free survival benefit from tamoxifen over placebo, in patients who already have had 5 years of taking tamoxifen without a recurrence.

Study design: Sequential randomised controlled study (National Surgical Adjuvant Breast and Bowel Project (NSABP) B-14) using O'Brien–Fleming stopping boundaries. Interim analyses were planned at intervals of approximately 1–1.5 years beginning in the fourth year of the study.

Outcome measure: Disease-free survival.

Planned sample size: To detect a 40% reduction in annual risk associated with tamoxifen (hazard ratio = 0.6), with 85% power and a one-sided tail area of 5%, 115 events were required. It had been planned that 624 patients were to be randomised, but eventually 1172 were recruited due to a lower than expected event rate.

Statistical model: Proportional hazards regression model, with summary using the approximate hazard ratio analysis. Following Section 2.4.2, if there are O_T events on treatment, and O_C events on control, then $2(O_T - O_C)/m$ is an approximate estimate of the log(hazard ratio) θ, with mean θ and variance $4/m$.

Prospective Bayesian analysis?: No, the DMC used conditional power and current data in order to make decisions.

Prior distribution: An 'enthusiastic' (or optimistic) prior was centred on a 40% hazard reduction and a 5% chance of a negative effect, *i.e.* HR > 1, equivalent on the log(HR) scale to a normal prior with mean -0.51 and standard deviation 0.31 ($\sigma = 2$, $n_0 = 41.4$). Also a sceptical prior was adopted with the same standard deviation as the enthusiastic prior but centred on 0, thus displaying a 5% chance of the true difference exceeding the alternative hypothesis of 40% hazard reduction.

Loss function or demands: No explicit loss function or range of equivalence.

Computation/software: Conjugate normal analysis.

Evidence from study: The DMC was presented with the data in Table 6.6. Unexpectedly, the results favoured the control treatment. At the third analysis in June 1995, there was a nominal two-sided $P = 0.01$ using the full survival data; this was not sufficient to cross the O'Brien–Fleming stopping boundary which demands two-sided $P < 0.003\,46$. Eighty-eight of the planned 115 events had been observed, and the DMC calculated that even if all 27 remaining events occurred in the control arm, the final results would still not 'significantly' favour tamoxifen. The DMC also considered the conditional power if the trial was extended until 229 events were observed – this was less than 50% for HR $= 0.5$ in favour of tamoxifen, and 15% for HR $= 0.6$. Since these hazard ratios were implausible in the light of the current data, the DMC recommended stopping the trial since the data favoured the control treatment and there

Table 6.6 Summary data from B-14 trial, with hazard ratios and *P*-values estimated using approximate normal analysis based only on the total number of events.

Date	No. events (O_C) on placebo	No. events (O_T) on tamoxifen	Estimated log(HR) (SD)	Estimated hazard ratio (95% CI)	Two-sided *P*-value
Sept. 1993	18	28	0.435 (0.295)	1.54 (0.87 to 2.75)	0.140
Sept. 1994	24	43	0.567 (0.244)	1.76 (1.09 to 2.85)	0.020
June 1995	32	56	0.545 (0.213)	1.72 (1.14 to 2.62)	0.010
Dec. 1995	36	66	0.588 (0.198)	1.80 (1.22 to 2.65)	0.003
Dec. 1996	50	85	0.519 (0.172)	1.68 (1.20 to 2.35)	0.003

was negligible chance of the conclusions being reversed. Further events were subsequently observed and are shown in Table 6.6.

Bayesian interpretation: Figure 6.10 shows the consequences of assuming the sceptical and enthusiastic (optimistic) priors considered by Dignam

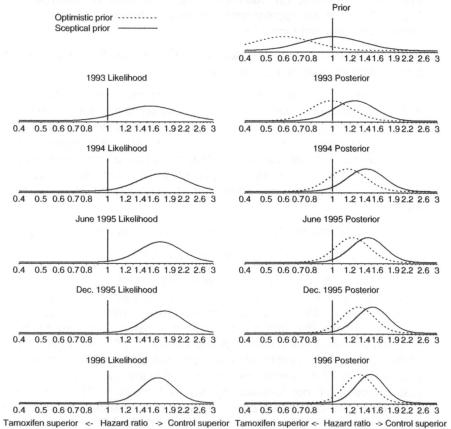

Figure 6.10 Sceptical and 'optimistic' prior distributions, likelihoods and posterior distributions at meetings of the DMC for the B-14 trial. The strong likelihood brings sceptics and enthusiasts into agreement.

et al. (1998). At the first interim analysis the evidence against tamoxifen is sufficient to bring an 'optimist' into a situation of equipoise, with a posterior mean of almost exactly 0. It is clear that by the end of the trial the likelihood is sufficiently in favour of control to bring the two extremes of opinion substantially into agreement.

We may use the results in Section 6.6.3 to calculate the predictive probability of the consequences of continuing the trial up to 115 events, based on the data observed at each of the five interim analyses. We first consider the situation after the first interim analysis in 1993 when 46 events had been observed. Three prior assumptions are examined: a reference analysis (essentially a classical analysis with no adjustment for repeated looks at the data), and sceptical and 'optimistic' analyses using the priors derived above. Each column in Figure 6.11 is headed by the posterior distribution under each assumption, and below are shown the conditional probability of obtaining different conclusions at the planned end of the trial, *i.e.* after a further $115 - 46 = 69$ events have occurred. The conclusions are: 'tamoxifen superior', defined as a 95% posterior interval for the hazard ratio lying wholly below 1; 'equivocal', defined as a 95% posterior interval including 1; and 'control superior', defined as a 95% posterior interval lying wholly above 1. Conditional on each value of $\theta = \log(\mathrm{HR})$, the probabilities of these outcomes can be obtained from (6.17) by substituting the appropriate values for the prior distribution.

Under the reference analyses, the chance of concluding in favour of control is fairly substantial for true hazard ratios greater than 1.5, and such values are supported by the current posterior distribution. The chance of finding in favour of tamoxifen is negligible unless the true hazard ratio is as low as 0.4, which is essentially ruled out by the reference posterior. Integrating the power curves with respect to the reference posterior provides the expected powers shown in the first column of Table 6.7. These probabilities can be obtained as follows. The current z statistic in favour of control is $0.435/0.295 = 1.475$, the fraction of the trial completed is $f = 46/115 = 0.4$, and $\epsilon = 0.025$. From Figure 6.9 we can read off that the expected power is approximately 0.6, and substituting in (6.20) gives the exact value of 0.619. For the expected power to find in favour of tamoxifen, we can take one minus the expected power for control when $\epsilon = 0.975$, which is 0. The unconditional probability of finishing with an equivocal result is simply one minus the other expected powers.

The sceptical analysis has a greater tendency to find an equivocal result as the sceptical prior will be included in the final analysis, and this is reflected in both the conditional power curves and the expected powers

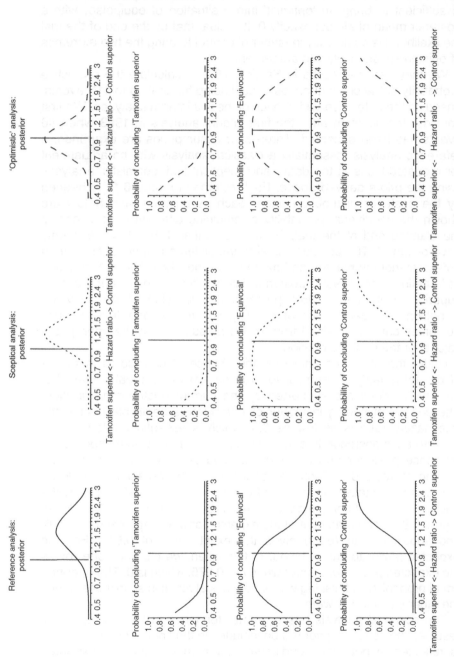

Figure 6.11 Alternative predictions that could be made at the first interim analysis in 1993. The reference analysis uses the data alone, and the power curves are the standard conditional power curves for each of the three possible conclusions after a further $115 - 46 = 69$ events are observed. The sceptical and optimistic analyses show the conditional power for each possible conclusion assuming the prior is to be used in the analysis.

Table 6.7 Probabilities of eventual conclusions for the B-14 trial after the first interim analysis in 1993. Three different prior assumptions are considered, first with the prior to be used in the analysis as well as the predictions, and then with the prior not being used in the final analysis.

Final conclusion	Reference	When using prior in analysis		When *not* using prior in analysis	
		Sceptical	'Optimistic'	Sceptical	'Optimistic'
'Tamoxifen superior'	0.000	0.000	0.017	0.000	0.003
'Equivocal'	0.380	0.724	0.972	0.610	0.846
'Control superior'	0.619	0.276	0.011	0.390	0.151

shown in Table 6.7. The optimistic analysis is even more reluctant to draw a firm conclusion given its current balanced opinion, and firmly (and wrongly, with hindsight) predicts an equivocal result at the end of the trial.

In practice it is likely that the final analysis of the trial would be classical, and therefore it is of interest to carry out a 'hybrid' or mixed prediction in which the prior is used for prediction but not for analysis. This essentially means that the classical conditional power curves shown in the first column of Figure 6.11 are averaged with respect to the sceptical or optimistic posterior distributions. The results are shown in the last two columns of Table 6.7. The chance of finding a result in favour of control is strengthened.

The consequences of making mixed predictions at each interim analysis are shown in Figure 6.12; only the chances of obtaining a conclusion in favour of control are shown, as the chance of finding in favour of tamoxifen is less than 0.003 in all cases.

Sensitivity analysis: Dignam *et al.* (1998) considered a range of prior distributions with means varying between optimistic and sceptical – we have just illustrated the extremes of this range.

Comments: A predictive calculation suggests that continued follow-up would almost certainly not lead to evidence of benefit for tamoxifen. However, when the DMC recommended stopping at the third interim analysis, Figure 6.10 shows that an optimist could still have 13% belief in a benefit from tamoxifen, and therefore would not rule out further trials. Dignam *et al.* (1998) defend the decision to stop and state that 'even an advocate of continued testing of the question might argue that we should have closed and reported the B-14 study, if for no other reason than to make way for a confirmatory trial in which participants could be adequately consented'.

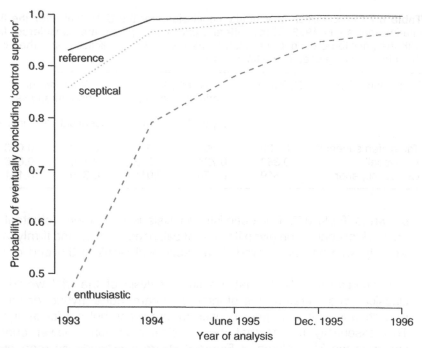

Figure 6.12 Predictive probability of reaching the conclusion 'control superior' at the end of the trial, under different prior assumptions but assuming a classical analysis. The predictive probability of a 'significant' result in favour of tamoxifen is negligibly small and is not shown. At the third interim analysis (June 1995), even an enthusiast would admit only a 16% chance of eventually drawing any conclusion except that control was superior.

6.6.4   Monitoring using a formal loss function

The full Bayesian decision-theoretic approach requires the specification of losses associated with all combinations of possible true underlying states and all possible actions. The decision whether to terminate a trial is then, in theory, based on whether termination has a lower expected loss than continuing, where the expectation is with respect to the current posterior distribution, and the consequences of continuing have to consider all possible future actions. This 'backwards induction' requires the computationally intensive technique of 'dynamic programming' and typically makes practical implementation troublesome. There is also an extensive theoretical literature on sequential trials designed from a non-Bayesian decision-theoretic perspective (Bather, 1985).

However, reasonably straightforward solutions can be found in some somewhat idealised circumstances. For example, Anscombe (1963) considers n pairs of patients randomised equally to two groups, a total patient horizon of N, a uniform prior on true treatment benefit, and a loss function proportional to the number of patients given the inferior treatment times the size of the inferiority. He concludes it is approximately optimal to stop and give the 'best to the rest' when the standard one-sided P-value is less than n/N – half the proportion of patients already randomised.

Berry and Pearson (1985) and others have extended such theory to allow for unequal stages and so on, while Carlin *et al.* (1998) claim backwards induction is computationally feasible using Markov chain Monte Carlo methods, in which forward sampling is used as an approximation to the optimal strategy.

As an illustrative (but retrospective) example, Berry *et al.* (1994) consider a trial of influenza vaccine for Navajo children. They construct a theoretical model consisting of priors for the effectiveness of the vaccine and the placebo treatment, the probability of obtaining regulatory approval and the time taken to obtain it, and the probability of a superior vaccine appearing in the next 20 years and the time taken for it to appear. After each month the expected number of cases of the strain amongst Navajo children in the next 20 years is calculated in the case of stopping the trial and of continuing the trial (the latter being calculated by dynamic programming). The trial is stopped when the former exceeds the latter.

As already discussed in Section 6.2, the level of detail required for such an analysis has been criticised as being unrealistic (Breslow, 1990), but it has been argued that trade-offs between benefits for patients within and outside the trial should be explicitly confronted (Etzioni and Kadane, 1995) and decision theory used to decide whether a trial is worth embarking on in the first place (Section 9.10).

6.6.5 Frequentist properties of sequential Bayesian methods

Although the long-run sampling behaviour of sequential Bayesian procedures is irrelevant from the strict Bayesian perspective, a number of investigations have taken place which generally show good sampling properties (Rosner and Berry, 1995). In particular, Grossman *et al.* (1994) explore the sampling properties of the boundaries described in (6.11) arising from assuming a sceptical prior (Section 5.5) centred on zero and with 'sample size' n_0, and a planned maximum experimental sample size n. They estimate by simulation and interpolation the values for the 'handicap' n_0/n that would give rise to an overall Type I error of 5% and 1% for different numbers of equally spaced interim analyses. The results in Table 6.8 show the required handicap is fairly stable over a range of designs: in particular, the boundaries displayed in Figure 6.6, based on an 'imaginary' prior trial of around 26% of the planned sample size, will have Type I error around 5% for five interim analyses. Grossman *et al.* (1994) also show this boundary has good power and expected sample size. Thus an 'off-the-shelf' Bayesian procedure assuming a sceptical prior essentially mirrors the conservative behaviour of the Neyman–Pearson approach. The sampling properties of Bayesian designs has been particularly investigated in the context of phase II trials (Section 6.12).

Table 6.8 Handicaps to fix Type I error rate when monitoring using a sceptical prior for different number of analyses: the handicap is n_0/n, the ratio of the prior 'sample size' to the maximum intended sample size.

Number of analyses	'Handicap' for two-sided $\alpha = 0.05$	'Handicap' for two-sided $\alpha = 0.01$
1	0	0
2	0.16	0.11
3	0.22	0.15
4	0.25	0.17
5	0.27	0.18
6	0.29	0.20
7	0.30	0.21
8	0.32	0.22
9	0.33	0.22
10	0.33	0.23

One contentious issue is 'sampling to a foregone conclusion' (Armitage *et al.* 1969). This mathematical result proves that repeated calculation of posterior tail areas will, even if the null hypothesis is true, eventually lead a Bayesian procedure to reject that null hypothesis. This does not, at first, seem an attractive frequentist property of a Bayesian procedure. Nevertheless, Cornfield (1966) argued that 'if one is seriously concerned about the probability that a stopping rule will certainly result in the rejection of a null hypothesis, it must be because some possibility of the truth of the hypothesis is being entertained', and if this is the case then one should be placing a lump of probability on it, as discussed in Section 5.5, and so fit within the Bayesian hypothesis-testing framework (Section 3.3). He shows that if such a lump, however small, is assumed then the problem disappears in the sense that the probability of rejecting a true null hypothesis does not tend to one. Armitage (1990) is not persuaded, claiming that even with a continuous prior distribution with no lump at the null hypothesis, one might still be interested in Type I error rates at the null as giving a bound to those at non-null values.

A somewhat more subtle objection, well described by Rosenbaum and Rubin (1984), is that the properties of a Bayesian stopping rule based on posterior tail areas may be over-dependent on the precise prior distribution (Jennison, 1990). A possible response is that Bayesian stopping should not be based on a strict rule derived from a single prior, and instead a variety of reasonable perspectives investigated and a trial stopped only if there is broad convergence of opinion.

6.6.6   Bayesian methods and data monitoring committees

A DMC is charged with both safeguarding the patients involved in a trial, and ensuring the quality of a trial's conduct and conclusions. The principles and practice of DMCs are fully discussed in Ellenberg *et al.* (2002), and here we restrict ourselves to the possible impact of Bayesian methods on a DMC's deliberations. Perhaps the most relevant elements are the ability to use external evidence as a basis for prior opinion in any analysis, and the formalisation through sceptical and enthusiastic priors of the wide range of clinical opinion that it may be necessary to convince before a trial's results have the appropriate impact. As outlined in Section 6.6.4, a full decision-theoretic approach would be attractive but difficult to put into practice in a convincing manner, although Kadane *et al.* (1998) report an intention to elicit prior distributions and utilities from members of the DMC for a large collaborative cancer trials group (NSABP), and use the forward sampling approach to solve the dynamic programming problem. Their success in this ambitious venture remains to be seen.

At an interim analysis of trial data, a DMC may be faced with a variety of possible recommendations that it can make concerning the future conduct of the trial. Using the structure of Altman *et al.* (2004), these may include the following:

- *The study should stop completely.* We have already seen in Example 6.6 how a DMC might use Bayesian methods in order to inform a recommendation whether to stop in favour of an apparent benefit of the new intervention on a primary outcome measure, possibly through using a sceptical prior to assess the degree to which the results would be convincing to a wide range of opinion. Similarly, in Example 6.7 we saw how an enthusiastic prior can be used to temper claims for apparent benefit in the control group. The DMC might also recommend stopping because of safety concerns on secondary outcomes, although these may not be so amenable to formal stopping procedures. A recommendation to stop could also be influenced by a 'futility' argument which assesses the chance of ever reaching a particular conclusion were the trial to continue, and this naturally falls into the framework outlined in Section 6.6.3. Finally, there may be convincing evidence of equivalence or non-inferiority: while a frequentist framework requires prespecification of this as an objective of the trial with pre-chosen limits, a Bayesian analysis allows the 'goalposts' to change as the trial progresses and hence a DMC can make such a recommendation on the basis of all currently available evidence. In all these deliberations the DMC is free to incorporate external evidence, such as recently published studies, into a prior opinion.

- *Part of the study should stop.* A recommendation could be made for randomisation to cease for a subgroup of patients or one of many arms in a multi-arm trial. Hierarchical models may be useful in these contexts: again stopping might be based on posterior tail areas to assess the extent to which available evidence would convince a wide body of clinical opinion.

- *The study should continue with modifications.* Design changes such as additional interim analysis, extending recruitment or extending follow-up time can have serious implications for frequentist designs that have pre-set criteria for assigning statistical significance based on pre-set design characteristics. A Bayesian analysis is completely unaffected by such decisions and so a DMC is given considerably more freedom to adapt trial designs.

Of course, a DMC that adopts a Bayesian approach must do so in full recognition of any regulatory issues, and in such a context it would currently be unwise not to carry out such an analysis in parallel with a traditional analysis – see Section 9.12 for future discussion of regulatory acceptance of Bayesian analyses.

6.7   THE ROLE OF 'SCEPTICISM' IN CONFIRMATORY STUDIES

After a clinical trial has given a positive result for a new therapy, there remains the problem of whether a confirmatory study is needed. Fletcher *et al.* (1993) argue that the first trial's results might be treated with scepticism, and Berry (1996b) claims that using a sceptical prior is a means of dealing with 'regression to the mean', in which early extreme results tend to return to the average over time. Example 6.8 illustrates the potential value of this approach.

Example 6.8   *CALGB: Assessing whether to perform a confirmatory randomised clinical trial*

Reference: Parmar *et al.* (1996).

Intervention: Adjunct chemotherapy for non-small-cell lung cancer.

Aim of study: To compare adjunct chemotherapy with radiotherapy alone.

Study design: A RCT conducted by the Cancer and Leukemia Group B (CALGB) between 1984 and 1987 planned to enrol 240 patients with locally advanced stage III non-small-cell lung cancer and to observe approximately $n = 190$ deaths. From (2.38), this design has 80% power to detect at the 5% level a log(hazard ratio) of $\theta_A = (z_{0.8} - z_{0.025})$ $\sigma/\sqrt{n}$ where $\sigma = 2$ (Section 2.4.2). Thus $\theta_A = 0.405$, corresponding to a hazard ratio (HR) of $\exp(-0.405) = 0.67$, where HR < 1 favours new over standard therapy.

Outcome measure: Full survival data were available, with results presented in terms of estimates of HR, the 2-year survival improvement, and the median improvement in survival in months. From Section 2.4.2, the relation between these quantities is as follows. Let the 2-year survival probability under the standard and new therapies be p_S and p_N, respectively. Then, assuming proportional hazards, HR $= \log(p_N)/\log(p_S)$. Further, let the median survival time under the standard and new therapies be s_S and s_N, respectively. If we assume an exponential survival distribution (constant hazard rate), then HR $= s_S/s_N$.

Statistical model: Proportional hazards model, providing an approximate normal likelihood for $\theta = \log(HR)$ (Section 2.4.2).

Prospective analysis?: The Bayesian analysis was carried out retrospectively.

Prior distribution: A default reference (uniform on the log(HR) scale) prior was termed 'enthusiastic' by Parmar *et al.* (1996). They also derived a sceptical prior by the method described in Section 5.5.2, with mean 0 and standard deviation $\sigma/\sqrt{n_0}$. The original alternative hypothesis was $\theta_A = \log(0.67) = -0.405$, and a prior centred at zero and with 5% chance of exceeding this value would have standard deviation $0.405/1.645 = 0.246$. Using $\sigma = 2$, this is equivalent to a 'prior sample' of size $n_0 = (2/0.246)^2 = 66$. Figure 6.13 shows this sceptical prior distribution with a median HR of 1, which is equivalent to an 'imaginary' trial in which 33 patients died on each treatment.

Loss function or demands: Parmar *et al.* (1996) argue that it might be reasonable to demand an improvement equal to the alternative hypothesis of a hazard ratio of 0.67, or an additional 5 months' median survival. The sceptical prior expresses a probability of 45% that the true benefit lies in the range of equivalence.

Evidence from study: The trial stopped early after enrolling 156 patients and observing the data shown in Table 6.9. These results suggested a substantial improvement – the two-sided *P*-value adjusted for covariates was 0.0075. The results show an estimated log (hazard ratio) $y_m = -0.489$ with standard error $(-0.489 + 0.846)/1.96 = 0.183$, which from the likelihood above is equivalent to $m = (\sigma/0.183)^2 = 120$ deaths.

Computation/software: Conjugate normal analysis.

Bayesian interpretation: The likelihood plot shows the inferences to be made from the reference prior, essentially equivalent to those in Table 6.9. The probability that the new treatment is actually inferior is 0.004 (equivalent to the one-sided *P*-value 0.0075/2.) The probability of clinical superiority is 68%, which might be considered sufficient to change treatment policy. The posterior plot shows the impact of the sceptical prior, in that the chance of clinical superiority is reduced to 27% – hardly sufficient to change practice.

Comments: In fact, Parmar *et al.* (1996) report that the NCI Intergroup Trial investigators were unconvinced by the CALGB trial due to their previous negative experience, and so carried out a further confirmatory study. They found a significant median improvement but of only 2.4 months, from 11.4 to 13.8 months. Under an exponential assumption this corresponds to a hazard ratio of 0.83, suggesting the sceptical approach might have given a more reasonable estimate than the likelihood based on the CALGB trial alone.

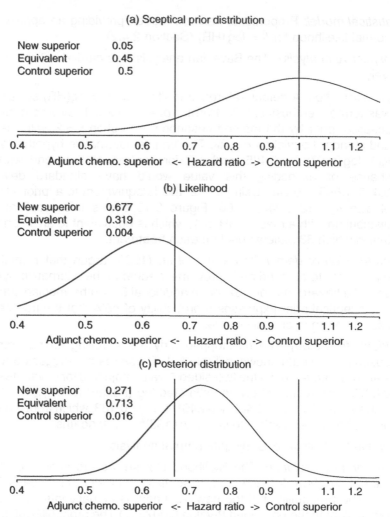

Figure 6.13 Prior, likelihood and posterior distributions arising from CALGB trial of standard radiotherapy versus additional chemotherapy in advanced lung cancer. The vertical lines give the boundaries of the range of clinical equivalence. Probabilities of lying below, within and above the range of equivalence are shown.

Table 6.9 Results of CALGB trial comparing adjunct chemotherapy with radiotherapy alone in advanced non-small-cell lung cancer.

Outcome	Estimate of improvement	95% CI
Median survival (mo)	6.3	1.4 to 13.3
2-year survival (%)	16	4 to 29
Hazard Ratio HR	0.61	0.43 to 0.88
$\theta = \log(HR)$	−0.489	−0.846 to −0.131

6.8   MULTIPLICITY IN RANDOMISED TRIALS

6.8.1   Subset analysis

The discussion on multiplicity in Section 3.17 has already described how multiple simultaneous inferences may be made by assuming a common prior distribution with unknown parameters, provided an assumption of exchangeability is appropriate, *i.e.* the prior does not depend on the units' identities. Within the context of clinical trials this has immediate relevance to the issue of estimating treatment effects in subgroups of patients.

A reasonable model might be to assign a reference (uniform) prior for the overall treatment effect, and then assume the subgroup-specific deviations from that overall effect have a common prior distribution with zero mean. This prior expresses scepticism about widely differing subgroup effects, although the variability allowed by the prior is usually estimated from the data: this procedure 'leads to 1) pooling subgroups if the differences among them appear small, 2) keeping them separate if differences appear large, and 3) providing intermediate results for intermediate situations.' (Cornfield, 1976). This specification avoids the need for detailed subjective input, which may be seen as an attractive feature. Many applications consider this an empirical Bayes procedure which gives rise to traditional confidence intervals which are not given a Bayesian interpretation. Donner (1982) sets out the basic ideas, and Dixon and Simon (1991), Simon (1994b) and Simon *et al.* (1996) have elaborated the techniques in a number of examples.

6.8.2   Multi-centre analysis

Methods for subset analysis (Section 6.8.1) naturally extend to multi-centre analysis, in which the centre-by-treatment interaction is considered as a random effect drawn from some common prior distribution with unknown parameters. Explicit estimation of individual institutional effects may be carried out, which in turn relates strongly to the methods used for institutional comparisons of patient outcomes (Section 7.4).

There have been numerous examples of this procedure (Section 6.13), generally adopting Markov chain Monte Carlo techniques due to the intractability of the analyses. Recent case studies include Gould (1998) who provides WinBUGS code (Section 3.19.3), and Jones *et al.* (1998) who compare estimation methods. Senn (1997b, p. 199) discusses when a random-effects model for centre-by-treatment interaction is appropriate, emphasising the possible difficulty of interpreting the conclusions particularly in view of the somewhat arbitrary definition of 'centre'.

6.8.3   Cluster randomisation

Rather than randomising individual patients, some trials randomise *clusters* of patients, grouped (say) by their general practitioner, both for administrative convenience and because some interventions, for example those involving education or organisation, are applied at the cluster level. A Bayesian approach to the analysis of such trials has been considered by Spiegelhalter (2001) with respect to continuous responses, and Turner *et al.* (2001) for binary responses. In each situation they assume exchangeable clusters, and discuss the appropriate choice of priors on between-cluster variances. Of particular interest is the growing body of empirical evidence on the magnitude of intra-class correlation coefficients observed in different clinical trial contexts, and its value in deriving appropriate prior distributions.

6.8.4   Multiple endpoints and treatments

Multiple endpoints in trials can often be of interest when dealing with, say, simultaneous concern with toxicity and efficacy. This tends to occur in early phase studies, and a Bayesian approach allows one to create a two-dimensional posterior distribution over toxicity and efficacy (Etzioni and Pepe, 1994; Dominici, 1998; Thall and Sung, 1998). General random-effects models for more complex situations can be constructed (Legler and Ryan, 1997). Naturally, a two-dimensional prior is required and particular care must be taken over the dependence assumptions.

A similar situation arises with many treatments: if one is willing to make exchangeability assumptions between treatment effects, then a hierarchical model can be constructed to deal with the multiple-comparison problem. This was proposed long ago by Waller and Duncan (1969). Brant *et al.* (1992) update this procedure by assuming exchangeable treatments and setting the critical values for the posterior probabilities of treatment effects by using a decision-theoretic argument based on specifying the relative losses for Type I to Type II error.

Both multiple endpoints and treatments are also common in meta-analysis of randomised controlled trials (Chapter 8).

6.9   USING HISTORICAL CONTROLS*

A Bayesian basis for the use of historical controls in clinical trials, generally in addition to some contemporaneous controls, is based on the idea that it is wasteful and inefficient to ignore all past information on control groups when making a new comparison. Pocock (1976) argued that careful use of historical controls may allow fewer controls in current studies and give more accurate effect estimates, and methods have since been developed particularly within the field of carcinogenicity studies (Ryan, 1993).

The crucial issue is the extent to which the historical information can be considered similar to contemporaneous data: Pocock (1976) suggests somewhat stringent criteria for use of historical controls, demanding that, in comparison to contemporaneous controls, they should have the same treatment, the same eligibility, the same evaluation, the same baseline characteristics, and the same organisation and investigators, and that there should be no reason to suspect systematic differences. These issues are essentially indistinguishable from those to be taken into account when using any historical evidence, such as when basing prior opinion on past data. We can therefore place the possible approaches within the structure laid out in Sections 3.16 and 5.4, keeping in mind that here we are concerned with past evidence concerning a single (control) arm of a trial, whereas in Section 5.4 we were concerned with past data on a treatment effect. However from an analytic perspective there is little difference between these two contexts. Possible approaches include the following:

(a) *Ignore the historical control data.* This is the standard option in which each trial uses only its own control group.
(b) *Assume the historical control groups are exchangeable with the current control group*, and hence build or assume a hierarchical model for the response within each group (Tarone, 1982; Dempster *et al.*, 1983). Pocock's criteria, described above, seem a natural basis for making a subjective judgement of exchangeability, and such an assumption leads to a degree of pooling between the control groups, depending on their observed or assumed heterogeneity – a classical random-effects formulation of this approach is also possible (Thall and Simon, 1990). Gould (1991) suggests using past trials to augment current control group information, assuming exchangeable control groups. Rather than directly producing a posterior distribution on the contrast of interest, he uses this historical information to derive predictive probabilities of obtaining a significant result were a full trial to have taken place (Section 6.5); his example is treated in Example 8.4.
(c) *Assume the historical controls are a biased sample.* With only one group of historical controls, Pocock (1976) adopts the model in Section 5.4 in which one assumes an additional bias with prior mean 0 – we shall give details of this method and illustrate its use in Example 6.9. Let y_t, y_c and y_h be the observed response in the randomised treated, randomised control and historical control groups respectively, where we assume

$$y_t \sim N[\theta_t, \sigma_t^2], \qquad (6.21)$$

$$y_c \sim N[\theta_c, \sigma_c^2], \qquad (6.22)$$

$$y_h \sim N[\theta_c + \delta, \sigma_h^2], \qquad (6.23)$$

and the degree of bias δ in the historical control evidence is assumed to be

$$\delta \sim N[0, \sigma_\delta^2]. \qquad (6.24)$$

From (6.23) and (6.24) we find the marginal distribution of y_h to be

$$y_h \sim N[\theta_c, \sigma_h^2 + \sigma_\delta^2].$$ (6.25)

Both (6.22) and (6.25) provide evidence concerning θ_c, and a combined likelihood for θ_c is obtained by weighting the two estimates of θ_c inversely by their variances:

$$\frac{y_c + W y_h}{1 + W} \sim N\left[\theta_c, \left(\frac{1}{\sigma_c^2} + \frac{1}{\sigma_h^2 + \sigma_\delta^2}\right)^{-1}\right],$$ (6.26)

where $W = \sigma_c^2/(\sigma_h^2 + \sigma_\delta^2)$. (6.26) can also be obtained in a somewhat convoluted way by assuming a uniform prior for θ_c, doing two Bayesian updates using the likelihoods (6.22) and (6.25), and then seeing what likelihood would have given rise to the resulting posterior.

The parameter of interest is the treatment effect $\theta = \theta_t - \theta_c$, and we can obtain a likelihood for θ from (6.21) and (6.26), giving

$$y_t - \frac{y_c + W y_h}{1 + W} \sim N\left[\theta, \sigma_t^2 + \left(\frac{1}{\sigma_c^2} + \frac{1}{\sigma_h^2 + \sigma_\delta^2}\right)^{-1}\right].$$ (6.27)

The likelihood (6.27) can then be combined with a prior for θ in the standard manner.

In addition to the assumptions above, values or estimates are also required for σ_e^2, σ_c^2 and σ_h^2. Finally, prior opinion regarding σ_δ^2 also has to be specified.

(d) *Discount the size of the historical control group.* This is essentially the 'power' prior described in Section 5.4, but applied solely to the control arm.

(e) *Functional dependence.* This would be relevant if, for example, the historical controls were considered entirely compatible with current controls, but needed to be adjusted for imbalance in covariates.

(f) *Assume the historical control individuals are exchangeable with those in the current control group,* which leads to a complete pooling of historical with experimental controls.

Various combinations of these assumptions are possible: Berry and Stangl (1996a) assume a parameter representing the probability that any past individual is exchangeable with current individuals, while Racine *et al.* (1986) assume a certain prior probability that the entire historical control group exactly matches the contemporaneous controls and hence can be pooled. It is also possible to use such models as a basis for designing future studies and deciding the number of patients to be allocated in each arm.

Example 6.9 *ECMO: incorporating historical controls*

Reference: Ware (1989) and the subsequent discussion.

Intervention: Extracorporeal membrane oxygenation (ECMO), an invasive technique for blood oxygenation in newborn babies.

Aim of study: Until the advent of ECMO, conventional medical therapy (CMT) for infants with severe persistent pulmonary hypertension of the newborn (PPHN) achieved less than a 20% survival rate. Early experiences with ECMO were promising, and by 1985 survival rates of over 80% were being reported. Following a review of the evidence of CMT prior to 1985, an RCT was undertaken at two hospitals at Harvard between 1986 and 1988, in order to evaluate the use of ECMO compared to CMT in this extremely poor prognosis patient population.

Study design: Adaptive two-phase RCT. Phase I randomised patients to either ECMO or CMT, while in phase II patients were to be allocated to whichever was the superior treatment in phase I. We consider here an evaluation of the effectiveness of ECMO based on the evidence from the first, randomised, phase of the trial, including information from historical control patients.

Outcome measure: Odds ratio (OR) of death (OR < 1 favours ECMO).

Planned sample size: The study was designed so that when stopped with at most four deaths in each arm, the study would have approximately 77% power to detect an odds ratio of 1/16 at the 5% significance level corresponding to mortality rates of 20% and 80% in the ECMO and CMT groups, respectively.

Statistical model: A normal likelihood based on the observed log(odds ratio) is adopted: more accurate methods would make use of the full binomial likelihood and MCMC methods (Section 3.19.2).

Prospective analysis?: No.

Prior distribution: Following the approach of Kass and Greenhouse (1989), we shall investigate the use of a sceptical prior distribution for the treatment effect, and historical evidence for survival in the control group. As prior evidence of survival under CMT, we shall follow Ware (1989) in restricting attention to cases of severe PPHN treated with CMT in the specific Harvard hospitals immediately preceding the trial: 13 patients were thus identified as 'historical controls', of whom 11 died. Table 6.10 shows the resulting estimated odds of death, log-odds of death and its variance (Section 2.4). Whilst the use of such historical data may be discounted totally or simply used at 'face-value', it may also be reasonable to discount it in some manner, such as assuming exchangeability,

Table 6.10 Historical and observed data for Harvard ECMO study showing notation for estimates and variances of log-odds of death.

Trial	ECMO deaths/ cases	CMT deaths/ cases	Odds	log(odds)	Variance of log(odds)
Historical data		11/13	4.60	1.53(y_h)	0.49 (σ_c^2)
Harvard phase I	0 / 9		0.05	−2.94(y_t)	2.11 (σ_t^2)
		4/10	0.69	−0.37(y_c)	0.38 (σ_c^2)

Table 6.11 Use of historical controls in assessing odds ratio of death for patients receiving ECMO compared to conventional treatment: OR < 1 favours ECMO. For example, a fourfold relative bias corresponds to a 95% chance that the odds ratio between historical and current control mortality lies between 0.25 and 4.

Potential relative bias assumed in historical controls	σ_δ	Posterior distribution of odds ratio			
		Mean	95% interval	P(OR<1)	P(OR<0.4)
0	0.000	0.033	0.0017 to 0.658	98.7%	94.9%
1.1	0.048	0.033	0.0017 to 0.659	98.7%	94.9%
1.5	0.207	0.035	0.0017 to 0.686	98.6%	94.6%
2	0.354	0.037	0.0018 to 0.741	97.7%	92.1%
4	0.707	0.045	0.0022 to 0.929	97.1%	90.3%
8	1.061	0.053	0.0025 to 1.113	96.8%	89.8%
16	1.415	0.055	0.0026 to 1.166	96.7%	89.4%
Not using historical controls		0.076	0.0035 to 1.673	94.9%	85.4%

bias or simply discounting its sample size (Section 6.9). For a single historical source, and assuming normal likelihoods, all these methods lead to essentially the same model (Section 5.4), and here we shall illustrate the use of the bias model (Pocock, 1976).

Assuming a model such as (6.27) requires prior opinion concerning the potential extent of the bias as measured by σ_δ. For example, if it were thought that in fact the historical controls may over- or underestimate the odds of death in the randomised controls by a factor of 2, then $\exp(1.96\sigma_\delta) = 2$, or $\sigma_\delta = (\log(2)/1.96) = 0.35$: this is similar to the analysis in Section 5.7.3 for interpreting the standard deviation of random effects. Table 6.11 gives a variety of values for σ_δ corresponding to beliefs which range from acceptance of the historical evidence at 'face value', *i.e.* $\sigma_\delta = 0$, to stating that the potential bias could be such that the historical controls could over- or underestimate the odds of death in the randomised controls by a factor of 16.

The choice of a suitable value for σ_δ will depend on the circumstances and the extent to which Pocock's criteria are met (Section 6.9). In this instance the historical controls seem reasonable in that they came from the same centre and were treated in a similar way, except they were not involved in a clinical trial which is known can have an impact on outcomes.

Loss function or demands: No, but an OR of 0.4 was taken to be of clinical importance by Kass and Greenhouse (1989).

Computation/software: Conjugate normal model.

Evidence from study: The results of phase I of the ECMO study are shown in Table 6.10: of the ten patients randomised to conventional therapy four died, whilst of the nine randomised to ECMO none died. The estimates and variances of the log-odds of death were obtained using the adjustments given in Section 2.4. We note the apparent contrast between the mortality rates under CMT before and during the trial: it is generally felt that all participants in a randomised trial get superior treatment. Using the randomised evidence alone, the treatment effect θ would be estimated by $-2.94 + 0.37 = -2.57$, with variance $2.11 + 0.38 = 2.49$. A traditional standardised test statistic, ignoring the sequential nature of the design, is therefore $-2.57/\sqrt{2.49} = 1.63$, corresponding to a one-sided P-value of 0.052; Fisher's exact test yields a one-sided P-value of 0.054 (Ware, 1989).

Bayesian interpretation: We first consider an analysis with a reference prior on the treatment effect. If the historical evidence is totally discounted ($\sigma_\delta = \infty$) then it can be seen from Table 6.11 that the posterior mean of the odds ratio is 0.076, and the posterior probablity of ECMO being inferior is 5.1%; the posterior probability of ECMO not being clinically superior, *i.e.* an odds ratio above 0.4, is 14.6%. However, treating the historical controls as exchangeable with the randomised controls, *i.e.* at 'face value' ($\sigma_\delta = 0$), gives a posterior mean for the odds ratio of 0.033, but now the probability of ECMO being inferior is only 1.3%, and of it not being clinically superior is 5.1%.

Sensitivity analysis: Table 6.11 displays a range of intermediate results between the extremes of totally accepting and totally ignoring the historical controls. A 95% posterior interval for the odds ratio will exclude 1 provided σ_δ is less than around 8, corresponding to a relative bias of around 5. The probability of the odds ratio being less than 0.4 is only around 95% provided that the historical controls are accepted at near face value.

We might also consider a sceptical prior on the treatment effect: the original alternative hypothesis in the Harvard trial was a reduction of the mortality rate from 80% to 20%, equivalent to an odds ratio of 1/16 or $\log(\mathrm{OR}) = -2.77$. Using the argument in Section 5.5.2, we might assume a prior centred on 0 and with 5% of its probability below this alternative of -2.77 – this corresponds to a prior standard deviation of

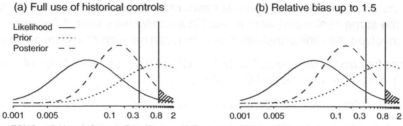

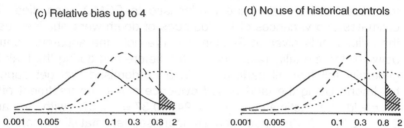

Figure 6.14 Sensitivity analysis of different choices of potential bias in historical controls in the ECMO trial, assuming a sceptical prior with mean 0 (on the log(OR) scale), and a 5% chance of an odds ratio less than 0.0625.

$-2.77/-1.64 = 1.69$. The consequences of using such a sceptical prior are shown in Figure 6.14 for a range of choices of potential bias in the historical controls. As Kass and Greenhouse (1989) conclude, a reasonable sceptic, even taking account of the historical data, is not going to be completely convinced by the ECMO trial.

Comments: This trial presents a number of interesting challenges which are fully argued in the discussion of Ware (1989) and in subsequent publications. For example, there are other historical data available, including some which show good survival on CMT, and there is a database of outcomes on ECMO. Other statistical models for this trial, including and discounting historical data, have been considered by Kass and Greenhouse (1989), Greenhouse and Wasserman (1995) and Berry and Stangl (1996a). Berry (1989b) also considers the inclusion of evidence from an RCT using a play-the-winner design which was also conducted before 1985. Such information could be included, if assumed to be exchangeable with the study reported by Ware (1989), using either a meta-analytic approach (Section 8.2) or by using this historical trial evidence to derive a prior distribution for the intervention effect (Section 5.4).

The discussants of Ware (1989) also have opposing views concerning the ethics of randomisation (Section 6.4): Royall and Berry (1989) say the trial should never have been started since it was unethical to randomise given the available evidence, whereas Begg (1989) takes the completely conflicting view that the Harvard trial was stopped too early since, as we have seen in the analysis above, the result was not convincing to a wide range of opinion.

It is notable that the evidence concerning ECMO was not considered sufficient to prevent a further large trial. After ECMO was introduced in the UK in 1989, it was agreed to organise a randomised trial involving 55 referral hospitals, in which patients randomised to ECMO were referred to one of five specialist centres (Field *et al.*, 1996). This pragmatic trial was designed to randomise 300 babies, but the DMC stopped the trial after 185 cases when the mortality rate was 30/93 on ECMO and 54/92 on CMT, with an odds ratio of 0.55 (95% interval from 0.39 to 0.77). Long-term follow-up of the patients over 4 years (Bennett *et al.*, 2001) revealed only one additional death (in the ECMO arm) but a high rate of disability and impairment: overall only 16% of survivors were without abnormal signs or disability, but with no significant excess in the ECMO group. Treatment was, however, confounded with hospital and the trial was of a referral service rather than ECMO being carried out in direct competition to conventional treatment.

6.10 DATA-DEPENDENT ALLOCATION

So far we have only covered standard randomisation designs in which patients are allocated 50:50 or in some other constant ratio to alternative treatments. However, a full decision-theoretic approach to trial design would consider data-dependent allocation so that, for example, in order to minimise the number of patients getting the inferior treatment, the proportion randomised to the apparently superior treatment could be increased as the trial proceeded. Such 'adaptive' designs are claimed to satisfy ethical considerations for the patients under study (Section 6.4). They can be called 'bandit' designs, as they are analogous in theory to a gambler deciding which arm of a two-armed bandit to pull in order to maximise the expected return: both Bayesian and non-Bayesian approaches are available. An extreme example is Zelen's (1969) 'play-the-winner' rule in which the next patient is given the currently superior treatment, and randomisation is dispensed with entirely; Palmer and Rosenberger (1999) review non-standard trial designs and suggest circumstances where they may be appropriate. Palmer (2002) claims that many of the current difficulties faced in carrying out trials could be relieved by using adaptive designs, and Berry (2001) provides a recent argument for their use.

Nevertheless, there has been considerable criticism of these ideas as not being practically rooted in the realities of clinical trials; see, for example, Byar *et al.* (1976), Simon (1977), Armitage (1985) and Peto (1985). Objections to adaptive allocation include the following:

1. Responses have to be observed without delay.
2. Adaption depends on a one-dimensional response.
3. Sample sizes may have to be bigger.
4. Patients may not be homogeneous throughout the trial.
5. Clinicians may be unhappy with adaptive randomisation.
6. Informed consent may be more difficult to obtain.
7. The trial will be complex and may deter recruitment.
8. Estimation of the treatment contrast will lose efficiency.
9. Potential inflation of Type I error.
10. Treatment assignments may be biased as clinicians may guess which treatment is 'in the lead'.

A careful analysis of two-armed trials has been carried out by Berry and Eick (1995), who conclude that balanced allocation is appropriate if the condition is reasonably common, but adaptive designs may yield a substantial improvement in the expected number of successful treatments when a large proportion of patients with the disease are likely to be in the trial. This is echoed by Senn (1997b, p. 88), who points out that future patients, who in general will greatly outnumber those in the trial, would value a more precise treatment estimate and therefore would prefer large trials with balanced allocation. The ECMO studies discussed in Example 6.9 provide one of the few examples of adaptive allocation, and the subsequent controversy did little to encourage the use of such designs; other examples include an adaptive trial in patients with depressive disorder (Tamura *et al.*, 1994), while the trial described in Kadane (1996) also adapts its allocation rules, in a somewhat complex way, to the current evidence.

A recent example has proved, however, that it is possible to carry out a large and complex adaptive trial. Berry *et al.* (2001a) describe the design of a phase II/III dose-finding study in acute stroke, in which 15 different doses were to be given at random at the start of randomisation, with steady adaptation to the range of doses around the ED95, i.e. the minimum dose that provides 95% of the maximum efficacy. This trial has now been completed. Various characteristics may have contributed to the success of the methodology: only short-term (90-day) outcomes were considered, modern communication technology was used to ensure rapid updating of the current posterior distribution of the dose–response curve, a minimum of 15% of patients given placebo dose ensured that the imbalance did not become too acute, the ability to completely blind clinicians as to the dose provided, the replacement of the original decision-theoretic stopping criterion with one based on posterior tail areas being less than a certain value, and classical estimation of the size and power of the study based on pre-trial simulations.

We may conclude that adaptive designs, which are not a specifically Bayesian issue, may be better accepted when there are many arms in the trial and not just an imbalanced randomisation between two arms. In addition, formulation of a trial as a decision rather than an inference problem leads to many objections (Section 6.2), and adaptation may be better based on posterior distributions.

6.11 TRIAL DESIGNS OTHER THAN TWO PARALLEL GROUPS

Equivalence trials. There is a large statistical literature on trials designed to establish equivalence between therapies. From a Bayesian perspective the solution is straightforward: define a region of equivalence (Section 6.3) and calculate the posterior probability that the treatment difference lies in this range – a threshold of 95% or 90% might be chosen to represent strong belief in equivalence. Several examples of this remarkably intuitive approach have been reported (Section 6.13), which tend to give similar results to traditional analysis. In contrast, Lindley (1998) explores a decision-theoretic formulation that can give radically different conclusions.

Crossover trials. The Bayesian approach to crossover designs, in which each patient is given two or more treatments in an order selected at random, is fully reviewed by Grieve (1994a). More recent references concentrate on Gibbs sampling approaches (Forster, 1994) – see Section 6.13 for other relevant papers.

N-of-1 trials. N-of-1 studies can be thought of as repeated within-person crossover trials in which interest focuses on the response of an individual patient: such trials may be appropriate in chronic conditions in which short-term symptom relief is of interest. A natural approach to combining such studies is to assume patients are exchangeable (perhaps conditional on covariates), and adopt a hierarchical model – an example based on Zucker *et al.* (1997) is given in Example 6.10. This can be thought of as an extreme example of the subset procedure described in Section 6.8.1, in which the subsets have been reduced to individual patients.

Example 6.10   *N of 1: pooling individual response studies*

Reference: Zucker *et al.* (1997).

Intervention: Amitriptyline for treatment of fibromyalgia to be compared with placebo.

Aim of study: To estimate population treatment effects and evaluate individual patient responses.

Study design: Each individual had an *N*-of-1 study in which they were treated in a number of periods (3 to 6 per patient), and in each period both amitriptyline and placebo were administered in random order. All trials were carried out by a single physician at a single centre.

Outcome measure: Each measurement comprised a difference (amitriptyline minus placebo) in response to a symptom questionnaire in each paired crossover period. Higher scores indicated fewer negative symptoms, and so a positive difference indicated amitriptyline as the superior treatment.

Statistical model: If y_{kj} is the *j*th measurement on the *k*th individual, we assume

$$y_{kj} \sim N[\theta_k, \sigma_k^2].$$

We then assume that both θ_ks and σ_k^2s are exchangeable, as it may not be reasonable to assume common between-period variability for all individuals. We make the specific distributional assumption that

$$\theta_k \sim N[\mu_\theta, \tau_\theta^2],$$
$$\log(\sigma_k^2) \sim N[\mu_\sigma, \tau_\sigma^2].$$

A normal distribution for the log-variances is equivalent to a log-normal distribution for the variances (Section 2.6.8).

Prospective analysis?: No.

Prior distribution:

Independence model. In order to reproduce the classical analysis, we may assume each θ_k has a uniform distribution, and each σ_k^{-2} has a Gamma[0.001,0.001] distribution. The latter is essentially equivalent to $\log(\sigma_k^2)$ having a uniform distribution and hence leads to the classical *t* distribution as a basis for testing for an effect in an individual (Sections 5.5.1 and 5.7.3).

Exchangeable model. We initially adopt uniform priors for μ_θ, τ_θ, μ_σ and τ_σ. Other prior distributions for the between-individual variation τ_θ are considered as part of a sensitivity analysis.

Loss function or demands: Zucker *et al.* (1997) suggest that a difference of 0.5 might be considered as important.

Computation/software: Markov chain Monte Carlo in WinBUGS software.

Evidence from study: The raw data are shown in Figure 6.15, ordered in terms of the observed sample mean. Seven out of 23 experienced

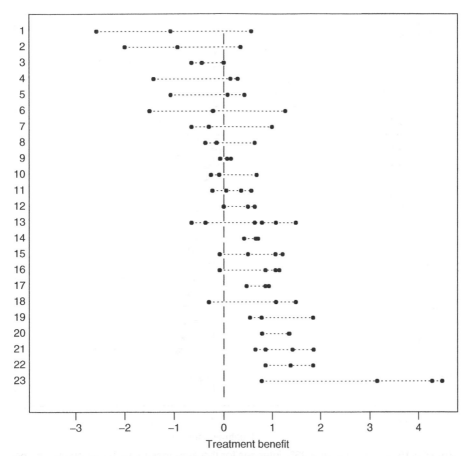

Figure 6.15 Raw data from *N*-of-1 clinical trials on 23 patients, ordered by their mean response. Each dot represents the difference in responses (amitriptyline minus placebo) in a single period in which both treatments have been tried in random order.

benefit from the new treatments in all their periods. There appears to be substantial variability both in the average response and within patients, justifying the statistical model adopted.

Bayesian interpretation: The independent and exchangeable estimates of the individual and overall treatment effects are shown in Figure 6.16. The independent estimates closely follow the raw data, exhibiting substantial uncertainty. In only six patients do the 95% intervals exclude 0, although Zucker *et al*. (1997) report that patients 11–23 were all advised to continue on the active treatment, while patients 1–10 were advised to stop active treatment.

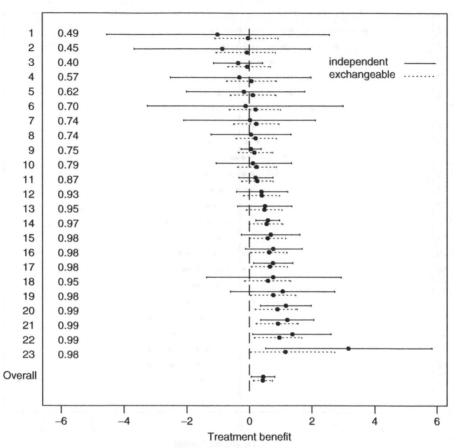

Figure 6.16 Estimates and 95% intervals for the response in each person, assuming both independent and exchangeable individuals. The vertical lines represent the null hypothesis of no treatment difference. $P(\theta_k > 0)$, the posterior probability that each individual's effect lies above 0, is given on the left.

Table 6.12 Summary of posterior distributions of parameters in exchangeable analysis.

Parameter		Median / estimate	95% interval
Overall mean	μ_θ	0.42	0.13 to 0.73
Prob. overall positive effect	$P(\mu_\theta > 0)$	0.997	
Prob. overall important effect	$P(\mu_\theta > 0.5)$	0.29	
Between-patient sd	τ_θ	0.50	0.20 to 0.92
Between-patient variability in log-variances	τ_σ	1.03	0.42 to 1.77
Mean within-patient variance	$\exp(\mu_\tau + \tau_\theta^2)/2$	0.94	0.49 to 3.05

The parameter estimates resulting from the exchangeable analysis are shown in Table 6.12. There is a clear overall positive effect in the population which is estimated to be 0.42, although the chance that it is an important effect (*i.e.* greater than 0.5) is only 29%. There is also strong evidence of patient heterogeneity in their response, with an estimated between-patient standard deviation of 0.50, suggesting that individual patient effects might vary between roughly −0.5 and 1.5.

There is also clear evidence of between-patient heterogeneity in their variability in responses, as shown by τ_σ being substantially away from 0. Transforming from a log-variance to a variance scale (Section 2.6.8) reveals a mean within-patient variance of 0.94.

From the individual estimates shown in Figure 6.16 it is clear that the exchangeable model brings about substantial shrinkage in the extreme patients, reflecting the limited information from each individual. For example, patient 23, with four positive measurements, three of which are extreme, has a posterior mean of 0.55, less than its minimum observation! It might be felt that the model is exercising undue influence in this situation, and some possible alternatives are discussed below. In spite of the shrinkage, the narrower intervals mean that the number of patients with 95% intervals excluding 0 rises to nine, compared to six with the independent analysis. We note one consequence of allowing exchangeable within-patient variances: patient 9, whose observations were remarkably close together and who hence has a very tight independent interval, obtains an exchangeable interval that is *wider* due to their within-patient variance being pulled towards the population mean of around 0.94.

Sensitivity analysis: Changing the prior distribution for τ_θ to the alternatives listed in Section 5.7.3 makes negligible difference to the conclusions, due to the considerable evidence available concerning τ_θ.

Comments: As pointed out by Zucker *et al.* (1997), it is straightforward to include patient-level covariates in such a model, and they illustrate this by including dose as a predictor. However, this can be shown to have minimal influence. It might be reasonable to carry out further analysis of sensitivity to the shape of both the sampling and the random-effects distribution: assuming t distributions (Section 2.6.9) for either may result, for example, in substantially less shrinkage for patient 23.

Factorial designs: Factorial trials, in which multiple treatments are given simultaneously to patients in a structured design, can be seen as another example of multiplicity and hence a candidate for hierarchical models. Simon and Freedman (1997) and Miller and Seaman (1998) suggest suitable prior assumptions that avoid the need to decide whether interactions do or do not exist.

6.12   OTHER ASPECTS OF DRUG DEVELOPMENT

Pharmacokinetics. The 'population' approach to pharmacokinetics, in which the parameters underlying each individual's drug clearance curve are viewed as being drawn from some population, is well established and is essentially an empirical Bayes procedure (Sheiner and Wakefield, 1999). Proper Bayesian analysis of this problem is described in Racine-Poon and Wakefield (1996) and Wakefield and Bennett (1996), emphasising MCMC methods for estimating both population and individual parameters, as well as individualising dose selection (Wakefield and Walker, 1997).

Phase I trials. Phase I trials are conducted to determine that dosage of a new treatment which produces a level of risk of a toxic response which is deemed to be acceptable. The primary Bayesian contribution to the development of methodology for phase I trials has been the continual reassessment method (CRM) originally proposed by O'Quigley *et al.* (1990). In CRM a parameter underlying a dose–toxicity curve is given a proper prior which is updated sequentially and used to find the current 'best' estimate of the dosage which would produce the acceptable risk of a toxic event if given to the next subject, as well as giving the probability of a toxic response at the recommended dose at the end of the trial (O'Quigley, 1992). High sensitivity of the posterior to the prior distribution (Gatsonis and Greenhouse, 1992) has been reported in a similar procedure. Numerous simulations and modifications of the method have been proposed (Section 6.13); Dougherty *et al.* (2000) report a practical application described in Example 6.11.

Example 6.11   *CRM: An application of the continual reassessment method*

Dougherty *et al.* (2000) provide the following application of the continual reassessment method, in which they wish to establish the maximum tolerated dose of the opioid antagonist nalmefene. Lack of tolerability is measured by reversal of anaesthesia. They are interested in establishing the maximum dose with probability p of reversal of anaesthesia nearest to 0.20. The available doses are 0.25, 0.50, 0.75 and 1.00, which are given labels 1 to 4. They adopt a one-parameter logistic response model in which, for dose i,

$$\text{logit}(p_i) = 3 + \alpha d_i, \tag{6.28}$$

where α is an unknown parameter with prior set as an exponential distribution with mean 1 (*i.e.* Gamma[1,1]), and the d_i are transformations of the dose to enable this logistic curve to fit the prior judgements of p_i, denoted p_i^0. Hence the d_i are calculated by setting α equal to its prior mean of 1, and inverting (6.28) to give $d_i = \text{logit}(p_i^0) - 3$.

Table 6.13 Summary of prior and posterior distributions of parameters in CRM experiment.

	Prior		Observed data		Posterior	
Dose	p_i^0: prior guess at p_i	d_i	No. patients	No. not tolerating	Mean	SD
1	0.10	−5.20	4	0	0.10	0.05
2	0.20	−4.39	18	3	0.19	0.08
3	0.40	−3.41	3	2	0.38	0.09
4	0.80	−1.61	0	0	0.79	0.03

Table 6.13 shows the prior judgements, the observed data and consequent posterior distributions. The analysis is straightforward to carry out in Win-BUGS.

We can make a number of observations concerning this analysis. First, the posterior means for the p_i show strong agreement with the prior, perhaps suggesting undue influence. Second, the actual doses used do not enter into the model. Third, a tolerability for dose 4 is estimated with considerable accuracy, even though no one was ever given this dose. Finally, the implied prior distributions for the p_i are actually bimodal. These all suggest that the basic CRM procedure should be used with great caution.

Etzioni and Pepe (1994) suggest monitoring a phase I trial with two possible adverse outcomes via the joint posterior distribution of the probabilities of the two outcomes with frequentist inference at the end of the trial.

Phase II trials. Phase II clinical trials are carried out in order to discover whether a new treatment is promising enough (in terms of efficacy) to be submitted to a controlled phase III trial, and often a number of doses may be compared. Bayesian work has focused on monitoring, sample-size determination and adaptive design. Monitoring on the basis of posterior probability of exceeding a desired threshold response rate was first recommended by Mehta and Cain (1984), while Heitjan (1997), Cronin *et al.* (1999) and Weiss *et al.* (2001) adapt the proposed use of sceptical and enthusiastic priors (Section 6.6.2) in phase III studies.

With regard to design, Herson (1979) used predictive probability calculations to select among designs with high power in regions of high prior probability. Thall and co-workers have also developed stopping boundaries for sequential phase II studies based on posterior probabilities of clinically important events, but where the designs are selected from the frequentist properties derived from extensive simulation studies: see Section 6.13 for references. However Stallard

(1998) has criticised this approach as being demonstrably sub-optimal when evaluated using a full decision-theoretic model with a monetary loss function.

Finally, John Whitehead and colleagues have taken a full decision-theoretic approach to allocating subjects between phase II and phase III studies. For example, Brunier and Whitehead (1994) consider the case where a single treatment with a dichotomous outcome is being evaluated for a possible phase III trial, and use Bayesian decision theory to determine the number of subjects needed. They place a prior on the probability of success and calculate the expected cost of performing or not performing a phase III trial, using a cost function which includes consideration of the costs to future patients if the inferior treatment is eventually used, the power of the possible phase III trial (which they assume will be carried out by frequentist methods), and the costs of experimentation. They show how to determine, for given parameter values, the expected cost of performing a phase II trial of any particular size, and thus the optimal size for a trial.

When faced with selecting among a list of treatments and allocating patients, Pepple and Choi (1997) have considered two-stage designs, Yao *et al.* (1996) deal with screening multiple compounds and allocating patients within a programme, while Strauss and Simon (1995) use a prior distribution and horizon. The successful adaptive study of Berry *et al.* (2001a) discussed in Section 6.10 can also be considered as a phase II dose-finding study monitored using posterior tail areas.

Phase IV – safety monitoring. A considerable literature exists on Bayesian causality assessment in adverse drug reactions: see, for example, Lanctot and Naranjo (1995).

6.13   FURTHER READING

There is a huge literature on Bayesian appraches to trials, which is reviewed in Spiegelhalter *et al.* (2000). General discussion papers include tutorial introductions at a non-technical (Lewis and Wears, 1993) and slightly more technical level (Abrams *et al.*, 1994). Pocock and Hughes (1990) provide a non-mathematical discussion concentrating on estimation issues, while Armitage (1989) attempts a balanced view of the competing methodologies. A special issue of *Statistics in Medicine* has been devoted to 'Methodological and Ethical Issues in Clinical Trials', containing papers both for (Berry, 1993; Urbach, 1993; Spiegelhalter *et al.*, 1993) and against (Whitehead, 1993) the Bayesian perspective, and featuring incisive discussion by Armitage, Cox and others. Particular emphasis has been placed on the ability of Bayesian methods to take full advantage of the accumulating evidence provided by small trials (Lilford *et al.*, 1995; Matthews, 1995).

Somewhat more technical reviews are given by Spiegelhalter *et al.* (1993, 1994). Berry (1991, 1995) has long argued for a Bayesian decision-theoretic basis for clinical trial design, and has described in detail methods for elicitation, monitoring, decision-making and using historical controls. Proponents of a decision-theoretic choice of sample size include Claxton and Posnett (1996), Hornberger and Eghtesady (1998) and Hornberger (2001).

Pocock (1992), O'Brien (1998) and Whitehead (1997b) provide good reviews on sequential trials, and applications of monitoring using posterior intervals include Berger and Berry (1988), Brophy and Joseph (1997), Carlin *et al.* (1993), DerSimonian (1996), George *et al.* (1994) and Rosner and Berry (1995). Papers investigating monitoring using predictions include Choi and Pepple (1989), Qian *et al.* (1996) and Spiegelhalter *et al.* (1986).

Empirical Bayes analyses of subsets are provided by Louis (1991) and Pocock and Hughes (1990), which give rise to traditional confidence intervals that are not given a Bayesian interpretation. Bayesian techniques for subsets are elaborated in Dixon and Simon (1991), Simon (1994b) and Simon *et al.* (1996). Hierarchical models for multicentre analysis have been considered by Gray (1994), Stangl (1996) and Stangl and Greenhouse (1998), while Matsuyama *et al.* (1998) allow a random centre effect on both baseline hazard and treatment, and examine the centres for outliers using a Student's t prior distribution for the random effects.

Examples of the Bayesian approach to equivalence trials have been reported by Selwyn *et al.* (1981), Fluehler *et al.* (1983), Selwyn and Hall (1984), Breslow (1990), Grieve (1991) and Baudoin and O'Quigley (1994). Bayesian approaches to crossover trials include Grieve (1985, 1995), Albert and Chib (1996) and Grieve and Senn (1998).

The continuous reassessment method for phase I studies has been developed by Goodman *et al.* (1995), Whitehead and Brunier (1995), and Gasparini and Eisele (2000). For phase II studies, Korn *et al.* (1993) consider a phase II study which was stopped after three out of four patients exhibited toxicity; Bring (1995) and Greenhouse and Wasserman (1995) re-examine their problem from a Bayesian perspective. See also Thall and Estey (1993), Thall *et al.* (1996), Thall and Russell (1998) and Whitehead (1986, 1997a).

6.14   KEY POINTS

Table 6.14 briefly summarises some major distinctions between the Bayesian and the frequentist approach to trial design and analysis.

1. The Bayesian approach provides a framework for considering the ethics of randomisation.
2. Prior information can be incorporated in power calculations, which should warn against conditioning on optimistic alternative hypothesis. 'Average' power may give a more realistic assessment of the chances of a trial reaching a positive conclusion.

Table 6.14   A brief comparison of Bayesian and frequentist methods in clinical trials.

Issue	Frequentist	Bayesian
Information other than that in the study being analysed	Informally used in design	Used formally by specifying a prior probability distribution
Interpretation of the parameter of interest	A fixed state of nature	An unknown quantity which can have a probability distribution
Basic question	How likely are the data given a particular value of the parameter?	How likely is a particular value of the parameter, given the data?
Presentation of results	Likelihood functions, *P*-values, confidence intervals	Plots of posterior distributions of the parameter, calculation of specific posterior probabilities of interest, and use of the posterior distribution in formal decision analysis
Interim analyses	*P*-values and estimates adjusted for the number of analyses	Inference not affected by the number or timing of interim analyses
Interim predictions	Conditional power analyses	Predictive probability of getting a firm conclusion
Dealing with subsets in trials	Adjusted *p*-values (*e.g.* Bonferroni)	Subset effects shrunk towards zero by a 'sceptical' prior

3. Monitoring trials with a sceptical and other priors may provide a unified approach to assessing whether a trial's results would be convincing to a wide range of reasonable opinion, and could provide a formal tool for data monitoring committees.
4. Predictions of the consequences of continuing a trial provide a useful adjunct to current posterior distributions, but should not be used as a formal monitoring tool.
5. Various sources of multiplicity can be dealt with in a unified and coherent way using hierarchical models.
6. A variety of models exist for incorporating historical controls, analogous to those for using historical data as a basis for a prior distribution.
7. Adaptive studies that change the randomisation ratio dependent on outcomes may be appropriate when a large proportion of available patients are taking part in the trial, or when many treatment arms are being simultaneously investigated.
8. It is generally unrealistic to formulate a phase III trial as a decision problem, except in circumstances where future treatments can be reasonably predicted. Earlier phase studies may be more amenable to this approach.

EXERCISES

6.1. Prove (6.4), (6.6) and (6.7).

6.2. In Example 6.2, calculate the expected power given that the treatment is effective. [Hint: There are two possible methods. You could generate the joint distribution of θ and the power, and only count those iterations for which $\theta > 0$. Alternatively, generate θ from its prior distribution constrained to be positive, using the I(0,) construct in WinBUGS.]

6.3. Consider the prior beliefs for the MRC neutron therapy RCT introduced in Exercise 5.2. The actual trial results at an interim analysis produced a hazard ratio of 0.66 (95% CI from 0.40 to 1.10) in favour of the control group. For each of the prior distributions in Exercise 5.2, update these priors in the light of the observed results.

6.4. Ben-Shlomo *et al.* (1998) report the results of the UK Parkinson's Disease Research Group RCT of the evaluation of levodopa, levodopa and selegiline, and bromocriptine in the treatment of early stage Parkinson's disease; we focus on the comparison of levodopa against levodopa and selegiline in terms of mortality. At a second interim analysis 44 deaths were observed out of 249 patients in the levodopa alone arm and 76 out of 271 patients in the levodopa and selegiline arm, producing a hazard ratio of 1.57 (95% CI from 1.09 to 2.03) for levodopa and selegiline vs. levodopa alone. At this point the trial was terminated, but follow-up continued and a subsequent analysis reported 73 and 103 deaths, producing a hazard ratio of 1.32 (95% CI from 0.98 to 1.79).

   (a) Use the credibility analysis of Section 3.11 to establish the degree of scepticism that would be required not to have found the interim results convincing of benefit.

   (b) In a trial in which $m = 120$ events were to be observed, what alternative log(hazard ratio) could be detected with 80% power?

   (c) What sceptical prior would express 5% belief that the effect would be as large as this alternative hypothesis?

   (d) Discuss whether, on the evidence provided, it was reasonable to stop the trial early.

6.5. Table 6.15, adapted from Wheatley and Clayton (2003), shows the accumulating data in a trial of five vs. four treament courses in the MRC Acute Myeloid Leukaemia trial. An unexpectedly large treatment effect in favour of five courses was observed early in the trial, which disappeared as the trial progressed.

   (a) Plot the likelihoods for the log(hazard ratio) at each timepoint, and calculate the two-sided *P*-values.

   (b) If the trial were planned to observe 300 events, what might a reasonable sceptical prior distribution be?

   (c) What would have been the effect had this prior been used to monitor the trial?

Table 6.15   Mortality in MRC Acute Myeloid Leukaemia RCT.

Timepoint	5 courses		4 courses		$O - E$	$V[O - E]$
	deaths	total	deaths	total		
1997	7	102	15	100	−4.6	5.5
1998(1)	23	171	42	169	−12.0	15.9
1998(2)	41	240	66	240	−16.0	26.7
1999	51	312	69	309	−11.9	30.0
2000	79	349	91	345	−9.5	42.4
2001	106	431	113	432	−6.2	53.7
2002	157	537	140	541	+6.7	74.0

6.6. Prove (6.17) and (6.18).

6.7. Consider the situation in which the Parkinson's disease trial was stopped in Exercise 6.4, and the predictions that could have been made concerning the status of the trial at its eventual publication when 176 events had occurred (an additional 56).

   (a) What would have been the expected power, given the data so far, of rejecting the hypothesis that the log(hazard ratio) was 0, *i.e.* the probability that the final 95% interval will lie wholly above 0, with and without the inclusion of the sceptical prior?

   (b) Was there evidence of conflict between the data in the first part of the trial and that collected in the second part, *i.e.* after the decision was made to stop? [Hint: One way to do this is to calculate the predictive distribution for the observed log(hazard ratio) arising in the second part and use Box's measure of conflict to compare it to that actually observed.]

6.8. (a) Derive the results given in the ECMO study in Example 6.9. (b) Reanalyse the ECMO study assuming the historical data are to be discounted using the 'power prior' model explored in Example 5.2, with prior weights 0, 10%, 50% and 100%.

6.9. Reanalyse the ECMO study in Example 6.9 with full binomial likelihoods instead of normal approximations and using WinBUGS for the analysis. You will need to select a prior distribution for the mortality rates in the control and ECMO groups ignoring both historical and trial data: compare the use of (a) independent uniform distributions in each group, (b) independent Beta[0.5,0.5] distributions, (c) a uniform distribution for the control group mortality and a sceptical prior for the treatment effect on the log(odds ratio) scale.

6.10. Consider Exercise 2.1, repeating the study with the *other* hand. Using a subjectively chosen sceptical prior distribution for the log(odds ratio) for the difference between hands, conduct the second 12 tosses, and update the prior beliefs in the light of the evidence that you have collected.

Table 6.16 Estimates of log(hazard ratio) and standard errors for disease-free survival comparing tamoxifen with control for women with breast cancer within subgroups defined by œstrogen receptor status, nodal status and postmenopausal status.

Oestrogen receptor +ve	Node + ve	Postmenopausal	Total	Tamoxifen	Control	log (HR)	SE [log(HR)]
			No. patients				
1	0	0	183	72	111	−0.520	0.207
1	1	0	57	27	30	−0.096	0.319
1	0	1	262	101	161	−0.551	0.190
1	1	1	92	44	48	+0.040	0.278
0	0	0	493	210	283	−0.061	0.152
0	1	0	128	52	76	−0.256	0.242
0	0	1	583	280	303	−0.287	0.131
0	1	1	161	72	89	−0.275	0.205

6.11. Table 6.16 displays estimates of log(hazard ratio) for disease-free survival comparing tamoxifen with control for women with breast cancer for eight mutually exclusive subgroups of women defined by three binary factors: œstrogen receptor status, nodal status and postmenopausal status. Assuming exchangeable subgroups, obtain the posterior estimates of the hazard ratio for each subgroup, and thus assess the evidence for specific subgroup–treatment interactions. [Hint: You could use the empirical Bayes methodology of Example 3.13, or the full Bayes approach using WinBUGS shown in Example 8.1.]. Do you think the exchangeability assumption is reasonable?

7

Observational Studies

7.1 INTRODUCTION

The RCT is generally considered the 'gold-standard' methodology in evaluating health-care interventions, but there are circumstances in which randomisation is either impossible or unethical (*e.g.* evaluating the health effects of smoking) or where there is substantial valuable information available in non-randomised or 'observational' data (Concato *et al.*, 2000). In many circumstances such observational data would form part of an evidence synthesis, which is dealt with in Chapter 8.

It is important to understand that the probability models used in Bayesian analysis are expressions of personal or group uncertainty and so do not need to be based on randomisation. Therefore in principle non-randomised studies can be analysed in exactly the same manner as randomised comparisons. In Section 7.2 we describe how both case–control and cohort designs provide a likelihood which can be combined with prior information using standard Bayesian methods, perhaps with extra attention to adjusting for covariates in an attempt to control for possible baseline differences in the treatment groups with respect to uncontrolled risk factors or exposures.

Of course, the dangers associated with the use of observational studies in evaluating health-care interventions have been well described in the medical literature (Byar *et al.*, 1976). For example, Dunn *et al.* (2002) compare randomised and non-randomised evidence collected according to a common protocol, and find a potentially misleading treatment comparison based on the observational data. Essentially, randomised studies should provide an unbiased likelihood for the parameter of interest, while observational studies may have a degree of systematic bias. In this book we do not argue the case for or against the use of non-randomised studies, but suggest that *if* observational studies are to be used, then their analysis falls naturally into a Bayesian framework. Specifically, the possibility of bias leads inevitably to a degree of subjective judgement about the comparability of studies, and this fits well into the

Bayesian Approaches to Clinical Trials and Health-Care Evaluation D. J. Spiegelhalter, K. R. Abrams and J. P. Myles
© 2004 John Wiley & Sons, Ltd ISBN: 0-471-49975-7

acknowledged judgement underlying all Bayesian reasoning. Hence, in Section 7.3 we consider the explicit modelling of potential biases, building on the structure developed in the context of evidence-based priors (Section 5.4) and using historical controls (Section 6.9), in each of which a range of methods are possible for 'downweighting' studies to allow for doubts about their degree of relevance.

Finally, in Section 7.4 we consider the specific issue of making institutional comparisons, also known as 'profiling'. This fits naturally into a hierarchical modelling framework, and we also show how a Bayesian approach allows direct probability statements about the rank of an institution.

7.2 ALTERNATIVE STUDY DESIGNS

Case–control studies involve retrospective investigation of risk factors for a sample of cases and controls, possibly matched for known risk factors. Inference is generally on the odds ratio, which is directly estimable from this design. Bayesian approaches have generally relied on analytic approximations in order to obtain reasonably simple analyses (Zelen and Parker, 1986; Marshall, 1988; Nurminen and Mutanen, 1989; Zelen, 1990); for example, Ashby *et al.* (1993) examine two case–control studies studying leukaemia following chemo- therapy treatment for Hodgkin's disease, and consider the consequences of various prior distributions based on a cohort study. However, all the techniques for analysing clinical trials can be adopted, with the additional complication in relation to judgements on the potential for bias and appropriateness of the prior. Example 7.1 describes the analysis of Lilford and Braunholtz (1996) concerning potential side-effects of oral contraceptives using a likelihood arising from case– control studies.

A large *cohort* study or *registry* database may provide observational evidence on the 'natural history' of a disease, which might be used to model the conse- quences of an intervention; for example, Craig *et al.* (1999) describe an analysis of a population-based cohort of patients with diabetic retinopathy in order to evaluate different screening policies. It is, of course, possible to directly estimate apparent effects of different interventions from registry data, although again the potential for bias should be acknowledged: Example 9.3 illustrates one technique for downweighting registry and single cohort data in an evidence synthesis.

There is also a substantial literature on Bayesian methods for complex epi- demiological modelling, particularly spatial correlation (Heisterkamp *et al.*, 1993; Bernardinelli *et al.*, 1995; Richardson *et al.*, 1995; Ashby and Hutton, 1996), measurement error (Richardson and Gilks, 1993) and missing covariate data (Raghunathan and Siscovick, 1996).

7.3 EXPLICIT MODELLING OF BIASES

Bayesian techniques for explicitly modelling potential bias, both within studies and in the attempt to generalise studies outside their target population, were pioneered by Eddy *et al.* (1992) under their general title of the 'confidence profile method' (Section 8.1).

Biases to internal validity mean that the effect of interest is not being appropriately estimated within the circumstances of the study. For example, suppose we suspect that a proportion p of patients in a study did not comply with the intended treatment, although we do not know who these patients are. If we are interested in estimating the treatment effect θ_t in those who actually received the treatment, then the overall underlying treatment effect in the trial will be $\theta = (1 - p)\theta_t + p\theta_0$, where θ_0 is the effect in non-compliers. A likelihood for θ can thus be transformed into a likelihood for θ_t, provided there is other evidence or prior opinion concerning p and θ_0. The likelihood therefore provides information on a *function* of the parameters of interest, and a fairly complex example is provided in Example 8.7.

Eddy *et al.* (1992) identify a range of potential biases that can be modelled in this manner: these include dilution and contamination due to those who are offered a treatment not receiving it, errors in measurement of outcomes, errors in ascertainment of exposure to an intervention, loss to follow-up, and patient selection and confounding in which the groups differ with respect to measurable features. These biases may occur singly or in combination.

Biases to external validity concern the ability of a study to generalise to defined populations or to be combined with studies carried out on different groups, and may be relevant even if a study has been meticulously carried out and has obtained an unbiased assessment of the treatment effect within its own study population. These include 'population bias' in which the study and general population differ with respect to known characteristics, 'intensity bias' in which the 'dose' of the intervention is varied when generalised, and differences in lengths of follow-up.

We have previously discussed the use of historical data as a basis for prior opinion (Section 5.4) or as historical controls in clinical trials (Section 6.9), and in each case examined ways of 'discounting' the data from their face-value interpretation. In each of these contexts it has been assumed that the current observed data, for example in a randomised trial, directly depend on the parameter of interest. The potential biases, whether internal or external, in observational studies can be modelled using similar techniques, but in this context the current likelihood may be adjusted.

As a simple example, we assume a normal likelihood

$$y_m \sim \mathrm{N}[\theta_\mathrm{Int}, \sigma^2/m],$$

where θ_{Int} represents an 'internal' parameter that is being estimated in the current study. Following the development in Sections 5.4 and 6.9, we might assume a bias δ so that $\theta_{\text{Int}} = \theta + \delta$, where θ is the parameter of real interest. Options then include the following:

1. Assuming δ is known.
2. Assuming δ has a known distribution with mean 0, indicating a non-systematic bias. If we assume $\delta \sim \text{N}[0, \sigma^2/n_\delta]$, from (2.25) we obtain a likelihood for the parameter of interest θ,

$$y_m \sim \text{N}\left[\theta, \; \sigma^2 \left(\frac{1}{m} + \frac{1}{n_\delta}\right)\right],$$

   i.e. the sample variance is inflated to allow for the potential bias.
3. If we suspect systematic bias in one direction, we might take δ to have a known distribution with non-zero mean, say $\delta \sim \text{N}[\mu_\delta, \sigma^2/n_\delta]$. We then obtain a likelihood

$$y_m \sim \text{N}\left[\theta + \mu_\delta, \; \sigma^2 \left(\frac{1}{m} + \frac{1}{n_\delta}\right)\right],$$

or equivalently

$$y_m - \mu_\delta \sim \text{N}\left[\theta, \; \sigma^2 \left(\frac{1}{m} + \frac{1}{n_\delta}\right)\right]. \tag{7.1}$$

Hence, after subtracting the assumed mean bias μ_δ from the observation y_m, (7.1) provides a likelihood for the parameter of interest that can be combined with an appropriate prior distribution for θ.

Each of these approaches is illustrated in Example 7.1.

In practice, analytic solutions will rarely be possible and MCMC techniques will be necessary. More serious are the assumptions required concerning the extent of the biases, since although data may be available on which to base accurate estimates, there is likely to be considerable judgemental input. Any unknown quantity can, of course, be given a prior distribution, and Eddy *et al.* (1992) claim this obviates the need for sensitivity analysis. They also argue strongly against simple downweighting using the 'power prior' model (Section 5.4) in which the effective sample size is reduced: they claim this is an arbitrary technique and that potential biases should be explicitly modelled. In fact, as we showed in Section 5.4, the models are effectively equivalent when handling a single study. We also note the increasing pace of research concerning the quantitative bias of observational studies: see, for example, Kunz and Oxman (1998), Britton *et al.* (1998), Benson and Hartz (2000), Ioannidis *et al.* (2001), Reeves *et al.* (2001) and Sanderson *et al.* (2001).

Example 7.1 *OC: interpreting case–control studies in pharmacoepidemiology*

Reference: Lilford and Braunholtz (1996).

Intervention: Third-generation oral contraceptives (OCs).

Aim of study: Suspicions had been raised as to whether 'third-generation' OCs increased the risk of venous thromboembolism compared to second-generation OCs. The aim of Lilford and Braunholtz (1996) was to assess the evidence from a Bayesian perspective.

Study design: Interpretation of a meta-analysis of four case–control studies.

Outcome measure: Odds ratio for venous thromboembolism, OR < 1 being in favour of 3rd-generation OCs.

Planned sample size: Not applicable.

Statistical model: Normal likelihood for pooled estimate of log(OR) derived from the meta-analysis of case–control studies, discounted for potential biases according to the methods described in Section 7.3. Lilford and Braunholtz (1996) consider a potential bias δ in the meta-analysis with a normal distrtribution: in the notation of Section 7.3, $\delta \sim N[\mu_\delta, \ \sigma^2/n_\delta]$. They examine the effect of both a non-systematic and a systematic bias, as detailed below under 'Sensitivity analysis'.

Prospective analysis?: No.

Prior distribution: Prior beliefs were elicited from two gynaecologists with an interest in family planning. Expert 1 thought that a 20% risk reduction in venous thromboembolism would be associated with third-generation compared to second-generation OCs, i.e. OR $= 0.8$, but that the OR could be between 0.4 and 1.6. Assuming this corresponds to a 95% interval of a normal distribution, the true log(odds ratio), θ, can be assumed to have mean $\mu = \log(0.8) = -0.22$ and standard deviation $(\log(1.6) - \log(1.4))/(2 \times 1.96) = 0.35$. Equivalently, if we take $\sigma = 2$, we obtain a prior 'number of events' $n_0 = (\sigma/0.35)^2 = 31.9$.

Expert 2 thought that there was an equal chance of third-generation OCs reducing the OR of venous thromboembolism or increasing it, *i.e.* OR $= 1.0$, but was suitably uncertain as to think that the true OR was likely to be between 0.5 and 2.0. Using the same argument as for Expert 1, we assume an N[0, $\sigma^2/31.9$] prior for Expert 2.

Loss function or demands: No.

Computation/software: Conjugate normal model.

Evidence from study: The meta-analysis of case–control studies produced a pooled odds ratio of 2.0 with a 95% CI from 1.4 to 2.7. On a log(OR) scale, this provides a likelihood with mean $\log(2.0) = 0.69$ and standard deviation $(\log(2.7) - \log(1.4))/(2 \times 1.96) = 0.17$. Equivalently, taking $\sigma = 2$, we obtain a sample 'number of events' $m = (\sigma/0.17)^2 = 142.5$.

Bayesian interpretation: Combining the evidence from the meta-analysis with each expert's prior beliefs produced the posterior distributions seen in Figure 7.1(a). Given that both gynaecologists were a priori quite uncertain as to the true odds ratio, their corresponding posterior distributions are influenced considerably by the data, so that the posterior distributions for both experts indicate less than 0.02% probability that third-generation OCs reduce the OR of venous thromboembolism.

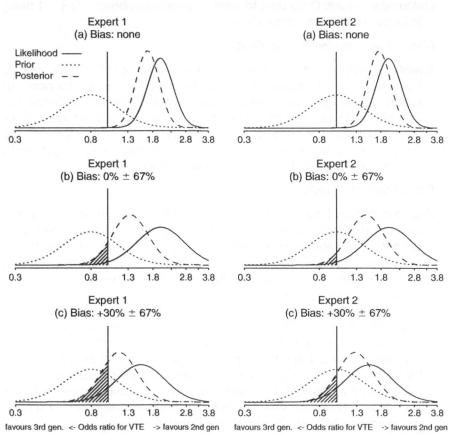

Figure 7.1 Likelihood, prior and posterior distributions for the oral contraceptive meta-analysis, showing the prior distributions for two experts and the results of (a) taking the meta-analysis at face value, (b) discounting the evidence by assuming the possibility of a random bias with standard deviation 30% on the HR scale, and (c) assuming an additional systematic bias of 30% on the HR scale.

Sensitivity analysis: It may be appropriate not to consider the evidence from such a meta-analysis at 'face value' since such retrospective epidemiological studies are known to be prone to various biases. Figure 7.1(b) shows an analysis in which the evidence from the meta-analysis is discounted using the non-systematic bias model described in Section 7.3.

Figure 7.1(b) shows the influence of a non-systematic ($\mu_\delta = 0$) bias such that the odds ratio θ_{int} being estimated may be between 60% and 167% of the true odds ratio θ, *i.e.* up to a 67% bias in either direction. This corresponds, on a log(OR) scale, to a bias with standard deviation $\log(1.67)/1.96 = 0.26$, equivalent, if we take $\sigma = 2$, to $n_\delta = (\sigma/0.26)^2 = 58.7$. The resulting posterior distributions for the two experts now give 11% and 5% probability to the notion that third-generation OCs may reduce the relative risk.

Figure 7.1(c) shows a further series of analyses in which the evidence from the meta-analysis is not only discounted, but also adjusted for the belief that case–control studies may have a systematic bias in which odds ratios are *overestimated* by a median of 30%: this is modelled by assuming $\mu_\delta = \log(1.3) = 0.26$, so that $\delta \sim N[0.26, 0.26^2]$. In this case the resulting posterior distributions show 27% and 15% probability that third-generation OCs may reduce the relative risk. Thus reasonable assumptions about the potential bias in the epidemiological studies, combined with a reasonably sceptical prior distribution, lead to substantial uncertainty as to the true effect of third-generation OCs.

Comments: There was great publicity surrounding the publication of this meta-analysis in 1995. Notification of family doctors in the UK was carried out in a 'panic' atmosphere, leading to a sudden drop in use of third-generation OCs, and reports of subsequent excess abortions. This Bayesian analysis suggests that such consternation may have been unfounded. A court case against the makers of third-generation OCs brought by 99 women who suffered strokes, deep vein thromboses and pulmonary embolisms was settled in July 2002 in the English courts, when the judge ruled that there was 'not, as a matter of probability, any increased relative risk' associated with the pills. It is notable that both sides in the case agreed that a doubling of risk had to be shown, in order that it was 'as likely as not' that any side-effect was caused by the third-generation OC. In view of this demand, it is hardly surprising the case against the companies failed.

Whilst this, and many other analyses have concentrated on the potentially negative effects of third-generation OCs, there has been evidence published that their use has been associated with a reduced relative risk of myocardial infarction compared to second-generation oral contraceptives. However, this example serves to illustrate the fact that in many situations in which there are numerous outcomes, both positive and negative, consideration of one in isolation is fraught with danger. It is also notable that policy decisions should depend on differences in expected utilities whch in turn depend on risk differences rather than odds ratio (Section 3.14), and hence this analysis, strictly speaking, is not in a suitable form for decision-making.

7.4   INSTITUTIONAL COMPARISONS

If we consider an individual clinician, a medical team or a hospital as representing a class of 'intervention', then the use of performance indicators to compare outcomes could be considered as a form of evaluation. There are many complex issues surrounding such 'profiling' of institutions, including risk adjustment, choice of indicator, frequency of analysis, public reporting and so on, but these are beyond the scope of this book. Bayesian approaches to institutional comparisons have been suggested by Goldstein and Spiegelhalter (1996), Normand *et al.* (1997) and Christiansen and Morris (1997a), while fully Bayesian methods have also been used in the analysis of panel agreement data on the appropriateness of coronary angiography (Ayanian *et al.*, 1998).

A popular method when comparing institutions is to plot the observed performance (possibly risk-adjusted) and 95% confidence interval; see, for example, the New York cardiac surgery indicators (New York State Department of Health, 1998). If the interval does not overlap a benchmark then attention focuses on that centre. However, by chance alone one can expect 2.5% of centres to be identified as 'significantly' below standard, even if they are actually performing at the benchmark level. This indicates the need for caution in interpreting 'statistically significant' results, as this is essentially testing the hypothesis that each surgeon has exactly the same underlying patient mortality rate, which is neither plausible nor particularly interesting. We can deal with this 'multiplicity' problem (Section 3.17) in an analogous way to subset estimation (Section 6.8.1) and meta-analysis (Section 8.2), in using hierarchical models to make inferences based on estimating a common prior distribution, leading to 'shrunken' estimates for each centre. Furthermore, *regression to the mean* describes the tendency for institutions that have been identified as 'extreme' to become less extreme when monitored in the future – put simply, part of the reason for their extremity was a run of good or bad luck. This simple phenomenon could lead to spurious claims being made about the benefit of interventions to 'rescue' failing institutions. Shrinkage estimation is intended to counter this difficulty (Christiansen and Morris, 1997a).

An additional benefit of using Markov chain Monte Carlo methods (Section 3.19) is the ability to derive uncertainty intervals around the rank order of each institution (Marshall and Spiegelhalter, 1998). Example 7.2 describes an analysis of success rates in *in vitro* clinics, in which Bayesian methods are used both to make inferences on the true rank of each clinic and to estimate the true underlying success rates with and without an exchangeability assumption.

Benefits of the Bayesian approach to institutional comparisons therefore include:

- methods for reporting probabilities that any specified centre's true rate exceeds any particular threshold of interest;
- a natural way of dealing with 'regression to the mean';
- explicit allowance for between-centre variability;
- an opportunity to incorporate covariates both at the patient and institutional level of the model;
- inferences on the true rank of the institution.

Example 7.2   *IVF: estimation and ranking of institutional performance*

Reference: Marshall and Spiegelhalter (1998).

Intervention: In vitro fertilisation (IVF).

Aim of study: The UK Human Fertilisation and Embryology Authority (HFEA) monitors clinics licensed to carry out donor insemination (DI) and IVF, and to help people who are considering fertility treatment to understand the services offered by licensed clinics and to decide which clinic is best for them (Human Fertilisation and Embryology Authority, 1996). They publish risk-adjusted live birth rates per treatment cycle started, and we are concerned with whether one can rank the institutions with any confidence.

Study design: Retrospective analysis of prospectively collected data on 52 clinics carrying out IVF treatment in the UK between April 1994 and March 1995.

Outcome measure: Estimated adjusted live birth rate $\hat{p}_k$, with 95% intervals, per treatment cycle started, where the case-mix adjustment is based on a pooled logistic regression of all IVF treatments.

Statistical model: If there are n_k treatments in the kth clinic, we calculate $r_k = \hat{p}_k n_k$ as the effective number of successful live births. The log-odds on success for each clinic are denoted y_k and estimated to be $y_k = \log\left[(r_k + 0.5)/(n_k - r_k + 0.5)\right]$, with estimated variance $s_k^2 = 1/(r_k + 0.5) + 1/(n_k - r_k + 0.5)$ (Section 2.4.1). Then we assume

$$y_k \sim N[\theta_k, s_k^2],$$

where θ_k is the true log-odds on success in the kth clinic; an exact likelihood based on the binomial distribution is possible but makes negligible difference in this example due to the substantial number of treatments.

Two models for the θ_ks are considered. First, that they are *independent*. Second, the clinics are assumed to be fully *exchangeable* (Section 3.4), with the true rates (on a logit scale) being drawn from a common normal distribution: if, after adjusting for case-mix, we can find no other contextually meaningful way to differentiate between the institutions, then the assumption of their exchangeability seems justified. Hence we assume

$$\theta_k \sim N[\mu, \tau^2].$$

Prior distributions:

Independence model. Originally assume the θ_k each have an independent uniform distribution: this is used for the ranking exercise.

Exchangeable model. Uniform priors are adopted for μ, τ.

Computation/software: MCMC techniques in the WINBUGS software are used to derive posterior distributions for the ranks of the institutions: this is done by calculating the current rank of each institution at each iteration of the simulation, and then summarising the distribution of these calculated ranks after many thousands of iterations.

Evidence from study: The raw data are shown in Figure 7.2.

Bayesian interpretation: It is clear from Figure 7.2 that there is substantial shrinkage towards the overall mean performance when assuming exchangeability, although there are still a number of clinics that would be considered 'significantly' above or below average. It can be argued that this adjustment is an appropriate means of dealing with the problem of multiple comparisons. In addition, this shrinkage should deal with 'regression to the mean', in which extreme institutions will tend back towards the overall average when they recover from their temporary run of good or bad luck.

Figure 7.3 shows that there is considerable uncertainty in the true rank of an institution, even when they show substantial differences in performance.

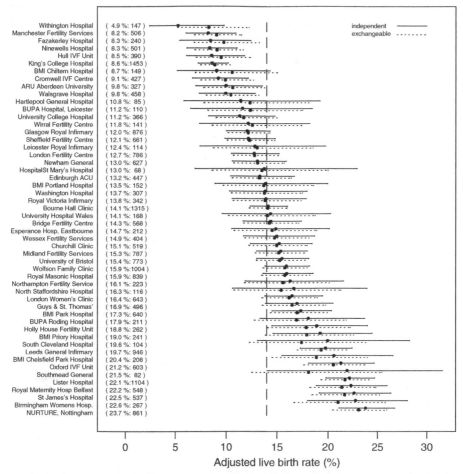

Figure 7.2   Estimates and 95% intervals for the adjusted live birth rate in each clinic, assuming both independent and exchangeable rates. The vertical lines represent the national average of 14%. The estimated adjusted live birth rate for each clinic is given in brackets, together with the number of treatment cycles started.

The consequence of assuming exchangeability is to reduce the differences between clinics and hence to make their ranks even more uncertain. Figure 7.3 shows this is the case to a limited extent, although since many of the extreme clinics are also fairly large, their rank is not unduly effected.

Sensitivity analysis: The results are extremely insensitive to the prior on τ and the use of a full binomial likelihood.

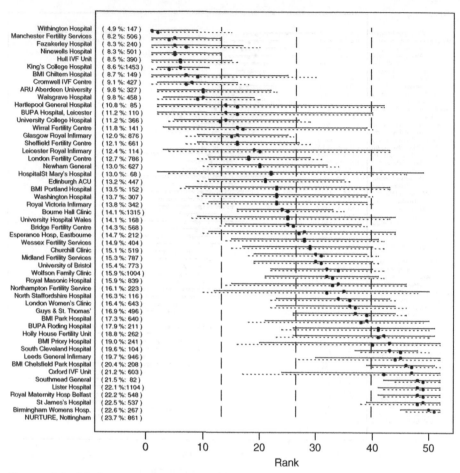

Withington Hospital	(4.9 %: 147)
Manchester Fertility Services	(8.2 %: 506)
Fazakerley Hospital	(8.3 %: 240)
Ninewells Hospital	(8.3 %: 501)
Hull IVF Unit	(8.5 %: 390)
King's College Hospital	(8.6 %:1453)
BMI Chiltern Hospital	(8.7 %: 149)
Cromwell IVF Centre	(9.1 %: 427)
ARU Aberdeen University	(9.8 %: 327)
Walsgrave Hospital	(9.8 %: 458)
Hartlepool General Hospital	(10.8 %: 85)
BUPA Hospital, Leicester	(11.2 %: 110)
University College Hospital	(11.2 %: 366)
Wirral Fertility Centre	(11.8 %: 141)
Glasgow Royal Infirmary	(12.0 %: 876)
Sheffield Fertility Centre	(12.1 %: 661)
Leicester Royal Infirmary	(12.4 %: 114)
London Fertility Centre	(12.7 %: 786)
Newham General	(13.0 %: 627)
HospitalSt Mary's Hospital	(13.0 %: 68)
Edinburgh ACU	(13.2 %: 447)
BMI Portland Hospital	(13.5 %: 152)
Washington Hospital	(13.7 %: 307)
Royal Victoria Infirmary	(13.8 %: 342)
Bourne Hall Clinic	(14.1 %:1315)
University Hospital Wales	(14.1 %: 168)
Bridge Fertility Centre	(14.3 %: 568)
Esperance Hosp, Eastbourne	(14.7 %: 212)
Wessex Fertility Services	(14.9 %: 404)
Churchill Clinic	(15.1 %: 519)
Midland Fertility Services	(15.3 %: 787)
University of Bristol	(15.4 %: 773)
Wolfson Family Clinic	(15.9 %:1004)
Royal Masonic Hospital	(15.9 %: 839)
Northampton Fertility Service	(16.1 %: 223)
North Staffordshire Hospital	(16.3 %: 116)
London Women's Clinic	(16.4 %: 643)
Guys & St. Thomas'	(16.9 %: 496)
BMI Park Hospital	(17.3 %: 640)
BUPA Roding Hospital	(17.9 %: 211)
Holly House Fertility Unit	(18.8 %: 262)
BMI Priory Hospital	(19.0 %: 241)
South Cleveland Hospital	(19.6 %: 104)
Leeds General Infirmary	(19.7 %: 946)
BMI Chelsfield Park Hospital	(20.4 %: 208)
Oxford IVF Unit	(21.2 %: 603)
Southmead General	(21.5 %: 82)
Lister Hospital	(22.1 %:1104)
Royal Maternity Hosp Belfast	(22.2 %: 548)
St James's Hospital	(22.5 %: 537)
Birmingham Womens Hosp.	(22.6 %: 267)
NURTURE, Nottingham	(23.7 %: 861)

Rank

Figure 7.3 Median and 95% intervals for the rank of each clinic, assuming both independent and exchangeable rates. The dashed vertical lines divide the clinics into quarters according to their rank.

7.5 KEY POINTS

1. Data from observational studies may, in principle, be analysed in exactly the same framework as for randomised trials.
2. Imperfections in the design and conduct, and generalisation to other populations, may be approached by adopting a more complex model.
3. There are likely to be increased demands for Bayesian analysis, particularly in areas such as institutional comparisons and gene–environment interactions.

4. The explicit modelling of potential biases in observational data may be widely applicable but needs some evidence base in order to be convincing.
5. Analysis of sensitivity to modelling and prior assumptions is even more important than in RCTs.

EXERCISES

7.1. Ashby *et al.* (1993) consider the association between treatment for Hodgkin's disease and the subsequent risk of leukaemia. An international case–control study reported data on 149 cases who had Hodgkin's disease followed by leukaemia and 411 matched controls who had Hodgkin's disease but no subsequent leukaemia. Table 7.1 displays cases and controls stratified according to treatment received.
   (a) Estimate the probability that cases with leukaemia had been treated with chemotherapy, *i.e.* $p(C|L)$, and compare this with the probability that controls without leukaemia had been treated with chemotherapy, *i.e.* $p(C|\bar{L})$.
   (b) Prove that from these quantities you can estimate the odds ratio associating leukaemia with treatment with chemotherapy, *i.e.* $[p(L|C)/p(\bar{L}|C)]/[p(L|\bar{C})/p(\bar{L}|\bar{C})]$.
   (c) Hence estimate the log(odds ratio) and its variance from the table.
   (d) Assuming a sceptical prior that doubts whether odds ratios as large as 10 are reasonable, how does this influence the conclusions?
7.2. Suppose that $r = 20$ people responded out of $n = 50$ given a particular drug. We then hear that $p = 20\%$ of individuals did not in fact take the drug. (a) Express the overall response rate θ in the experiment in terms of the true response rate θ_t of those who did take the drug, the proportion p of compliers, and the response rate θ_0 of those who did not take the drug. Assuming a uniform prior for θ_t, what inference would you make on θ_t, assuming (b) $\theta_0 = 0$, (c) a Beta[2,10] prior distribution for θ_0?
7.3. In Example 7.1, justify the statement that the bias is equivalent to a 'standard deviation of 30% on the HR scale'. How might you interrogate an expert concerning the potential size of a bias?

Table 7.1 Results from an international case–control study of leukaemia following treatment for Hodgkin's disease.

Treatment	Cases	Controls
No chemotherapy	11	160
Chemotherapy	138	251
Total	149	411

Table 7.2   Odds ratios and 95% CIs for venous thromboembolism in users of third-generation oral contraceptives compared to second-generation OCs.

Study	Odds ratio	95% CI
Farley *et al.*	2.6	1.4 to 4.8
Jick *et al.*	2.2	1.1 to 4.4
Bloemenkamp *et al.*	2.5	1.2 to 5.2
Spitzer *et al.*	1.5	1.1 to 2.2

7.4. Table 7.2 presents the results of the four case–control studies reported by Lilford and Braunholtz (1996) in Example 7.1. Estimate the log(odds ratio) assuming (a) a pooled-effects model and (b) a random-effects model, using the empirical Bayes methodology of Section 3.17. The analysis in Example 7.1 considers a conjugate normal analysis, using the results of a meta-analysis of the four studies to produce an approximate normal likelihood. (c) Examine the sensitivity of the conclusions to the assumptions underlying the meta-analysis.

7.5. In Example 7.2, investigate the claim that the findings are robust to the prior on τ and the use of a full binomial likelihood.

7.6. Goldstein and Spiegelhalter (1996) report the teenage conception rates shown in Table 7.3.

Table 7.3   Teenage conception rates (13–15-year-olds) in 1990–1992 for 15 health boards in Scotland.

Health Board	No. conceptions	Relevant population
Western Isles	6	1 935
Orkney	5	1 220
Highland	76	11 515
Borders	36	5 294
Lanark	230	31 944
Argyle	172	23 243
Forth	121	14 938
Glasgow	388	45 647
Shetland	13	1 512
Lothian	303	35 233
Dumfries	67	7 614
Grampian	267	27 526
Ayr	204	20 606
Fife	188	18 614
Tayside	208	20 000

(a) Calculate the observed conception rates per 10 000 population, and rank the health boards according to their rates.
(b) Assuming either Poisson or binomial responses, estimate the ranks of each health board in a 'league table', assuming both independent and exchangeable rates.
(c) What is the probability that Tayside truly has the highest rates?

8

Evidence Synthesis

8.1 INTRODUCTION

It is unusual for a policy question to be informed by a single study. Interest in more diffuse areas, such as health-care delivery or broad public health measures, means that health-care evaluations become more realistically complex and there is an inevitable demand to make use of the huge volume of published and unpublished evidence. A quantitative synthesis of multiple studies has become known as a *meta-analysis*, whose procedures for randomised trials have become increasingly formalised by the Cochrane Collaboration (Section A.2). This has led to parallel developments for observational studies (Stroup *et al.*, 2000), and in the context of social science by the Campbell Collaboration (Section A.2).

A Bayesian approach to such 'standard' meta-analyses is considered in Section 8.2, emphasising the additional flexibility that arises both from the use of prior information and the adoption of Markov chain Monte Carlo methods for dealing with more complex models (Section 8.2.2). In particular, Section 8.2.3 illustrates the ability to handle the tricky and controversial issue of dependence of the treatment effect on baseline risk. The basic meta-analysis procedure can be further extended to increasingly complex contexts. First, we examine the somewhat specific but useful issue of *indirect comparison* analyses (Section 8.3), which are required when multiple studies have been carried out in which multiple treatments have been compared in different combinations, and we wish to draw inferences about specific treatment contrasts. Second, we examine the broader topic of *generalised evidence synthesis* (Section 8.4), in which studies of possibly different designs are pooled in order to estimate quantities of interest – a wide range of alternative models for pooling are available, broadly following the structure outlined for handling historical data (Section 5.4).

Since the basic methodological procedures were established in Section 3.17, this chapter relies heavily on a series of quite detailed examples, featuring prediction from meta-analyses (Example 8.1), meta-analysis with rare events (Example 8.2), dependence on baseline risk (Example 8.3), indirect comparisons in drug trials (Example 8.4), synthesis of RCTs and observational studies

Bayesian Approaches to Clinical Trials and Health-Care Evaluation D. J. Spiegelhalter, K. R. Abrams and J. P. Myles
© 2004 John Wiley & Sons, Ltd ISBN: 0-471-49975-7

(Example 8.5), and two examples of the synthesis of multiple studies to estimate the effects of a screening programme (Examples 8.6 and 8.7).

Many of the ideas in this chapter were suggested by Eddy *et al.* (1992) under the general label 'confidence profile method', and promulgated with numerous worked examples and accompanying software (FAST*PRO). They used directed conditional independence graphs (Section 3.19.3) to represent the qualitative way in which multiple contributing sources of evidence relate to the quantity of interest, explicitly allowing the user to discount studies due to their potential internal bias or their limited generalisability (Section 7.3). Their analysis was essentially Bayesian, although it was possible to avoid specification of priors and use only the likelihoods. The need to make explicit subjective judgements concerning the existence and extent of possible biases, and the limited capacity and friendliness of the software, have perhaps limited the application of this technique. However, throughout this chapter we show that modern software can allow straightforward implementation of their ideas, and we fully acknowledge their foresight in promoting these concepts.

8.2 'STANDARD' META-ANALYSIS

8.2.1 A Bayesian perspective

A standard classical meta-analysis will comprise a series of K studies each estimating a treatment effect $\theta_k, k = 1, \ldots, K$, by means of a likelihood which can be expressed, possibly approximately, as

$$y_k \sim N[\theta_k, s_k^2], \tag{8.1}$$

whether the sample variances s_k^2 are generally considered known or estimated. Following the development in Section 3.17, individual estimates of the θ_k can be termed a *fixed-effects* analysis in which there is no pooling; at the other extreme an analysis in which all the θ_k are assumed equal may be termed *pooled-effect*. An intermediate *random-effects* analysis (DerSimonian and Laird, 1986) treats the θ_k as if they were drawn from a population distribution, generally taken as

$$\theta_k \sim N[\mu, \tau^2].$$

As mentioned in Section 3.17, a variety of classical techniques are available for estimating τ^2; see Sutton *et al.* (2000) and Whitehead (2002) for recent reviews.

From a Bayesian perspective, it is natural to treat meta-analysis as a standard problem of multiplicity (Section 3.17), and follow the approach taken in contexts such as subset analysis (Section 6.8.1), multi-centre trials (Section 6.8.2), multiple N-of-1 studies (Section 6.11) and institutional comparisons (Section 7.4). Thus, if we are willing to treat the trials as exchangeable, the 'true' treatment effect in each trial is considered a random quantity drawn from some population distribution, in exactly the same manner as the standard random-effects approach to meta-analysis. However, the latter tends to focus on estimating an overall treatment effect, while a full Bayesian approach also concentrates on estimating trial-specific effects and, as we shall see below, permits a variety of useful extensions. A simple 'empirical Bayes' meta-analysis has already been presented in Example 3.13.

The Bayesian approach requires prior distributions to be specified for the mean effect size μ, the between-studies standard deviation τ, and possibly the within-study variances; as in other hierarchical models, specifying default 'reference' priors for τ is not straightforward (Section 5.7.3).

Some of the potential advantages of the Bayesian approach to meta-analysis are rather briefly summarised below (Sutton *et al.*, 2000); of course, many of these issues can also be tackled from a classical perspective, but perhaps with less flexibility.

1. *Unified modelling.* The conflict between fixed- and random-effects meta-analysis is overcome by explicitly modelling between-trial variability (which could be assumed to be small). The 'random-effects' distribution can also be much more flexible than the standard normal assumption, for example partitioned into subgroups within which studies might be assumed equal or exchangeable.
2. *Borrowing strength.* As in all areas of Bayesian hierarchical modelling, an exchangeability assumption leads to each experimental unit 'borrowing' information from the other units, leading to a shrinkage of the estimate towards the overall mean, and a reduction in the width of the interval estimate. This degree of pooling depends on the empirical similarity of the estimates from the individual units.
3. *Exact likelihoods.* It is not necessary to adopt approximate normal likelihoods, although care may be required in dealing with nuisance parameters (Section 8.2.2).
4. *Allowing for uncertainty in all parameters.* The full uncertainty from all the parameters is reflected in the widths of the intervals for the parameter estimates; these will therefore tend to be wider than those from a classical random-effects analysis.
5. *Allowing for other sources of evidence.* Other sources of evidence can be reflected in the prior distributions for parameters, or in pooling multiple types of study (Section 8.4).
6. *Allowing direct probability statements on different scales.* Quantities of interest can be directly addressed, such as the probability that the true treatment effect in a typical trial is greater than 0. It is also possible to make inferences on a variety of scales, such as risk difference, risk ratio and odds ratio (Carlin, 2000; Warn *et al.*, 2002).

7. *Predictions.* The ease of making predictions within a Bayesian framework allows, for example, current meta-analyses to be used in designing future studies. For example, we may use the basic normal model to predict the treatment effect θ^{new} in a new trial by

$$\theta^{new} \sim N[\mu, \tau^2]. \tag{8.2}$$

Rather than making predictions based on the 'plug-in' random-effects distribution $p(\theta^{new}|\hat{\mu}, \hat{\tau})$, we can use the full predictive distribution

$$p(\theta^{new}|\text{data}) = \int p(\theta^{new}|\mu, \tau) \, p(\mu, \tau|\text{data}) \, d\mu \, d\tau, \tag{8.3}$$

which fully takes into account the uncertainty concerning μ and τ. This may be easily achieved when using MCMC methods by simulating a value θ^{new} at each iteration; the simulated values form a sample from the full predictive distribution (8.3).

It could be argued that this predictive distribution is a more appropriate summary of the treatment than conclusions regarding the mean effect μ. Such a predictive distribution may also be valuable as the basis for power calculations for confirmatory clinical trials (Section 6.5), and could also act as a prior distribution in their analysis. Predictions of effects in future populations are also required if the analysis is to contribute to a policy model, and these may need to be adjusted for different patient characteristics.

8. *Assessing compatibility between meta-analyses and individual clinical trials.* Suppose we have observed data y^{obs} in a new trial and we wish to assess their compatibility with a meta-analysis. We may consider y^{obs} as providing a likelihood term for a new treatment effect θ^{new}, and the issue becomes one of assessing compatibility between a likelihood and a prior $p(\theta^{new}|\text{data})$ obtained from (8.3). We have already considered such comparisons in Section 5.8, where Box's method was outlined. This compares y^{obs} with the predictive distribution of new data Y^{new}, given by

$$p(Y^{new}|\text{data}) = \int p(Y^{new}|\theta^{new}) \, p(\theta^{new}|\text{data}) \, d\theta^{new}.$$

Specifically, as a form of two-sided P-value, we calculate twice the minimum tail area $2\min(p(Y^{new} < y^{obs}|\text{data}), p(Y^{new} > y^{obs}|\text{data}))$. This is easily achieved when using MCMC by generating θ^{new}, then generating Y^{new} from $p(Y^{new}|\theta^{new})$, and counting the proportion of simulated Y^{new}s that exceed or are less than y^{obs}.

Suppose both prior $p(\theta^{new}|\text{data})$ and likelihood $p(y^{obs}|\theta^{new})$ can be assumed approximately normal with distributions $N[\hat{\theta}^{new}, \sigma^2/m]$ and $N[\theta^{new}, \sigma^2/n]$ respectively. Then Box's procedure is equivalent to a two-sided test based on a standardised comparison

$$Z = \frac{y^{\text{obs}} - \hat{\theta}^{\text{new}}}{\sigma\sqrt{m^{-1} + n^{-1}}}.$$

Example 8.1 illustrates the comparison of predictions Y^{new} from meta-analyses with observed y^{obs} in new trials, to show the conflict may not be as great as is often claimed – see also Berry (2000).

9. *Cumulative meta-analysis.* It is natural to use a cumulative meta-analysis as external evidence when monitoring a clinical trial (Henderson *et al.*, 1995), and cumulative meta-analysis can also be given a Bayesian interpretation as providing a prior distribution (Lau *et al.*, 1995; see also Section 5.4): in this situation the Bayesian approach relies on the assumption of exchangeability of trials but avoids concerns with retaining Type I error over the entire course of the cumulative meta-analysis.

10. *'Meta-regression'.* It is reasonably straightforward to investigate the relationship between treatment effect and study-level factors. For example, suppose we have measured a covariate x_k on each study. Then we could fit the model

$$\theta_k = \theta_k^{\text{adj}} + \beta(x_k - \overline{x}), \tag{8.4}$$

where θ_k^{adj} is the treatment effect adjusted for the covariate and might be assumed to have a population distribution $\theta_k^{\text{adj}} \sim N[\mu, \tau^2]$. However, particular care is required for examining the relationship with baseline rates (Section 8.2.3).

11. *Publication bias.* It is feasible to model the effects of different degrees of publication bias, although any conclusions must necessarily be somewhat dependent on uncheckable assumptions (Silliman, 1997; Begg *et al.*, 1997; Givens *et al.*, 1997; Smith *et al.*, 2000).

These methods are not restricted to randomised trials and may equally be applied to meta-analyses of case–control and other observational studies, with the usual caveats about adjustment for potential bias.

Example 8.1   *ISIS: Prediction after meta-analyses*

Reference: Higgins and Spiegelhalter (2002).

Background: Example 3.13 described a meta-analysis carried out in 1993 which showed an apparent survival benefit from magnesium sulphate following myocardial infarction. When the ISIS-4 'megatrial' announced its result of no benefit from magnesium, the apparent conflict with the meta-analysis led to a long-running argument – see Higgins and Spiegelhalter (2002) for a recent analysis. Here we derive a predictive distribution for the effect expected in a new trial based on the data available in the meta-analysis and presented in Example 3.13, and see whether that prediction is really in conflict with the results observed in ISIS-4. We carry out a full Bayesian analysis on all the parameters, and check sensitivity to prior assumptions.

Statistical model: The normal approximation for the log(odds ratios) described in Section 2.4.1 is adopted.

Prior distribution: As a baseline analysis, μ and τ, the between-study mean and standard deviation, are given uniform priors.

Computation/software: MCMC methods implemented using WinBUGS.

Evidence from study: The data contributing to the meta-analysis were given in Table 3.8. In ISIS-4 2216/29 011 (7.6%) deaths were observed in the magnesium arm, slightly in excess of the 2103/29 039 (7.2%) deaths observed under placebo. This corresponds to a log(OR) of $y^{obs} = 0.06$, with standard deviation 0.03.

Bayesian interpretation: Summaries of the simulated values of μ and τ are given in Table 8.1 under the uniform prior assumptions. It can be seen that the between-trial heterogeneity is poorly estimated from these data in that the 95% interval is extremely wide, and therefore some prior sensitivity might be expected. Nevertheless the 95% interval for the overall odds ratio does exclude 1. The predicted log(OR) θ^{new} in a new trial has an extremely wide interval, and this is reflected in the predictive distribution of the observed log(OR) Y^{new} in a trial of the size of ISIS-4, which has a point prediction of 0.56 but a 95% prediction interval from 0.10 to 2.43. We note that the huge sample size of ISIS-4 means that the distribution of Y^{new} is essentially the same as θ^{new}. The observed log(OR) of $y^{obs} = 0.06$ lies well within this interval with a one-sided tail area of 0.12; Box's compatibility measure is the probability of observing such an extreme result,

Table 8.1 Comparison of meta-analysis with megatrial. Y^{new} are the results from a further trial that would be predicted from the meta-analysis. The observed data y^{obs} from ISIS-4 are well within the 95% prediction interval.

Parameter		Median	95% interval	Median OR	95% interval for OR
μ:	mean effect	−0.59	−1.35 to −0.01	0.56	0.26 to 0.99
τ:	between-trial SD	0.55	0.02 to 1.62		
θ^{new}:	prediction of effect in new trial	−0.58	−2.28 to 0.89	0.56	0.10 to 2.43
Y^{new}:	prediction of log(OR) to be observed in new trial	−0.59	−2.29 to 0.88	0.56	0.10 to 2.43
y^{obs}:	observed log(OR) in ISIS-4	0.06	0.00 to 0.12	1.06	1.00 to 1.13

$2 \times 0.12 = 0.24$. This analysis does not therefore indicate strong conflict between the meta-analysis and the megatrial.

Sensitivity analyses: Six alternative prior distributions for τ give predictive distributions for Y^{new} shown in Figure 8.1. As expected from the discussion in Section 5.7.3, the Gamma(0.001,0.001) (a) (equivalent to a root-inverse-gamma on τ), DuMouchel (e) and half-normal with $\tau_u = 1.0$ (f) tend to support smaller values of τ and hence produce narrower posterior intervals, while the uniform on τ^2 (b) leads to very wide intervals. We note that $s_0 = 0.36$, roughly corresponding to an average of 31 events per trial (in fact a total of 286 events are recorded in Table 3.8, or an average of 36 events per trial).

The resulting one-sided P-values $P(Y^{\text{new}} < y^{\text{obs}})$ ranged from 0.06 (for (a) and (f)) to 0.18 (for (b)), so under no assumption was there particularly strong evidence of incompatibility.

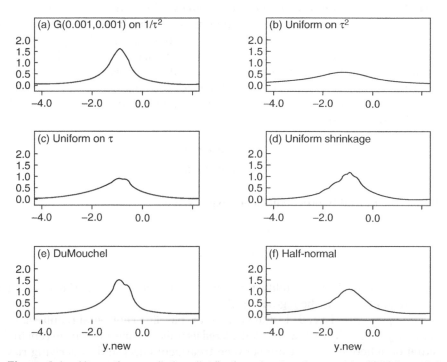

Figure 8.1 Alternative predictive distributions for the observed log(OR) in a trial the size of ISIS-4, arising from six different prior distributions on τ. The actual observed log(OR) was 0.058, and hence was not seriously in conflict with any of the predictive distributions.

8.2.2   Some delicate issues in Bayesian meta-analysis

The Bayesian approach to meta-analysis promises additional flexibility but raises some tricky issues, some of which are generic to hierarchical models and some more specific to this context. These include the following:

The between-study standard deviation τ. Comparative studies show that when there are few studies and hence τ cannot be accurately estimated from the data alone, the prior for this parameter may become important and the empirical Bayes approach, in which the uncertainty about the between-study variability is ignored, tends to provide intervals that are too narrow. Priors on the heterogeneity parameter have already been discussed in Section 5.7.3, in which it was noted that Higgins and Whitehead (1996) use proper priors derived from a series of meta-analyses. It is important to check the sensitivity to the prior on τ – see Example 8.1.

Exact likelihoods and nuisance parameters. The standard normal approximation given in (8.1) may not be appropriate when the studies are small or their results extreme, as the resulting likelihoods may not be approximately normal. For example, suppose in the kth trial there are n_{tk} and n_{ck} in the treatment and control groups respectively, and we observe r_{tk} and r_{ck} deaths. If either n_{tk} and n_{ck} is small, or mortality rates are near 0% or 100%, we may adopt a full binomial model instead of the normal approximation of Section 2.4. Specifically, we assume

$$r_{tk} \sim \text{Bin}[\,p_{tk}, n_{tk}],$$
$$r_{ck} \sim \text{Bin}[\,p_{ck}, n_{ck}],$$

where the mortality probabilities are expressed as

$$\text{logit}(p_{tk}) = \phi_k + \theta_k,$$
$$\text{logit}(p_{ck}) = \phi_k. \tag{8.5}$$

Hence ϕ_k is the logit(mortality rate) in the control group of trial k, and the treatment effect θ_k is the log(odds ratio).

The ϕ_k can also be called 'study effects' or 'baseline rates' and require careful handling. Generally they will be considered as nuisance parameters, except in the situation where a relationship between treatment effect and underlying risk is suspected (Section 8.2.3). Eliminating such nuisance parameters is a problem within all schools of statistical inference: see Section 3.18 for a brief review.

In the context of meta-analysis the following methods have been adopted:

- '*Approximate pivotal quantity*'. The standard normal approximation in (8.1) has a distribution which does not depend on the baseline ϕ_k.

- *'Conditional likelihood'*. By conditioning on the value of a statistic we derive a likelihood which depends only on the parameter of interest: see Liao (1999) for a Bayesian application of this procedure in meta-analysis.

- *Prior distributions.* The appropriate joint prior distribution for the ϕ_k and the θ_k presents a particular problem. The 'study effects' ϕ_k might be given independent uniform priors, but a choice must be made between the logit (ϕ_k) and probability (p_{ck}) scale. Random study effects can be assumed if the control group risks are considered exchangeable, but a normal distribution may not be appropriate. Finally, it may be reasonable to assume the ϕ_k and the θ_k are correlated, and hence carry out a 'bivariate meta-analysis' (van Houwelingen *et al.*, 1993). This is essential if one is explicitly investigating the relationship between effect and baseline risk (Section 8.2.3), but it has been argued that it would be appropriate in any situation in which one assumes random ϕ_k. The reasoning is as follows: if the ϕ_k and the θ_k are assumed independent, (8.5) shows that the variance of the treatment risks is forced to be greater than the variance of the control risks. Of course this may be a reasonable assumption, but it should be explicitly acknowledged.

Example 8.2 examines a meta-analysis of trials with rare events, and explores the sensitivity of conclusions to a range of these modelling options.

Example 8.2   *EFM: meta-analyses of trials with rare events*

References: Sutton and Abrams (2001), Sutton *et al.* (2002).

Intervention: Electronic foetal heart rate monitoring (EFM) in labour, with the aim of early detection of altered heart-rate pattern and hence a potential benefit in perinatal mortality.

Aim of study: EFM was gradually introduced in the early 1970s, and early evaluation of its impact in terms of perinatal death was in terms of either non-randomised comparative studies or before–after studies. A large body of evidence was collected which suggested that EFM was indeed clinically effective in reducing the risk of perinatal death. Despite this body of evidence a number of randomised trials were conducted, which were much smaller in terms of sample sizes, but which suggested that there was little benefit, if any, from the use of EFM. Here we consider the evidence from the randomised trials, with emphasis on the difficulties associated with rare events.

Study design: Meta-analysis of nine randomised trials.

Outcome measure: Perinatal mortality, as measured by the odds ratio in deaths per 1000 births, odds ratios less than 1 favouring EFM. We note that Sutton and Abrams (2001) consider the risk difference, which is directly related to the number needed to treat (NNT) and hence a policy decision (Section 3.14).

Statistical model: There are a number of options for dealing with the nuisance parameters in this model, *i.e.* the control group risks (Section 3.18), acknowledging that the standard normal approximation for the log (odds ratio) likelihood within each study may be inappropriate due to the rarity of perinatal deaths.

(a) *Fixed effects.* A normal approximation to the likelihood for the observed log(odds ratio) (Section 2.4), with the log(odds ratios) θ_k assumed to be independent.

(b) *Approximate normal likelihood, random effects.* A normal approximation to the likelihood for the observed log(odds ratio) (Section 2.4), with the log(odds ratios) assumed to have the distribution $\theta_k \sim N[\mu, \tau^2]$.

(c,d) *Binomial likelihood, random effects.* An exact binomial model (8.5), with the log(odds ratios) assumed to have the distribution $\theta_k \sim N[\mu, \tau^2]$. The control group risks are assumed independent, with options (c) and (d) representing different assumptions (see below).

An exchangeable model for the control group risks could also have been adopted.

Prior distribution: μ and τ, the between-study mean and standard deviation, are given uniform priors. For the full binomial models (c) and (d), two alternative priors for each study's control group mortality p_{ck} are considered: (c) p_{ck} is given an independent uniform prior, and (d) $\phi_k = \text{logit}(p_{ck})$ is given an independent uniform prior.

Computation/software: MCMC methods implemented using WinBUGS.

Evidence from study: The randomised data are presented in Figure 8.2. We note that trial 8 has a high mortality rate in the control group, which would cast doubt on a simplistic normal assumption for exchangeable control groups risks. The 0s in trials 3 and 6 also suggest that conclusions may be sensitive to ways of dealing with the nuisance parameters.

Bayesian interpretation and sensitivity analyses: Figure 8.2 shows the estimated odds ratios for each trial and for the population, for each of the four models (a) to (d). The approximate normal random-effects model (b) is consistently more conservative in its estimate than the models using a binomial likelihood, and also more precise. The binomial model (d) with a uniform prior on the logit of the control risks is more conservative than model (c) with a uniform prior on the control risks – this is presumably because model (d) will tend to estimate smaller control risks than model (c) and hence will reduce any apparent benefit of EFM.

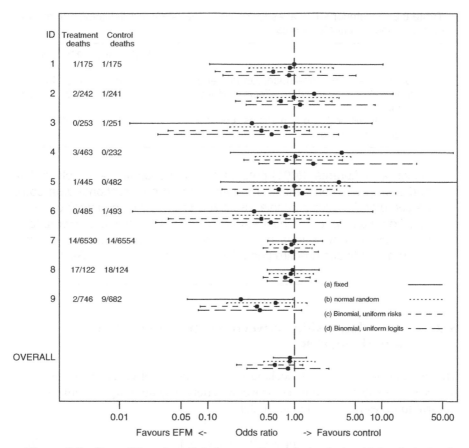

Figure 8.2 Four different models for a meta-analysis of nine trials of electronic foetal monitoring. The rare events lead to considerable sensitivity of conclusions to assumptions concerning the form of the likelihood and prior distributions on nuisance parameters.

Table 8.2 shows that the three random-effects models also give rise to different estimates of τ, although each has a wide interval with the bulk of the density near 0. There is likely to be considerable additional sensitivity to prior assumptions concerning τ.

Comments: This example shows there can be sensitivity to likelihood assumptions as well as prior distributions, and that analyses with rare events have to be handled with care. In particular, the traditional normal approximation, used in so many of our examples, would lead to excessive confidence in the conclusion, whereas the RCTs provide little evidence of efficacy on their own.

Table 8.2 Posterior summaries for between-trial standard deviation τ from three different random-effects models.

Model	Median of τ	95% interval
(b) Approximate normal likelihood, random effects:	0.32	0.01 to 1.50
(c) Binomial likelihood, random effects, uniform on control risks:	0.54	0.01 to 2.25
(d) Binomial likelihood, random effects, uniform on logit control risks:	0.75	0.09 to 2.82

Sutton and Abrams (2001) present both case–control and cohort data addressing this comparison: randomised and observational data could be combined by, for example, using the (possibly discounted) observational data as a prior for the meta-analysis presented above (Hornbuckle *et al.*, 2000), or by conducting a generalised evidence synthesis in which different study designs are pooled in a hierarchical model (Section 8.4).

8.2.3   The relationship between treatment effect and underlying risk

The appropriate means of modelling the dependence of effect on baseline risk has been the subject of some controversy. There is general agreement that it is natural to investigate the linear model

$$\theta_k = \theta_k^{\text{adj}} + \beta(\phi_k - \overline{\phi}), \tag{8.6}$$

where θ_k^{adj} is now the treatment effect adjusted for a measure of baseline risk ϕ_k, also known as a 'study effect'. θ_k^{adj} might be assumed to have a distribution

$$\theta_k^{\text{adj}} \sim \text{N}[\mu, \tau^2]. \tag{8.7}$$

We note from (8.6) and (8.7) that the treatment effect θ_k has distribution

$$\theta_k \sim \text{N}[\mu + \beta(\phi_k - \overline{\phi}), \tau^2], \tag{8.8}$$

and hence the treatment effect in any future trial with true baseline risk ϕ can be obtained by substitution in (8.8). In particular, the effect is expected to be 0 when ϕ obeys

$$\phi_0 = \frac{-\mu}{\beta} + \overline{\phi};$$

the solution to this equation is known as the 'breakeven' point. MCMC methods allow inferences to be drawn about this quantity, as demonstrated in Example 8.3. Such models have been investigated by McIntosh (1996), Thompson *et al.* (1997), Sharp and Thompson (2000) and Arends *et al.* (2000).

The controversy arises in the specification of a prior for the 'study effects' ϕ_k. Thompson *et al.* (1997) assume independent priors and hence fixed study effects, but this is strongly criticised by Houwelingen and Senn (1999), who argue that since this introduces an additional nuisance parameter for each trial, the procedure will be 'inconsistent' in the sense that under broad assumptions it will, as the number of trials grows, not tend to give the correct underlying relationship. In their reply the authors claim that fixed study effects are standard methodology, for example in using logistic regression, and will only give misleading conclusions in extreme situations. These alternative approaches are investigated in Example 8.3.

Van Houwelingen and Senn (1999) also make the important point that there will always, in a sense, be dependence between effect and baseline, since if there is no relationship on a logit scale, there would be on an absolute risk scale. An important aim may therefore be to find a scale on which the effect is most independent of baseline.

Example 8.3   *Hyper: Meta-analyses of trials adjusting for baseline rates*

References: Hoes *et al.* (1995) and Arends *et al.* (2000).

Intervention: Drug treatment in mild to moderate hypertension.

Aim of study: To determine whether drug treatment reduced mortality and to see whether the size of the treatment effect depended on the event rate in the control group.

Study design: Meta-analysis of 12 randomised trials with considerable variability in baseline risk.

Outcome measure: All-cause mortality per 1000 patient-years of follow-up.

Statistical model: A random-effects Poisson regression model was assumed. In a similar manner to Section 3.18, for the ith study the numbers of deaths r_{ti} and r_{ci} in treatment and control groups are assumed

$$r_{ti} \sim \text{Poisson}(m_{ti}),$$
$$r_{ci} \sim \text{Poisson}(m_{ci}),$$

using the notation of Section 2.6.2. The Poisson means are expressed as

$$m_{ti} = \log(n_{ti}/1000) + \phi_i + \theta_i,$$
$$m_{ci} = \log(n_{ci}/1000) + \phi_i,$$

where n_{ti} and n_{ci} are the patient-years of follow-up in the treatment and control groups. Hence ϕ_i is the log of the rate per 1000 patient-years in the control group of trial i, and the treatment effect θ_i is the log(rate ratio).

The dependence of treatment effect on baseline rate is then modelled exactly as described in Section 8.2.3.

Prior distribution: For the baseline analysis, μ and τ, the between-study mean and standard deviation, are given uniform priors. Following the discussion in Section 8.2.3, two priors are considered for each study's control log(event rate) ϕ_i: independent uniform priors, and exchangeable with a normal distribution

$$\phi_i \sim N\left[\mu_\phi, \tau_\phi^2\right],$$

where μ_ϕ, τ_ϕ are given uniform priors.

Computation/software: MCMC methods implemented using WinBUGS.

Evidence from study: The data are given in Table 8.3. Figure 8.3(a) shows the observed rate ratios from Table 8.3 plotted against the observed control group rates. There is a clear suggestion of a relationship.

Bayesian interpretation: Figure 8.3(b) shows the estimated rate ratios e^{θ_i} plotted against the estimated control group rates e^{ϕ_i} when adjusting for baseline, assuming independent uniform priors for the ϕ_i. There is clear shrinkage towards the assumed straight line, with the control group rate for centre 2 estimated to be even smaller than that observed. The intersection

Table 8.3 Data from 12 randomised trials of drug treatment for mild-to-moderate hypertension: r is the number of deaths, n is the patient-years of follow-up, and rates are events per 1000 patient-years.

Treatment group			Control group		
r_t	n_t	rate$_t$	r_c	n_c	rate$_c$
10	595.2	16.8	21	640.2	32.8
2	762.0	2.6	0	756.0	0.0
54	5 635.0	9.6	70	5 600.0	12.5
47	5 135.0	9.2	63	4 960.0	12.7
53	3 760.0	14.1	62	4 210.0	14.7
10	2 233.0	4.5	9	2 084.5	4.3
25	7 056.1	3.5	35	6 824.0	5.1
47	8 099.0	5.8	31	8 267.0	3.7
43	5 810.0	7.4	39	5 922.0	6.6
25	5 397.0	4.6	45	5 173.0	8.7
157	22 162.7	7.1	182	22 172.5	8.2
92	20 885.0	4.4	72	20 645.0	3.5

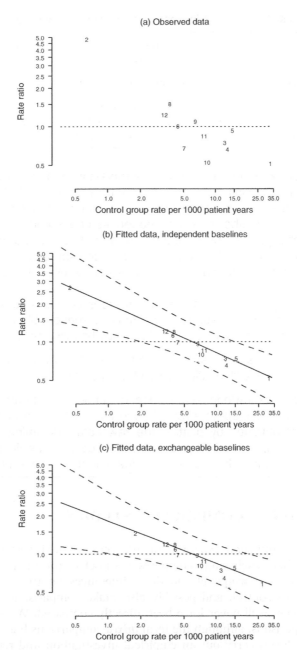

Figure 8.3 Estimated control group rates and rate ratios in 12 studies under different assumptions. (a) can be considered as fixed-effect estimates of control rate and treatment effects. In (b), the treatment effect is assumed linearly related to independent log(control group rates), whereas in (c) the log(control group rates) are assumed exchangeable and hence shrunk towards a common value.

Table 8.4 Results from fitting independent and exchangeable control group rates.

		Independent control rates		Exchangeable control rates	
	Parameter	Median	95% interval	Median	95% interval
β	Dependence on baseline	−0.38	−0.57 to −0.17	−0.33	−0.55 to −0.09
e^{ϕ_0}	'Breakeven' control rate	6.00	3.67 to 8.01	6.06	2.73 to 8.80
τ	Residual SD	0.10	0.01 to 0.28	0.10	0.01 to 0.30

of the upper and lower prediction intervals with the null rate ratio 1 corresponds to the interval for e^{ϕ_0}, the control group rate at which there is no treatment effect. The corresponding estimates are shown in Table 8.4.

Figure 8.3(c) shows the consequences of assuming the control rates are exchangeable: the estimates are shrunk towards a common value, particularly the smaller study 2. The reduced spread in the control group rates with the exchangeable analysis has resulted in increased uncertainty.

After adjusting for baseline risk, there is very little residual between-study heterogeneity suggesting it may be reasonable to set $\tau = 0$ and assume all heterogeneity is explained by baseline risk.

Sensitivity analyses: Alternative priors for the between-study standard deviation τ have little influence on this analysis.

Comments: Acknowledging functional dependence of treatment and baseline rates brings about a reduction in the apparent gradient, compared with that obtained by plotting the raw data. Assuming exchangeable control group rates brings some shrinkage but has little influence on the conclusions. There is little residual variability around the fitted line.

8.3   INDIRECT COMPARISON STUDIES

Suppose that a number of experimental interventions are investigated in a series of studies, where each study compares a subset of the interventions with a control group. We would like to draw inferences on the treatment effects compared with control, and possibly also make comparisons between treatments that may well never have been directly compared. We shall call these *indirect* comparisons, although the term *mixed* comparisons has also been used. Song *et al.* (2003) carry out an empirical investigation and report that such comparisons arrive at essentially the same conclusions as 'head-to-head' comparisons.

A specific application arises in the context of 'active control' studies. Suppose an established treatment C exists for a condition, and a new intervention T is being evaluated. The efficacy of T would ideally be estimated in randomised trial with a placebo P as the control group, but because of the existence of C this may be considered unethical. Hence C may be used as an 'active control' in a head-to-head clinical trial, and inferences about the efficacy of T may have to be estimated indirectly, using past data on comparisons between C and P.

Let ϕ_{jk} represent the expected response (on an appropriate scale) of treatment j being given in study k, where the control is labelled as $j = 0$. A simple model might express ϕ_{jk} as

$$\phi_{jk} = \phi_k + \theta_{jk}, \qquad (8.9)$$

where ϕ_k denotes a 'study effect' and θ_{jk} a treatment effect in the kth study. It is often convenient to set $\theta_{0k} = 0$, so that we can interpret ϕ_k as the response in the control group. Equation (8.9) needs to be further constrained in order to estimate parameters: we might assume a common treatment effect across all studies $\theta_{jk} = \theta_j$, or a random effect in which the θ_{jk} are assumed drawn from some population distribution, say, $\theta_{jk} \sim N[\theta_j, \tau_j^2]$ (Higgins and Whitehead, 1996; Hasselblad, 1998). A variety of models are possible for the distributions of the ϕ_k and θ_{jk}: Higgins and Whitehead (1996) point out that if we wish the contrasts between all possible treatment pairs (including control) to have the same distribution, then we need to assume a multivariate normal distribution for the θ_{jk} with a particular correlation structure. Example 8.4 re-examines a published example of such an analysis.

Example 8.4 *Blood pressure: Estimating effects that have never been directly measured*

Reference: Gould (1991).

Intervention: Alternative therapies for lowering blood pressure.

Aim of study: To estimate the contrast between two therapies that have never been compared head-to-head. Gould (1991) suggests such an inference could then be used to design a direct comparison study.

Available evidence: Table 8.5 displays the results from a set of eight crossover experiments comprising randomised comparisons and single-arm studies (Gould, 1991), showing mean and standard deviation of change in blood pressure, and sample size in each group. Four treatments (control, A, B and C) have been given, but there has been no direct comparison between treatments A and B and it is this contrast that is of particular interest.

Statistical model: Let y_{jk} be the mean response recorded in Table 8.5 for the jth treatment in the kth study. We assume

Table 8.5 Sample sizes m, mean and standard deviation of responses under each treatment given in eight studies: e.g. study 1 compared A with C, while study 2 randomised between control and B in a 1:2 ratio. The problem is to compare treatments A and B.

	Control ($j=0$)			$A(j=1)$			$B(j=2)$			$C(j=3)$		
Study	m	Mean	SD	m	Mean	SD	m	Mean	SD	m	Mean	SD
1				41	8.90	7.49				39	6.05	10.28
2	47	5.51	8.72				100	6.21	8.02			
3	53	3.75	7.07	54	10.20	9.39						
4	47	3.04	9.20	44	8.43	8.17						
5	30	2.97	7.69				32	6.53	7.80			
6	69	3.99	8.04									
7	68	5.28	7.58									
8	67	3.34	8.01									

$$y_{jk} \sim N\left[\phi_{jk}, \frac{\sigma^2}{m_{jk}}\right],$$

and assume $\phi_{jk} = \phi_k + \theta_j$ (8.9), where $\theta_0 = 0$ so that ϕ_k is the response in the control group in study k (although there was not necessarily an actual control in the kth study) and θ_1, θ_2, θ_3 measure the mean effects of A, B, C over placebo, respectively. Some of the studies have only a single arm, and if we assume fixed study effects then these will contribute no information (except in contributing to the estimate of σ^2). Since all the studies were carried out in a common research programme by the same investigators, it may be reasonable to adopt exchangeable study effects ϕ_k, with

$$\phi_k \sim N\left[\mu_\phi, \tau_\phi^2\right].$$

The treatment effects θ_1, θ_2, θ_3 are taken as independent fixed effects. We may use the following distribution theory to obtain a likelihood for σ (Section 2.6.5). The observed standard deviations s_{jk} have the property

$$\frac{(m_{jk} - 1)s_{jk}^2}{\sigma^2} \sim \chi_{m_{jk}-1}^2,$$

and hence $(m_{jk} - 1)s_{jk}^2 \sim \Gamma((m_{jk} - 1)/2,\ 1/(2\sigma^2))$.

Prospective analysis?: No.

Table 8.6   Posterior summaries.

	Parameter	Median	SD	95% interval
μ_δ	Control mean	4.01	0.50	3.00 to 4.98
θ_1	A	9.37	0.79	7.87 to 10.98
θ_2	B	6.10	0.87	4.28 to 7.73
θ_3	C	6.92	1.08	4.83 to 9.07
$\theta_1 - \theta_2$	A vs. B	3.28	1.16	1.08 to 5.68
σ	sampling sd	8.18	0.22	7.79 to 8.63
τ_ϕ	between-study sd	0.46	0.48	0.02 to 1.78

Prior distribution: Uniform distributions are given to $\log(\sigma)$, μ_ϕ, τ_ϕ and each of the θ_j.

Loss function or demands: None specified.

Computation/software: MCMC implemented in WinBUGS, with inferences based on 10 000 iterations after a burn-in of 1000.

Bayesian interpretation: The results are shown in Table 8.6, revealing the between-study standard deviation τ_ϕ to have a wide interval. The indirect analysis allows a posterior distribution to be obtained for $\theta_1 - \theta_2$ which might be used in designing a suitable trial for a direct comparison of A and B.

8.4   GENERALISED EVIDENCE SYNTHESIS

As noted when discussing observational studies in Chapter 7, in some circumstances randomised evidence will be less than adequate due to economic, organisational or ethical considerations (Black, 1996). Considering all the available evidence, including that from non-randomised studies, may then be necessary or advantageous. Droitcour *et al.* (1993) describe the limitations of using either RCTs or databases alone, in that RCTs may be rigorous but restricted, whereas databases have a wider range but may be biased. They introduce what they term *cross-design synthesis*, an approach for synthesising evidence from different sources, with the aim 'not to eliminate studies of overall low quality from the synthesis, but rather to provide the information needed to compensate for specific weaknesses'. Although not a strictly Bayesian approach, they are essentially explicitly modelling potential biases (Section 7.3), and then attempting to generalise the results of clinical trials for broader populations. Rubin (1992) emphasises pooling evidence through modelling in order to 'build and extrapolate a response surface', which models the true treatment effect conditional on both the design of the study and subgroup factors.

Cross-design synthesis was outlined in a report from the US General Accounting Office (General Accounting Office, 1992), but a *Lancet* (1992) editorial was critical of this approach, suggesting it would deflect attention from carrying out serious controlled trials: this was denied in a subsequent reply by Chelimsky *et al.* (1993). A commentary by Begg (1992) suggested they had underestimated the difficulty of the task, and appeared to assume that randomised trials and databases could be reconciled by statistical adjustments, whereas selection biases and differences in experimental rigour could not be eliminated so easily. A non-Bayesian case study is provided by Belin *et al.* (1995) who combine observational databases in order to evaluate interventions to increase screening rates, but need to impute missing data in some studies.

One must clearly be very cautious in such an endeavour, balancing the desire to make use of all available evidence with due acknowledgement of potential weaknesses. It is not a purely technical exercise, and must be carried out in loose collaboration with subject-matter experts. Nevertheless, it is natural to take a Bayesian approach to the synthesis of multiple study designs, in which relationships are assumed between some underlying parameters of the different studies. Such relationships may involve a huge variety of both deterministic models and probabilistic dependence, and again fall naturally into the taxonomy of relationships already explored in the use of historical data (Section 5.4)

(a) *Irrelevance.* It is always an option, possibly on purely subjective grounds, to declare certain studies irrelevant to the issue under study.
(b) *Exchangeable.* Typically we may be able to classify our studies according to a 'type', say randomised, case–control or cohort: this naturally leads to hierarchical exchangeability assumptions, which can specifically allow for the quantitative within- and between-study-type heterogeneity, and incorporate prior beliefs regarding qualitative differences between the various sources of evidence. Figure 8.4 shows a stylised graphical representation of a possible model, in which treatment effects are assumed exchangeability within study type, and also that mean study effects are exchangeable. Examples of this approach include Prevost *et al.* (2000) who pool randomised and non-randomised studies on breast cancer screening (Example 8.5), Larose and Dey (1997) who similarly assume open and closed studies are exchangeable, and Dominici *et al.* (1999) who examine migraine trials and pool open and closed studies of a variety of designs in a four-level hierarchical model. There is a clearly a difficulty in making such exchangeability assumptions, since there are few study types and hence little information on the variance component. Prior assumptions may be very important, and priors for the degree of 'similarity' between alternative designs might be empirically informed by studies comparing the results of RCTs and observational data, such as listed in Section 7.3.
(c) *Potential biases* and (d) *Equal but discounted.* Both biases and discounting can be incorporated into a model for between- and within-study-type variation such as that shown in Figure 8.4.

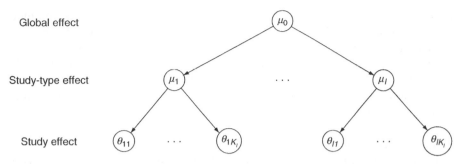

Global effect

Study-type effect

Study effect

Figure 8.4   Hierarchical model in which the effects θ_{ij} in studies of type i are assumed exchangeable with mean μ_i, and the study-type effects μ_i are assumed exchangeable with mean μ_0.

(e) *Functional dependence.* Suppose we are interested in drawing inferences on a quantity f about which no direct evidence exists, but where f can be expressed as a deterministic function of a set of 'fundamental' parameters $\theta = \theta_1, \ldots, \theta_N$. For example, f might be the response rate in a new population made up of subgroups about which we do have some evidence. More generally, we might assume we have available a set of K studies in which we have observed data $y_1, \ldots, y_K$ which depend on parameters $\psi_1, \ldots, \psi_K$, where each ψ_k is itself a function of the fundamental parameters θ. This structure is represented graphically in Figure 8.5. This situation sounds very complex but in fact is rather common, when we have a lot of studies, each of which informs part of a jigsaw, and which need to be put together to answer the question of interest. See Example 8.6 for a case where the fundamental parameters have directly relevant evidence, and Example 8.7 in which the fundamental parameters have only indirect evidence.

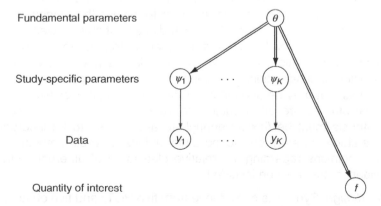

Fundamental parameters

Study-specific parameters

Data

Quantity of interest

Figure 8.5   Data y_k in each of K studies depend on parameters ψ_k, which are known functions of fundamental parameters θ. We are interested in some other function f of θ, and so need to propagate evidence from the y_k.

(f) *Equal.* It is of course possible to assume the treatment effect is common across studies of different designs. For example, Li and Begg (1994) present a non-Bayesian analysis of pooling controlled and single-arm studies, in which each is assumed to have a common treatment effect but the study effect is taken as random – this is essentially an application of the indirect comparison models considered in Section 8.3, in which some of the studies are non-comparative since only one treatment is given.

Such models allow enormous room for imagination and complexity, and graphical representations (Spiegelhalter, 1998) have been found to be very useful in clarifying the underlying structure. There is also considerable flexibility in the logical and stochastic assumptions: for example, Dominici *et al.* (1999) assume that between-study variability follows a 'mixture of normals' distribution to allow for skewness. Nevertheless, such analyses may be controversial, since there may be strong dependence on assumptions and there is concern that including studies with 'poor' designs will weaken the analysis. Careful sensitivity analyses are clearly vital, and perhaps one reason for the limited uptake of such syntheses is that they are not seen as 'clean' methods, with each analysis being context-specific, less easy to set quality markers for, easier to criticise as subjective and so on.

Example 8.5   *Screen: generalised evidence synthesis*

Reference: Prevost *et al.* (2000).

Intervention: Mammographic screening for breast cancer.

Aim of study: Breast cancer has the potential to be particularly amenable to screening in that RCTs and observational studies clearly indicate that prognosis is extremely good for early stage tumours, especially in women over 50 years of age. In order to assess the magnitude of this potential benefit, a number of RCTs and observational studies have been conducted world-wide. Whilst it is accepted that RCTs provide a 'gold standard' by which to assess efficacy, it has been argued that the inclusion of observational evidence may help in the estimation of effectiveness that may be seen in a potential population. However, observational studies are often subject to various biases and therefore any synthesis must be flexible enough to allow these to be incorporated. This study therefore developed a hierarchical Bayesian model in which prior opinions regarding the relative plausibility of different sources of evidence may also be included.

Study design: Synthesis of evidence from five RCTs and five observational studies which evaluated screening in women over 50.

Outcome measure: Breast cancer mortality per 1000 patient-years.

Statistical model: The three-level model follows that shown in Figure 8.4. Let y_{ik} be the observed log(risk ratio) in the ith study of type k, where $k = 1$ (RCT), 2 (observational), and σ^2_{ik} its associated variance. Then we assume

$$y_{ik} \sim N[\theta_{ik}, \sigma^2_{ik}],$$
$$\theta_{ik} \sim N[\mu_k, v^2_k], \qquad \qquad (8.10)$$
$$\mu_k \sim N[\mu_0, \tau^2].$$

The θ_{ik} represent the underlying effect, on the log(risk ratio) scale, in the ith study of type k. The θ_{ik} are distributed about an overall effect for the kth type of study, μ_k, with v^2_k representing the between-study variability for those studies of type k. At the third level of the model the study-type effects are distributed about an overall population effect, μ_0, with τ^2 representing the between-study-type variability. As with many other meta-analytic models the level 1 variances, σ^2_{ik}, can be replaced by the estimated sample variances s^2_{ik}, derived in this case using the methods described in Section 2.4.3. In this case prior distributions are required for μ_0, τ^2 and the v^2_k.

Prospective analysis?: No.

Prior distribution: A prior distribution for each of the v^2_k is derived using the techniques described in Section 5.7.3. We assume we are 95% sure that the true underlying risk ratio for a study of a particular type will be within a range from four times to a quarter the overall risk ratio of that type, which means that the upper 95% point of the prior distribution for each v_k is $\log(16)/(2 \times 1.96) = 0.71$. A half-normal distribution (Section 2.6.7) $v_k \sim HN[0.36^2]$ has this property.

In a similar manner a prior for the between-type variance, τ^2, can be derived from assuming 95% belief that the underlying risk ratio for a particular study type will be less than double or more than half the overall population effect. On this basis, a half-normal prior distribution $\tau \sim HN[0.18^2]$ is obtained.

For μ_0, the overall population effect, a relatively vague prior distribution is specified on the basis that the overall relative risk is unlikely to exceed 500 in favour of either screening or control, and therefore a prior distribution for μ_0 has standard deviation $\log(500)/1.96 = 3.17$, or $\mu_0 \sim N[0,10]$.

Loss function or demands: None used.

Computation/software: MCMC in WinBUGS.

Evidence from study: Figure 8.6 displays the observed risk ratios (together with 95% confidence intervals) for the five RCTs and five observational studies.

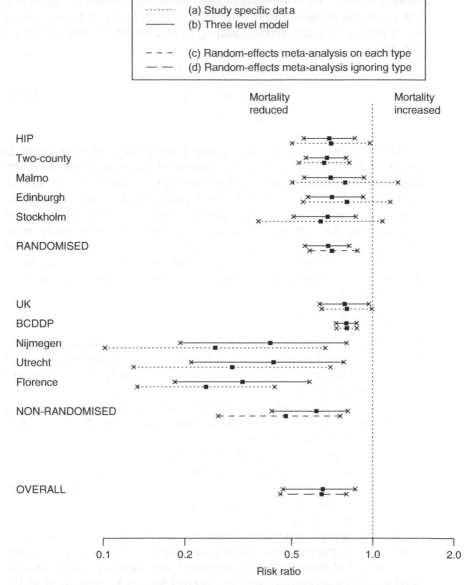

Figure 8.6 Observed risk ratio of breast cancer mortality in RCTs and observational studies in women over 50, together with Bayesian estimates of overall synthesis.

Bayesian interpretation: Figure 8.6 also displays the results, in terms of estimates and 95% intervals, of applying model (8.10) using the prior distributions derived above. In terms of the individual study estimates there is the usual shrinkage towards the overall study-type estimates, the degree of shrinkage dependent upon the within-study variances, and towards the overall population estimate for the study-type overall estimates. The overall population estimate is very little different from the overall RCT estimate, but the 95% interval for the population effect is considerably larger than that for the RCTs. The key point is that the effect of synthesising both RCT and observational evidence has not been to change our overall estimate of the effectiveness of breast cancer screening, but rather to be less certain about this estimate.

Sensitivity analysis: Table 8.7 shows the results of changing the prior distributions for the variance parameters used in the analysis above, together with that for μ_0, the overall population effect. As an alternative to the prior distributions described above for the variance parameters, uniform distributions over the range 0 to 5 are assumed on a standard deviation scale, and the prior distribution for μ_0 is made even more diffuse. The prior distribution for τ has the largest effect on the estimates for μ_0, μ_1 and μ_2, which is due to the fact that there are only two study types in this example, and therefore relatively little data on which to estimate τ^2.

A further sensitivity analysis was undertaken by Prevost *et al.* (2000) regarding the plausibility of introducing the observational evidence at all into the analysis. In a manner similar to the discounting of historical evidence (Section 5.4), they considered letting v_2, the between-study standard deviation for the observational studies, be a function of v_1 the between-study standard deviation of the RCTs, *i.e.* $v_2 = a \times v_1$. In this

Table 8.7 Sensitivity analysis of estimates of population risk ratio, e^{μ_0}, pooled risk ratio for randomised studies, e^{θ_1}, and pooled risk ratio for observational studies, e^{θ_2} (95% credible interval), under different prior distributions.

Prior for τ	Prior for $v_j (j = 1, 2)$	Prior for μ_0	
		N(0,10)	N(0,10 000)
HN(0.033)	HN(0.125)	e^{μ_0}: 0.65 (0.46, 0.86)	0.65 (0.47, 0.90)
		e^{θ_1}: 0.68 (0.56, 0.82)	0.68 (0.56, 0.83)
		e^{θ_2}: 0.62 (0.42, 0.81)	0.61 (0.41, 0.84)
	U(0,5)	e^{μ_0}: 0.65 (0.44, 0.92)	0.65 (0.44, 0.92)
		e^{θ_1}: 0.69 (0.53, 0.85)	0.69 (0.53, 0.85)
		e^{θ_2}: 0.62 (0.39, 0.88)	0.62 (0.39, 0.88)
U(0,5)	HN(0.125)	e^{μ_0}: 0.61 (0.24, 1.47)	0.80 (0.19, 13.15)
		e^{θ_1}: 0.70 (0.57, 0.88)	0.70 (0.56, 0.87)
		e^{θ_2}: 0.52 (0.30, 0.80)	0.49 (0.26, 0.80)
	U(0,5)	e^{μ_0}: 0.59 (0.15, 1.47)	0.67 (0.28, 3.64)
		e^{θ_1}: 0.70 (0.57, 0.85)	0.70 (0.58, 0.86)
		e^{θ_2}: 0.50 (0.22, 1.00)	0.52 (0.21, 0.99)

case *a* can be used to represent beliefs about the relative credibility of the two types of evidence. As an illustration they consider placing a N[3,1] prior distribution on *a*, which corresponds to prior beliefs that the RCTs could be 'valued' three times as highly as the observational studies, but that is also consistent with them being valued as much as five times the observational studies or in fact on an equal basis with the RCTs. Re-estimating the overall population relative risk incorporating this prior distribution yields an estimate of 0.66 with 95% credible interval from 0.47 to 0.92. As with the main three-level analysis above, the point estimate is similar to the overall population relative risk, but the uncertainty surrounding this estimate is now greater than both one based on only the RCTs and a full Bayesian three-level model.

Comments: A wide range of models could be applied to these data. For example, an alternative approach would be to use the observational evidence as a prior distribution for a likelihood based on only the RCT evidence. The model could also be extended to include covariates, and allow prediction on new populations. Nevertheless, there may be difficulties in overcoming suspicion of non-randomised studies, in spite of downweighting and sensitivity analysis.

Example 8.6   *Maple: estimating complex functions of parameters*

Reference: This example forms Chapter 27 of Eddy *et al.* (1992).

Intervention: Neonatal screening for maple syrup urine disease (MSUD), an inborn error in amino acid metabolism, the early detection of which should lead to reduced rates of retardation.

Aim of study: To estimate the probability of retardation without screening, and the change in retardation rate associated with screening. The latter is denoted $e_d = \theta_n - \theta_s$, where θ_n is the retardation rate in those not screened, and θ_s is the rate in those screened.

Study design: Modelling exercise using results from multiple epidemiological cohort studies.

Outcome measure: Expected retardations.

Statistical model: The data described above are all assumed to arise from binomial distributions with the appropriate parameters. The functional relationships shown in Table 8.8 then exist.

The graphical model is shown in Figure 8.7, using the graphical tool for WinBUGS.

Table 8.8 Model and notation for maple syrup urine disease example.

Factor	Notation	Derivation
Probability of MSUD	r	
Prob. of early detection with screening	ϕ_s	
Prob. of early detection without screening	ϕ_n	
Prob. of retardation with early detection	θ_{em}	
Prob. of retardation without early detection	θ_{lm}	
Prob. of retardation for a case of MSUD who is screened	θ_{sm}	$\phi_s\theta_{em} + (1 - \phi_s)\theta_{lm}$
Prob. of retardation for a case of MSUD who is *not* screened	θ_{nm}	$\phi_n\theta_{em} + (1 - \phi_n)\theta_{lm}$
Expected retardations per 100 000 newborns who are screened	$100\,000\theta_s$	$\theta_{sm}\ r$
Expected retardations per 100 000 newborns who are *not* screened	$100\,000\theta_n$	$\theta_{nm}\ r$
Change in retardations due to screening 100 000 newborns	e_d	$\theta_s - \theta_n$

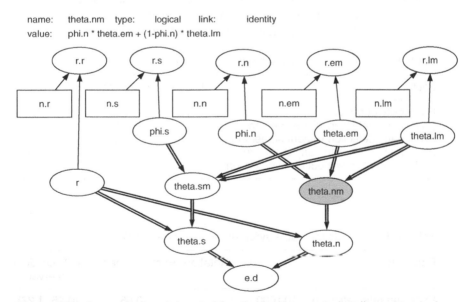

Figure 8.7 A graphical model underlying the maple syrup urine disease example. The observed data at the top of the graph depend on denominators and unknown proportions. The quantities of interest are functions of those proportions, where a double arrow corresponds to a deterministic function. This illustration is taken from WinBUGS, and shows the logical definition of node θ_{nm}, the probability of retardation for a case of a MSUD patient who is not screened.

Prospective analysis?: No.

Prior distribution: The prior distributions for all the binomial parameters used by Eddy *et al.* are the 'non-informative' Jeffreys priors, *i.e.* Beta[0.5, 0.5] (Section 5.5.1).

Loss function or demands: None.

Computation/software: MCMC analysis using WinBUGS; 100 000 iterations were carried out.

Evidence from study: There was no direct evidence on the change in retardation rate in screened and unscreened populations. The data shown in Table 8.9 were used, as provided by Eddy *et al.* (1992).

Bayesian interpretation: The posterior distribution of e_d had the properties shown in Table 8.10. Eddy *et al.* display a normal approximation to the posterior distribution for e_d, with an estimate of -0.35 (95% interval from -0.69 to -0.19). Our wider interval accurately reflects the skewed posterior distribution.

Comments: This example illustrates the synthesis of evidence from multiple studies, with appropriate allowance for the uncertainty of the parameter estimates. Further extensions could include allowance for various biases and uncertainty on the inputs to the model.

Table 8.9   Data used in maple syrup urine disease example.

Factor	Notation	Outcomes	Observations
Probability of MSUD	r	7	724 262
Prob. early detection with screening	ϕ_s	253	276
Prob. early detection without screening	ϕ_n	8	18
Prob. retardation with early detection	θ_{em}	2	10
Prob. retardation without early detection	θ_{lm}	10	10

Table 8.10   Results for maple syrup urine disease example.

Parameter	Notation	Posterior mean	95% credible interval
Expected retardations per 100 000 newborns who are *not* screened	θ_n	0.65	(0.25, 1.27)
Change in expected retardations due to screening 100 000 newborns	e_d	-0.35	$(-0.77, -0.11)$

Example 8.7 *HIV: synthesising evidence from multiple sources and identifying discordant information*

Reference: Ades and Cliffe (2002).

Intervention: Alternative strategies for screening for HIV in pre-natal clinics: *universal* screening of all women, or *targeted* screening of current intravenous drug users (IDUs) or women born in sub-Saharan Africa (SSA).

Aim of study: To determine the optimal policy, taking into account the costs and benefits. However, Ades and Cliffe (2002) point out that the formulation is not wholly realistic as the decision to screen universally throughout England has now been taken, and in any case a strategy of targeted testing may not be politically acceptable.

Study design: Synthesis of multiple sources of evidence to estimate parameters of the epidemiological model shown in Figure 8.8. The relevant fundamental parameters are described in Table 8.11. However, direct evidence is only available for a limited number of these parameters.

Outcome measure: SSA and IDU women will be screened under both universal and targeted strategies, and hence the only difference between the strategies comprises the additional tests and additional cases detected

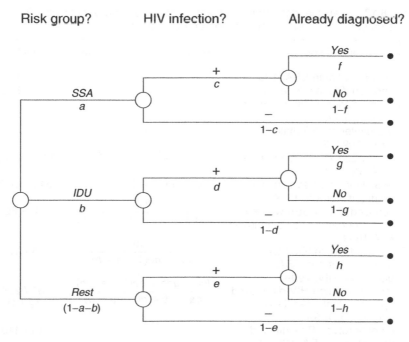

Figure 8.8 Probability tree showing how the proportions of women in different risk groups can be constructed.

Table 8.11 Definition of fundamental parameters in HIV model.

Label	Parameter
a	Proportion of women born in sub-Saharan Africa
b	Proportion of women who are intravenous drug users
c	HIV infection rate in SSA
d	HIV infection rate in IDUs
e	HIV infection rate in non-SSA, non-IDUs
f	Proportion HIV already diagnosed in SSA
g	Proportion HIV already diagnosed in IDUs
h	Proportion HIV already diagnosed in non-SSA, non-IDUs

in the non-SSA, non-IDU group. Additional tests per 10 000 women comprise those on non-SSA, non-IDU women who are not already diagnosed, and so the rate is given by $10\,000(1 - a - b)(1 - eh)$. The rate of new HIV cases detected is $10\,000(1 - a - b)e(1 - h)$.

Statistical model and evidence from study: Table 8.12 summarises the data sources available – full details and references are provided by Ades and Cliffe (2002) who also describe their efforts to select sources which are as 'independent' as possible.

Table 8.12 Available data from relevant studies, generally only allowing direct estimation of functions of fundamental parameters of interest.

Data items and sources	Parameter being estimated	Data
1 Proportion born in SSA, 1999	a	11 044 / 104 577
2 Proportion IDU last 5 years	b	12 / 882
3 HIV prevalence, women born in SSA, 1997–8	c	252 / 15428
4 HIV prevalence in female IDUs, 1997–9	d	10 / 473
5 HIV prevalence, women not born in SSA, 1997–8	$\dfrac{db + e(1 - a - b)}{1 - a}$	74 / 136 139
6 Overall HIV seroprevalence in pregnant women, 1999	$ca + db + e(1 - a - b)$	254 / 102 287
7 Diagnosed HIV in SSA women as a proportion of all diagnosed HIV, 1999	$\dfrac{fca}{fca + gdb + he(1 - a - b)}$	43 / 60
8 Diagnosed HIV in IDUs as a proportion of non-SSA diagnosed HIV, 1999	$\dfrac{gdb}{gdb + he(1 - a - b)}$	4 / 17
9 Overall proportion HIV diagnosed	$\dfrac{fca + gdb + he(1 - a - b)}{ca + db + e(1 - a - b)}$	87 / 254
10 Proportion of infected IDUs diagnosed, 1999	g	12 / 15
11 Prop of serotype B in infected women from SSA, 1997–8	w	14 / 118
12 Prop of serotype B in infected women not from SSA, 1997–8	$\dfrac{db + we(1 - a - b)}{db + e(1 - a - b)}$	5 / 31

The crucial aspect is that there is no direct evidence concerning the vital parameters *e* and *h* for the low-risk group, and hence their value must be inferred indirectly from other studies. For this reason the parameter *w* is introduced which is not part of the epidemiological model: the assumption that the low-risk group has the same prevalence of subtype B as SSA women, and that all IDU women are subtype B, allows use of data source 12 on non-SSA women.

Prior distribution: Uniform priors for all proportions are adopted.

Computation/software: MCMC methods implemented using WinBUGS.

Bayesian interpretation: The posterior estimates and intervals for the proportions underlying the studies are given in Table 8.13, together with the quantities of interest.

Sensitivity analyses: Here we focus on the consistency of data sources rather than the usual analysis of sensitivity to model assumptions. We have synthesised all available data, but the results may be misleading if we have included data that do not fit our assumed model. A simple way of assessing possible conflict is to compare the observed proportion in the 12 sources with that fitted by the model, and it is apparent that the observation for source 4 is only just included in the 95% interval, while the data for source 12 lie wholly outside its estimated interval. This is only a crude method, since a source may strongly influence its estimate, so a better procedure is to leave each source out in turn, re-estimate the model, and then predict the data we would expect in a source of that

Table 8.13 Estimates of parameters underlying the available data. Estimates of quantities of interest in selecting a screening strategy are also shown.

	Quantity	Observed proportion	Estimate	95% interval	*P*-value (excl 4)
1	Proportion SSA	0.106	0.106	0.104 to 0.108	0.47
2	Proportion IDUs	0.0137	0.0088	0.0047 to 0.149	0.46
3	HIV prevalence in SSA	0.0163	0.0172	0.0155 to 0.0189	0.27
4	HIV prevalence in IDUs	0.0211	0.0120	0.0062 to 0.0219	0.004
5	HIV prevalence non-SSA	0.000544	0.000594	0.000478 to 0.000729	0.35
6	Overall HIV prevalence	0.00248	0.00235	0.00217 to 0.00254	0.21
7	SSA as proportion of all diagnoses	0.717	0.691	0.580 to 0.788	0.50
8	IDU as proportion of non-SSA diagnoses	0.235	0.298	0.167 to 0.473	0.40
9	Proportion HIV diagnosed	0.343	0.350	0.296 to 0.408	0.47
10	Proportion IDU already diagnosed	0.800	0.747	0.517 to 0.913	0.44
11	Prop subtype B in SSA	0.119	0.111	0.065 to 0.171	0.43
12	Prop subtype B in non-SSA, 1997–8	0.161	0.285	0.201 to 0.392	0.23
	Additional tests per 10 000, $10\,000(1 - a - b)(1 - eh)$		8856	8789 to 8898	
	Additional HIV cases detected, $10\,000(1 - a - b)e(1 - h)$		2.49	1.09 to 3.87	

size. This predictive distribution, easily obtained using MCMC methods, is then compared to the observed data and a *P*-value calculated in a parallel manner to Box's test of prior/data compatibility described in Section 5.8 (although here we seek to criticise the data rather than the 'prior' based on the remaining studies). We may term these 'cross-validatory *P*-values'.

Removing data source 4 from the analyis leads to the cross-validatory *P*-values shown in Table 8.13. The small *P*-value for source 4 shows its lack of consistency with the remaining data, whereas the predictions for the remaining data seem quite reasonable. Removing source 4 from the analysis leads to an estimate of 8810 (8717 to 8872) for additional tests per 10 000, and 2.73 (1.31 to 4.12) for additional HIV cases detected, so the removal of this divergent source does not in fact have much influence on the conclusions. The estimates for the fundamental parameters are presented in Table 8.14.

Comments: Example 9.5 extends this example to include cost-effectiveness analysis.

Table 8.14 Estimates of fundamental parameters in HIV model, ignoring evidence from source 4.

Label	Parameter	Median	95% interval
a	Proportion of women born in SSA	0.106	0.104 to 0.108
b	Proportion of women who are IDUs	0.013	0.007 to 0.022
c	HIV infection rate in SSA	0.0172	0.0156 to 0.0189
d	HIV infection rate in IDUs	0.0046	0.0015 to 0.012
e	HIV infection rate in non-SSA, non-IDUs	0.00051	0.00039 to 0.00065
f	Proportion HIV already diagnosed in SSA	0.32	0.24 to 0.40
g	Proportion HIV already diagnosed in IDUs	0.78	0.55 to 0.93
h	Proportion HIV already diagnosed in non-SSA, non-IDUs	0.40	0.22 to 0.67

8.5   FURTHER READING

Sutton *et al.* (2000) review the whole area of meta-analysis and Bayesian methods in particular; other reviews are provided by Jones (1995), Normand (1999) and Hedges (1998). See also the book edited by Stangl and Berry (2000).

Empirical Bayes approaches for meta-analysis have received most attention in the literature until recently, largely because of computational difficulties in the use of fully Bayesian modelling (Raudenbush and Bryk, 1985; Stijnen and van Houwelingen, 1990). However, the full Bayesian hierarchical model has been investigated extensively by DuMouchel and Harris (1983), DuMouchel (1990),

DuMouchel and Waternaux (1992) and Abrams and Sansó (1998) using analytic approximations, and also using MCMC methods (Morris and Normand, 1992; Smith *et al.*, 1995). Carlin (1992), for example, considers meta-analyses of both clinical trials and case–control studies; he examines the sensitivity to choice of reference priors, and explores checking the assumption of normal random effects. There have been many comparative studies of the full Bayesian approach, including trials (Rogatko, 1992; Su and Po, 1996; Tunis *et al.*, 1997) and observational studies (Biggerstaff *et al.*, 1994; Su and Po, 1996; Tweedie *et al.*, 1996).

Tutorial articles on the confidence profile method include Eddy (1989), Eddy *et al.* (1990a, 1990b) and Shachter *et al.* (1990). The method has been used in meta-analysis of the benefits of antibiotic therapy (Baraff *et al.*, 1993), mammography in women aged under 50 (Eddy *et al.*, 1988) and angioplasty (Adar *et al.*, 1989).

8.6   KEY POINTS

1. A unified Bayesian approach appears to be applicable to a wide range of problems concerned with evidence synthesis.
2. The Bayesian approach provides a natural structure for many subtle issues that arise in meta-analyses, such as adjusting for baseline risk.
3. Priors on nuisance parameters can be important when there is limited evidence, such as when there are rare events or few studies.
4. 'Indirect' comparisons enable one to infer comparisons where there is limited or no head-to-head evidence.
5. Generalised evidence synthesis is likely to become increasingly important as evidence from disparate studies is used in the construction of health-policy models.
6. Complex synthesis models make extensive use of assumptions, only some of which can be empirically checked, and careful sensitivity analysis is vital.

EXERCISES

8.1. Repeat the analysis in Example 3.13 but using a full Bayesian analysis as in Section 8.2, using WinBUGS. Given the relatively small number of studies, it is important to consider the sensitivity of the posterior results to the prior distribution for the between-study variability (Section 5.7.3): explore the options illustrated in Example 8.1.

8.2. Table 8.15 is adapted from Berry (2000) and presents the results of six RCTs which evaluated cholesterol reduction compared to control in terms of coronary deaths in patients who had previously suffered a myocardial infarction.

Table 8.15   RCTs evaluating cholesterol reduction compared to control in terms of coronary deaths in patients who had previously suffered a myocardial infarction.

	Intervention		Control	
Study	Deaths	Total	Deaths	Total
CDP	398	2224	535	2789
Newcastle	25	244	44	253
Edinburgh	34	350	35	367
Stockholm	47	279	73	276
Oslo	37	206	50	206
MRC	35	322	37	323

    (a) Obtain and compare the posterior distribution for the overall pooled odds ratio using a random-effects meta-analysis based on: (i) a normal approximation to the likelihood arising from the observed log(odds ratio) and standard error in each RCT; (ii) modelling the events in the two arms of each RCT using binomial distributions.

    (b) In each case assess the sensitivity of the results to the prior distribution assumed for the between-study variability, as in Example 8.1.

    (c) An additional large-scale RCT (4S) was reported after those in Table 8.15, in which 111 deaths occurred out of 2221 patients in the intervention arm, and 189 deaths occurred out of 2223 patients in the control arm. The observed effect in the 4S trial was considered to be in conflict with that of those in Table 8.15. Obtain the predictive distribution based on the six RCTs in Table 8.15 for a future RCT and therefore assess whether the assertion that there was a conflict was in fact warranted, and in particular whether the sensitivity analyses considered in (a) affect this assessment.

8.3.   Geddes *et al.* (2000) consider a meta-analysis of 23 RCTs which compared the use of atypical anti-psychotic drugs with haloperidol in patients with schizophrenia. The summary data are shown in Table 8.16 with the relevant dose. Evaluate whether there is evidence for an effect of dose on treatment effect.

8.4.   Using the techniques described in Section 8.2.3, investigate the extent to which the effect of diuretic therapy on risk of pre-eclampsia considered in Exercise 3.12 depends upon the baseline level of risk.

8.5.   In Example 8.2 a meta-analysis of nine RCTs evaluating the effect of electronic foetal heart rate monitoring on perinatal mortality was presented. In addition to the nine RCTs, Sutton and Abrams (2001) also considered evidence from the seven non-randomised comparative studies and ten before–after studies which are presented in Table 8.17 together with the results for the RCTs. Explore the effect that consideration of both randomised and non-randomised evidence has on the conclusions obtained in Example 8.2 when: (a) the non-randomised evidence is

Table 8.16 Standardised effect sizes and associated standard errors (SE) for 23 RCTs evaluating comparing atypical anti-psychotic drugs with haloperidol in patients with schizophrenia.

Study	Standardised effect size	SE	Dose
1	−0.014	0.158	12.0
2	−0.070	0.150	15.0
3	−0.191	0.136	15.0
4	−0.663	0.312	8.0
5	−0.488	0.320	20.0
6	+0.455	0.254	11.0
7	−0.273	0.250	20.0
8	+0.129	0.309	6.0
9	−0.109	0.142	10.0
10	−0.779	0.330	22.5
11	−0.765	0.225	7.6
12	−0.214	0.214	7.5
13	−0.775	0.437	13.5
14	+0.216	0.116	16.0
15	+0.018	0.105	10.0
16	−0.406	0.145	20.0
17	−0.234	0.146	17.5
18	−0.112	0.075	10.0
19	−0.294	0.147	16.0
20	−0.469	0.131	17.5
21	−0.903	0.365	20.0
22	−0.237	0.048	12.5
23	+0.049	0.099	9.4

considered as *prior* evidence, either at 'face value' or downweighted; and (b) when both the randomised and non-randomised sources of evidence are considered within a single hierarchical model following the methods of Section 8.4 and Example 8.5. You will need to make some explicit prior assumptions about the size of the potential bias of the non-randomised studies, and conduct suitable sensitivity analysis.

8.6. In addition to the 17 single-arm studies evaluating either radiotherapy alone (RTx) or radiotherapy together with adjuvant chemotherapy (RTx+Chm) following surgery for childhood medulloblastoma reported in Table 5.7, Sutton *et al.* (2000) also considered six RCTs comparing the two interventions and summarised in Table 8.18. Using the prior distribution for the difference in 5-year survival rates between the two therapies in Exercise 5.6, together with the RCT evidence in Table 8.18, obtain a posterior distribution for the difference: (a) using the evidence from the single-arm studies at 'face value'; (b) possibly downweighting the uncontrolled evidence or allowing for bias; (c) modelling both the randomised and non-randomised sources of evidence within a single model following the methods of Section 8.4.

Table 8.17 RCTs, non-randomised comparative studies and before–after studies evaluating electronic foetal heart rate monitoring (EFM) in terms of perinatal mortality.

Study	Year of publication	EFM Deaths	EFM Total	Control Deaths	Control Total
RCTs					
1	1976	1	175	1	175
2	1976	2	242	1	241
3	1978	0	253	1	251
4	1979	3	463	0	232
5	1981	1	445	0	482
6	1985	0	485	1	493
7	1985	14	6 530	14	6 554
8	1987	17	122	18	124
9	1993	2	746	9	682
Non-randomised					
1	1973	2	1 162	17	5 427
2	1973	0	150	15	6 836
3	1975	1	608	37	6 179
4	1977	1	4 210	9	2 923
5	1978	1	554	3	692
6	1979	0	4 978	2	8 634
7	1982	10	45 880	45	66 208
Before–after					
1	1975	4	991	0	1 024
2	1975	7	1 161	9	1 080
3	1975	14	11 599	1	1 950
4	1976	15	4 323	1	3 529
5	1977	53	4 114	21	3 852
6	1978	35	15 357	6	7 312
7	1980	19	4 240	2	4 503
8	1980	15	6 740	5	8 174
9	1984	13	7 582	2	7 911
10	1986	7	17 409	5	17 586

Table 8.18 Five-year survival rates and standard errors for RCTs comparing radiotherapy alone (RTx) with radiotherapy together with adjuvant chemotherapy (RTx+Chm) following surgery for childhood medulloblastoma.

Study	RTx+Chm S_5	RTx+Chm $SE(S_5)$	RTx S_5	RTx $SE(S_5)$
1	0.55	0.026	0.42	0.020
2	0.58	0.058	0.60	0.054
3	0.74	0.083	0.56	0.099
4	0.59	0.060	0.50	0.065
5	0.17	0.217	0.63	0.341
6	0.46	0.114	0.30	0.118

8.7. In Example 8.7, suppose an additional trial came to light which showed an HIV prevalence of 10/10 000 in non-SSA, non-IDU women.

(a) Does this study conflict with the available evidence?

(b) How would its inclusion alter the findings?

9

Cost-Effectiveness, Policy-Making and Regulation

9.1 INTRODUCTION

In this chapter we go beyond making inferences based on single or multiple studies in order to focus on the consequences of adopting particular health interventions. This broader perspective reflects the increasing attention given to the cost-effectiveness of new and existing treatments, leading to the development of technology-appraisal agencies, such as the National Institute of Clinical Excellence (NICE) in the UK, which are intended to give guidance to health providers and decide on treatments to be covered under relevant reimbursement schemes. We need, however, to take careful account of the context of the evaluation, particularly with regard to specification of prior distributions and loss functions, and a framework is outlined in Section 9.2.

As is clear from the name, cost-effectiveness analysis requires a focus on the dual outcomes of *costs* and *effectiveness*, and a typical formulation requires specification of a model for both, which will contain parameters whose plausible values will depend on both judgement and evidence. The 'standard' approach to cost-effectiveness analysis is outlined in Section 9.3, in which the value of concepts such as *incremental net benefit* and the *cost-effectiveness plane* are emphasised. In many circumstances randomised trial evidence may be lacking or limited to certain aspects of the model, leading naturally to the use of the generalised evidence synthesis techniques outlined in Chapter 8.4. In Section 9.4 we identify two alternative approaches to combining evidence synthesis with a cost-effectiveness model. The first approach is termed *two-stage*: in the first stage the evidence from multiple sources is synthesised and used as a basis for the distributions given to parameters; in the second stage, the effects of the

Bayesian Approaches to Clinical Trials and Health-Care Evaluation   D. J. Spiegelhalter, K. R. Abrams and J. P. Myles
© 2004 John Wiley & Sons, Ltd   ISBN: 0-471-49975-7

resulting uncertainty are propagated through the cost-effectiveness model. The second stage, in which distributions are placed on unknown parameters, has become known in the cost-effectiveness literature as *probabilistic sensitivity analysis*. The second, *integrated*, approach simultaneously carries out the synthesis and cost-effectiveness analysis. The two-stage approach is illustrated in Section 9.5, in which *cost-effectiveness acceptability curves* are introduced and shown to be easily handled in the Bayesian framework, illustrated using closed-form, Monte Carlo and MCMC approaches. The integrated approach is then demonstrated in Section 9.6.

In view of the potential complexity of the resulting models and analysis it is important that there is a clear description of the different components of uncertainty, and in Section 9.7 a taxonomy is provided. This is applicable to complex cost-effectiveness models, typically discrete-state, discrete-time Markov models, which are commonly used to make predictions of the longer-term consequences of a particular intervention. Section 9.8 describes their structure and the use of simulation methods both for micro-simulation of individual cases and probabilistic sensitivity analysis.

Since this chapter emphasises decisions as well as inferences, a strict decision-theoretic approach may be appropriate (see Sections 3.14 and 6.2). For example, Luce and Claxton (1999) point out that hypothesis testing is of limited relevance in economic studies, and when a cost-effectiveness analysis is being used as one of the inputs into a formal decision concerning drug regulation or health policy, they recommend a full decision-theoretic approach in which an explicit loss function of the decision-maker is assessed. Such a loss function can also be used as a basis for valuing the expected benefit from further evidence, and this *expected value of information* approach to deciding research priorities is discussed in Section 9.10; a brief critique of this approach is contained in Section 9.11. Finally, we briefly consider the role of regulatory authorities and the particular issues that arise in relation to Bayesian analysis (Section 9.12).

The combined literature on these topics is becoming large and only selected references will be provided: Briggs (2000) introduces many of these issues in a non-technical style, and we make extensive use of Spiegelhalter and Best (2003) although with some changes in notation. We also note a special issue on Bayesian methods of the *International Journal of Health Technology Assessment in Health Care* which features many relevant articles (Luce *et al.*, 2001), and the primer by O'Hagan and Luce (2003).

9.2   CONTEXTS

Throughout this book we have emphasised that it is vital to take into account the *context* in which a clinical trial is being either designed or analysed and interpreted, and more generally when evaluating any health-care intervention. The appropriate prior opinions, and the possibility of explicit loss functions, depend crucially on whose behalf any analysis is being reported or a decision is being made.

This becomes particularly important when considering the 'end stage' of an evaluation – predicting the effects of actually getting the intervention into practice. We can address this issue using the broad categories of *stakeholders* introduced in Section 3.1:

- *Sponsors*, e.g. pharmaceutical industry, medical charities or granting agencies. In deciding whether to fund studies, they will be concerned with the potential 'payback' from research (Section 9.10), which in industry takes the form of a portfolio of drug development programmes. For such 'internal' analyses it will be quite reasonable for prior distributions to be based on subjective judgements and for loss functions to be based, in industry, on profitability. Very different considerations apply for 'external' analyses done on behalf of others – see below.

- *Investigators*, *i.e.* those responsible for the conduct of a study, whether funded by industry or publicly. In previous chapters we have focused primarily on those carrying out a single study, whose main concern is with the accuracy of the inferences to be drawn from their work, although again they may carry out a cost-effectiveness analysis on behalf of others.

- *Reviewers*, e.g. regulatory bodies (Section 9.12). They will be concerned with the appropriateness of the inferences drawn from the studies, and so may adopt their own prior opinions and reporting standards (Section 3.21). Regulatory bodies will generally only be concerned with safety and efficacy issues, and cost-effectiveness analyses will be dealt with by health-policy agencies.

- *Policy-makers*, e.g. agencies or clinicians setting health policy. Health-care organisations may be concerned with the cost-effectiveness of an intervention, although the sponsor or investigator may carry out this analysis on their behalf. Any analysis is likely to be open to external scrutiny, and hence any prior distributions used at this stage would need to be evidence-based or subject to careful justification and sensitivity analysis. Values would be societally based such as quality measures based on surveys, and future costs and benefits may be discounted according to accepted criteria.

- *Consumers*, e.g. individual patients or clinicians acting on their behalf. These would ideally demand individualised prognostic predictions under available alternative interventions, which could be combined with the patient's own utility function. We shall not deal with such individualised decision-making here, although it has been recommended that clinical trial results are presented in such a form as to help such judgements to be made (Simes, 1986).

There is a large literature on the appropriate means of dealing with values, whether concerning utility measures, quality adjustments, discount rates for costs and benefits, and so on, but these important issues are beyond the scope of this book. See Claxton *et al.* (2000) for a brief overview from a health-economic perspective, including a contrast between the perspective of health-policy agencies and the wider society in general.

9.3 'STANDARD' COST-EFFECTIVENESS ANALYSIS WITHOUT UNCERTAINTY

Cost-effectiveness analyses aim to combine information regarding both clinical effectiveness and economic costs. Given known mean economic costs m_{c1} and m_{c2} under two different treatment options T1 and T2, and similar estimates of mean clinical effectiveness, m_{e1} and m_{e2}, define $\theta_c = m_{c2} - m_{c1}$, $\theta_e = m_{e2} - m_{e1}$ as the incremental mean costs and effectiveness. Then the *incremental cost-effectiveness ratio* (ICER) is defined by

$$\text{ICER} = \frac{\theta_c}{\theta_e} = \frac{m_{c2} - m_{c1}}{m_{e2} - m_{e1}}. \tag{9.1}$$

The ICER can be considered as the cost per unit increase in effectiveness by adopting treatment option T2 rather than T1.

Until recently almost all cost-effectiveness analyses reported findings in terms of the ICER. Nevertheless, whilst the ICER appears appealing, difficulties arise in both the calculation of confidence intervals and its interpretation when the denominator is negative or zero. Figure 9.1 (O'Hagan *et al.*, 2000) shows a *cost-effectiveness plane* divided into four quadrants corresponding to different signs of θ_c and θ_e, with the line $\theta_c = K\theta_e$ drawn, where K represents a maximum acceptable cost per unit of effectiveness; we shall discuss the specification of K at the end of this section.

A conceptual difficulty with the ICER is that its interpretation changes according to the sign of θ_e. Quadrants II and IV correspond to the 'domination' of T1 and T2 respectively, in that one treatment is both less costly and more effective; in these quadrants the ICER is negative and the interpretation is clear. In quadrant I, T2 is more costly but more effective: in area IA, T2 is an acceptable choice as the additional benefit is achieved at a smaller unit cost than K (here ICER $< K$), whereas in IB, T2 would be unacceptable. In quadrant III, T2 is less costly but less effective: in area IIIA, T2 would be considered unacceptable as insufficient gains in cost were being obtained for the effectiveness lost, the ICER being less than K, whereas in the area IIIB, where T2 is acceptable, the ICER is greater than K.

Thus, if there is any possibility that $\theta_e < 0$, it could be very misleading to base any conclusions on possible values of the ICER, since T2 is favoured by small values of the ICER when $\theta_e < 0$, and large values of the ICER when $\theta_e > 0$. In fact, the area where T2 is favoured corresponds to all the cost-effectiveness plane lying below the dashed line, which includes *all* possible values of the ICER. See O'Hagan *et al.* (2000) and Heitjan *et al.* (1999) for further discussion and illustrations.

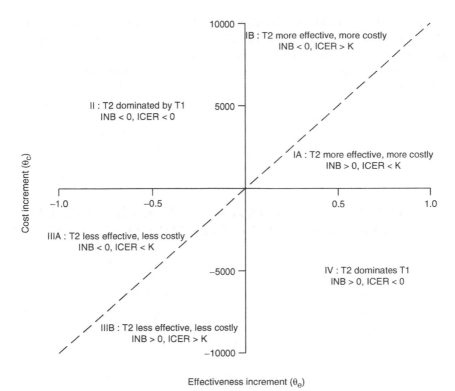

Figure 9.1 Interpretation of different segments of the incremental cost-effectiveness plane. The dashed line represents $\theta_c = K\theta_e$, where K is the willingness to pay for a unit of benefit. Since the incremental net benefit $\mathrm{INB} = K\theta_e - \theta_c$, the dashed line represents $\mathrm{INB} = 0$, the breakeven point. The incremental cost-effectiveness ratio $\mathrm{ICER} = \theta_c/\theta_e$.

The *incremental net benefit* (INB) function has been proposed as an alternative means of interpretation of cost-effectiveness analyses which avoids the problems associated with the ICER, and is defined by

$$\mathrm{INB}(K) = K\theta_e - \theta_c. \tag{9.2}$$

$\mathrm{INB}(K)$ as defined by (9.2) represents the incremental net monetary benefit in terms of economic costs, and provides a connection to classical cost–benefit analysis. INB can also be transformed to the incremental net *health* benefit, in which case $\mathrm{INB}^*(K)$ is given by

$$\mathrm{INB}(K)/K = \mathrm{INB}^*(K) = \theta_e - \theta_c/K. \tag{9.3}$$

It is straightforward to see that the regions in Figure 9.1 which correspond to $\mathrm{INB} > 0$, *i.e.* acceptability of T2, represent all the regions below the dashed line, *i.e.* IA, IV and IIIB.

Setting INB $= 0$ yields the 'breakeven' cost per unit effectiveness $K_0 = \theta_c/\theta_e$ which is numerically equal to the ICER, and this value can be subject to deterministic sensitivity analysis of alternative assumptions.

The value K must be handled with care. Taking the perspective of a health-care agency, it represents their 'willingness to pay' for the gain of a unit of effectiveness. Such a value would not usually be considered as fixed, nor as a random quantity. Instead it is natural to carry out an analysis of sensitivity to alternative values of K, with values of around \$50 000 perhaps being considered reasonable in the USA, and lower values such as £20 000 in the UK. See Claxton *et al.* (2000) for a recent discussion of this quantity.

9.4   'TWO-STAGE' AND INTEGRATED APPROACHES TO UNCERTAINTY IN COST-EFFECTIVENESS MODELLING

Let ψ represent state-of-the-world parameters in a cost-effectiveness model, for example the true mean cost and benefit of an intervention, and let X be a set of unknown generic outcomes of interest, both costs and benefits, taking on a value x. Suppose, for a specified value of ψ, we can specify a predictive distribution $p(x|\psi)$, the *chance variability* between outcomes on future patients. Our primary interest is in $E(X|\psi) = \int x \, p(x|\psi)dx = m_\psi$, the expected outcome in a homogeneous population. m_ψ will often be available in closed form, say when using discrete-time, discrete-state Markov models (Section 9.8).

Any uncertainty concerning ψ may be expressed as a distribution $p(\psi)$, from which we can obtain a joint distribution for m_ψ, the expected costs and benefits of the intervention. By considering different interventions we can thus obtain a joint distribution over the incremental expected costs and effectiveness from a new intervention, denoted θ_c and θ_e respectively, the quantities of interest in a cost-effectiveness analysis (Section 9.3). In practice this will generally require simulation of a value of ψ from $p(\psi)$, which is propagated through the cost-effectiveness model to obtain m_ψ, which in turn provides a value for θ_c, θ_e. Repeated simulations provide a joint distribution for θ_e, θ_c, and hence a distribution for any functions of θ_c, θ_e such as the INB. The construction and analysis of this joint distribution has been termed *probabilistic sensitivity analysis* in the cost-effectiveness literature, to distinguish it from *deterministic sensitivity analysis* in which parameters are varied systematically across ranges.

Two approaches are possible. The *two-stage* approach proceeds as follows. First, $p(\psi)$ is constructed as a closed-form distribution, based on subjective judgements, data analysis or a combination of the two: $p(\psi)$ can be thought of as a prior distribution even though it may be partly based on evidence. Generally the elements of ψ will be assumed independent and parametric distributions adopted. Values of ψ are then simulated from $p(\psi)$ and the cost-effectiveness model provides the relevant outcomes θ_e, θ_c. This is a natural application of Monte Carlo methods (Section 3.19.1) in homogeneous populations, which has become a standard tool in risk analysis to deal with 'second-order uncertainty', as opposed to first-order 'chance' uncertainty (Section 9.7). It is implementable as a Microsoft Excel® macro, either from commercial software such as @RISK (Palisade Europe, 2001) and Crystal Ball (Decisioneering, 2000), or self-written. Here, however, we use the freely available WinBUGS software (Section 3.19.3) in order to facilitate both approaches. A schematic representation is shown in Figure 9.2(a). Applications of the two-stage approach are demonstrated in Example 9.1 for the simple normal case, and Example 9.3 for a more complex model.

The *integrated* or *unified* approach unifies the two stages described above, in that $p(\psi)$ is taken to be a posterior distribution arising from a data analysis, which feeds directly into the cost-effectiveness model without an intermediate summary step. This corresponds to a full Bayesian probability model and

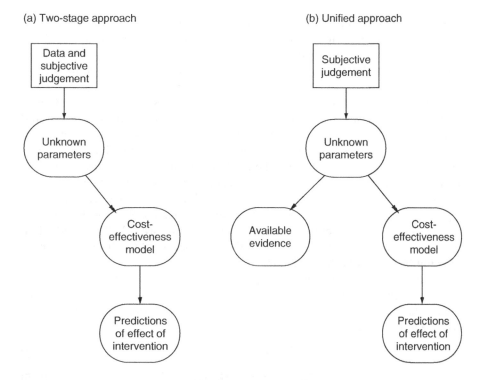

(a) Two-stage approach

(b) Unified approach

Figure 9.2   Schematic graph showing the two approaches to incorporating uncertainty about parameters into a cost-effectiveness analysis. (a) The *two-stage approach* subjectively synthesises data and judgement to produce a prior distribution on the parameters which is then propagated through the cost-effectiveness model. (b) The *unified* or *integrated approach* adopts a fully Bayesian analysis: after taking into account the available evidence, initial prior opinions on the parameters are revised by Bayes theorem to posterior distributions, the effects of which are propagated through the cost-effectiveness model in order to make predictions. An integrated Bayesian approach ensures that the full joint uncertainty concerning the parameters is taken into account.

requires MCMC rather than simply Monte Carlo techniques, since in effect the evidence from the data has to be propagated 'against the arrow' in order to give the uncertainty on the parameters, and then 'forwards' through the cost-effectiveness model; a schematic representation is shown in Figure 9.2(b). Implementation will generally be in a full MCMC program such as WinBUGS: see Examples 9.2 and 9.4. The potential advantages and disadvantages of this integrated approach over the two-stage process are discussed in Section 9.9.2.

9.5 PROBABILISTIC ANALYSIS OF SENSITIVITY TO UNCERTAINTY ABOUT PARAMETERS: TWO-STAGE APPROACH

From a strict decision-theoretic approach, any uncertainty about the parameters θ_c, θ_e is irrelevant to decision-making, and their expectations need only be placed in (9.2) for a specified K, and T2 chosen if INB > 0. Nevertheless, for reasons outlined in Sections 3.14 and 6.2, and discussed further in Section 9.11, it is generally considered appropriate to specify a measure of certainty that T2 is in fact an acceptable option. Confidence intervals for INB can be derived within the classical framework, but a Bayesian approach is natural and straightforward and allows the inclusion of additional prior information.

If we take the two-stage approach (Section 9.4) and assume that a joint prior distribution (θ_e, θ_c) is available based on judgment, data, or a mixture of the two, then this can be plotted on the cost-effectiveness plane shown in Figure 9.1 and the probability of specific conclusions may be obtained by integrating over the appropriate areas (Grieve, 1998). As mentioned in Section 9.4, this has become known as probabilistic sensitivity analysis (Briggs and Gray, 1999). In addition, Heitjan *et al.* (1999) suggest obtaining the distribution of the ICER conditional on being in each quadrant of Figure 9.1.

A joint distribution on (θ_e, θ_c) implies a distribution on INB. If we denote $E[\theta_e] = \mu_e, V[\theta_e] = \tau_e^2, E[\theta_c] = \mu_c, V[\theta_c] = \tau_c^2, \mathrm{Corr}[\theta_e, \theta_c] = \rho$, and similarly for costs, then without further distributional assumptions we have, for INB $= K\theta_e - \theta_c$, that

$$E[\mathrm{INB}] = K\mu_e - \mu_c, \qquad (9.4)$$

$$V[\mathrm{INB}] = K^2\tau_e^2 - 2K\rho\tau_e\tau_c + \tau_c^2. \qquad (9.5)$$

Thus we can plot $E[\mathrm{INB}]$ and, for example, its ± 2 standard deviation interval for different values of K. The breakeven point K_0 occurs at μ_c/μ_e.

In terms of decision-making it is natural to consider the probability that INB(K) in (9.2) is positive for any given value of K, i.e.

$$Q(K) = P(\mathrm{INB}(K) > 0). \qquad (9.6)$$

$Q(K)$ is referred to as the *cost-effectiveness acceptability curve* (CEAC); see van Hout *et al.* (1994). Although $Q(K)$ has been interpreted in frequentist terms, the CEAC is most naturally handled within a Bayesian approach.

It may be reasonable to make a normal approximation to the distribution of INB, and then the CEAC is given by

$$Q(K) = P(\text{INB} > 0) = \Phi\left(\frac{K\mu_e - \mu_c}{\sqrt{K^2\tau_e^2 - 2K\rho\tau_e\tau_c + \tau_c^2}}\right), \qquad (9.7)$$

and this expression is exact if we assume bivariate normality (Section 2.6.10) for θ_e, θ_c – it is also possible to solve (9.7) explicitly to find the value K at which, for example, $Q(K) = 0.95$ or some other desired level of 'significance'. O'Hagan *et al.* (2000) describe various closed-form approximations when normality is not assumed, but in this situation it seems preferable to move to the MCMC approaches as described in the next section.

Not all inferences of interest can be obtained in closed form even when assuming joint normality for θ_e, θ_c, and in this case it can be better computationally to model the joint distribution in two stages: from Section 2.6.10 we see that $\theta_e \sim N[\mu_e, \tau_e^2]$, and $\theta_c|\theta_e$ is normal with mean and variance

$$\begin{aligned} E[\theta_c|\theta_e] &= \mu_c + \frac{\rho\tau_c}{\tau_e}(\theta_e - \mu_e), \\ V[\theta_c|\theta_e] &= \tau_c^2(1 - \rho^2). \end{aligned} \qquad (9.8)$$

Thus we can simulate θ_e followed by $\theta_c|\theta_e$. This is illustrated in Example 9.1.

Example 9.1 *Anakinra: Two-stage approach to cost-effectiveness analysis*

Reference: van Hout *et al.* (1994).

Intervention: Human recombinant interleukin-1 receptor antagonist (anakinra) in the treatment of sepsis syndrome.

Aim of study: To assess the cost-effectiveness of anakinra compared to placebo.

Study design: RCT with 25 patients per arm.

Outcome measure: Effectiveness measured by survival (proportion surviving), and costs of treatment measured in Dutch guilders. The guilder, now replaced by the euro, was valued at around 2.2 to the US dollar.

Statistical model and evidence from study: Table 9.1 shows the data for one of the outcomes of the trial. There is clearly substantial evidence of a clinical benefit, but considerable uncertainty about increases in costs.

Table 9.1   Available data from anakinra study.

Quantity	Estimate	SD	Correlation
θ_e: Increase in effectiveness (survival)	0.28	0.123	
			0.34
θ_c: Increase in costs (guilders)	1380	5657	

Prior distribution: We may approximate a joint prior as having the same properties as the sample data shown in Table 9.1, so that $\mu_e = 0.28$, $\tau_e = 0.123$, $\mu_c = 1380$, $\tau_c = 5657$, $\rho = 0.34$. By further assuming joint normality, the contours for (θ_c, θ_e) may be plotted as in Figure 9.3.

Computation/software: The distribution of INB can be obtained exactly from (9.4) and (9.5), while the CEAC is given by (9.7). Other calculations, such as the distribution of the ICER and the probabilities of lying in each of the quadrants, are carried out by Monte Carlo methods implemented using WinBUGS, taking advantage of the conditional sampling scheme described in (9.8).

Bayesian interpretation: Figure 9.3(a) plots cost per extra survivor when $K = 5000$ and 35 000 guilders. The probabilities of lying in quadrants I, ..., IV are 59.3%, 0.3%, 0.9%, 39.6% respectively, so that there is around a 40% chance that anakinra dominates placebo in costs and benefits. The ICER has median 5146 and 95% interval −79 260 to +57 990. However, it is not clear whether the high values occur in quadrant I or III, which would have a completely different interpretation. Heitjan *et al.* (1999) report that *if* the ICER is in quadrant I, then it has an interval from 791 to 163 400 additional guilders per life saved, while *if* the ICER is in quadrant III, the interval is from 8400 to 4 580 000 guilders saved per life sacrificed. While these conditional statements reveal the different nature of the ICER in different quadrants, their interpretation is not straightforward.

Figure 9.3(b) plots the distribution of the incremental net benefit INB for $K = 5000, 35\,000, 100\,000$: for $K = 5000$ there appears to be almost complete indifference between the options, while the INB increases substantially as the willingness to pay per additional survivor increases. The mean and 95% intervals for the INB for a wide range of K are shown in Figure 9.3(c), while Figure 9.3(d) plots $Q(K) = P(\text{INB} > 0)$ against K: the analysis suggests, on balance, that anakinra is cost-effective provided K is greater than around 5000 guilders, and we can be 95% sure that anakinra is cost-effective provided K is greater than around 45 000 guilders. Whether this would provide an appropriate basis for recommendation of the treatment depends on the decision-maker.

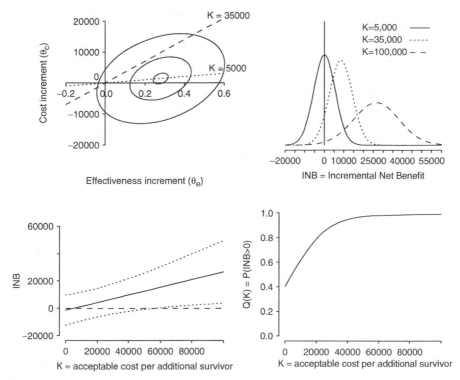

Figure 9.3   Results for anakinra study. (a) Joint distribution of (θ_e, θ_c), superimposed on lines representing maximum acceptable cost per additional survivor $K = 5000$, 35 000. (b) Distribution of incremental net benefit for $K = 5000$, 35 000, 100 000. (c) E[INB] and 95% intervals for a range of values of K. (d) Cost-effectiveness acceptability curve.

Sensitivity analyses: The primary sensitivity analysis concerns the specification of K.

9.6   COST-EFFECTIVENESS ANALYSES OF A SINGLE STUDY: INTEGRATED APPROACH

In the previous section we assumed $p(\theta_e, \theta_c)$ was a prior distribution based on a subjective synthesis of evidence and judgement. We now suppose we have data sources available from which to derive a posterior distribution $p(\theta_e, \theta_c | \text{data})$, and adopt the integrated approach outlined in Section 9.4. We emphasise that θ_e and θ_c must be population *mean* effectiveness and cost increments, in order to make measures additive across individuals. Hence, although cost data will generally have a highly skewed distribution, we must be careful to make inferences about their mean rather than some other measure of location.

Data sources available may include clinical trials, meta-analyses, observational studies and so on, and in later sections we shall consider how to exploit various sources of evidence. Here we shall only consider data from a single clinical trial, in which we assume we have observed pairs (e_{ij}, c_{ij}) representing the observed effect and cost when treatment i is given to patient j. The process of modelling the joint sampling distribution of (e_{ij}, c_{ij}) within each treatment group requires care and statistical insights which are beyond the scope of this book – we refer to O'Hagan and Stevens (2002a) for a variety of approaches in this context. An obvious starting point is to assume bivariate normality (O'Hagan *et al.*, 2001), although the skewness of the cost data will generally make this unreasonable and log-costs might better be assumed normal. Cost data are frequently bimodal and a mixture of distributions may be appropriate (O'Hagan and Stevens, 2001; Cooper *et al.*, 2003c). It is also natural to consider a two-stage approach in which we model effectiveness and then costs conditional on effectiveness: this is the approach taken in Example 9.2. In any of these situation the complexity of the necessary inferences makes MCMC the computational procedure of choice; Fryback *et al.* (2001a) provide a further example of a posterior distribution being used as a direct input to probabilistic sensitivity analysis using WinBUGS.

Example 9.2   *TACTIC: integrated cost-effectiveness analysis*

References: O'Hagan *et al.* (2001), O'Hagan and Stevens (2001, 2002a).

Intervention: Turbuhaler (treatment 2), a novel inhaler for asthmatics, compared to conventional CFC pressurised metered dose inhaler (pMDI, treatment 1).

Aim of study: To investigate whether asthmatic patients who were considered to be adequately treated using a conventional pMDI could be transferred to Turbuhaler without decrease in the effect of treatment, whilst reducing average costs.

Study design: RCT with prospective collection of costs: we use the data of O'Hagan *et al.* (2001) which comprise only the UK portion of the study.

Outcome measure: Number of days with exacerbation and total costs in pounds sterling.

Planned sample size: The original trial was designed to be able to detect a 10% improvement in the proportion of patients experiencing no exacerbations during the course of the trial, from 50% on pMDI to 60% on Turbuhaler.

Evidence from study and statistical model: The summary data are presented in Table 9.2. Turbuhaler patients suffered fewer exacerbations: the high proportion with no exacerbations suggests a normal distribution for

Table 9.2   Results from UK portion of TACTIC trial of Turbuhaler compared to pMDI: log-costs are given separately for patients with and without exacerbations.

Treatment		n	No. exacerbations	Log-costs (mean and SD)	
				With exac.	No exac.
T1	pMDI	58	26 (45%)	6.02 (1.11)	5.87 (1.47)
T2	Turbuhaler	62	36 (58%)	6.37 (0.98)	6.13 (0.85)

clinical outcome is unreasonable and instead we follow O'Hagan and Stevens (2001) in adopting a binary outcome to measure benefit: $e_{ij} = 0$ if exacerbation occurred, 1 otherwise, with proportion ϕ_i in treatment group i.

Figure 9.4 shows the distribution of log-costs in the two treatment groups and according to whether exacerbations were experienced: it is important to note that there were two extremely high costs of 19 871 and 26 201 in the pMDI group who suffered no exacerbations, which are extremely influential in a normal model for costs (O'Hagan *et al.*, 2001) and lead to a higher standard deviation for log-costs. Nevertheless, the empirical distributions in Figure 9.4 suggest adopting a dependent model in which log-costs are assumed normally distributed with mean and standard deviation dependent on treatment and exacerbation. We thus have a model

$$e_{ij} \sim \text{Bern}[\phi_i],$$
$$\log(c_{ij})|e_{ij} = 0 \sim N[\lambda_{i0}, \sigma_{i0}^2],$$
$$\log(c_{ij})|e_{ij} = 1 \sim N[\lambda_{i1}, \sigma_{i1}^2].$$

The mean costs m_{ci} in each treatment group are therefore a weighted average of the means in each exacerbation group and hence, from the known properties of the log-normal distribution (Section 2.6.8), are

$$m_{ci} = (1 - \phi_i)e^{\lambda_{i0} + \sigma_{i0}^2/2} + \phi_i e^{\lambda_{i1} + \sigma_{i1}^2/2},$$

from which we can derive the mean cost and effectiveness differences

$$\theta_c = m_{c2} - m_{c1},$$
$$\theta_e = \phi_2 - \phi_1,$$

which are the inputs to the cost-effectiveness analysis.

Prospective analysis?: No.

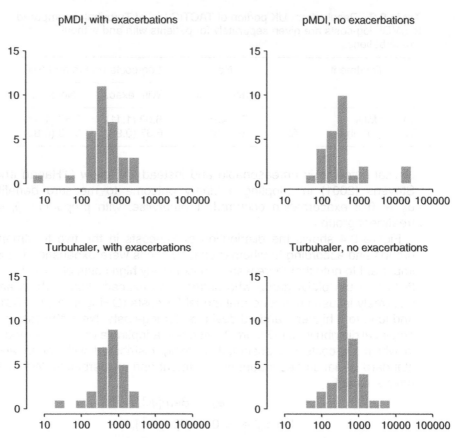

Figure 9.4 Costs for TACTIC data, broken down by treatment (pMDI or Turbuhaler) and whether exacerbations occurred or not.

Prior distribution: O'Hagan and Stevens (2001) use an informative prior for the clinical effectiveness (ϕ_1, ϕ_2), with a mean of 0.1 on $\phi_2 - \phi_1$ which matches the difference used in the power calculations. This initial bias may be considered unreasonable by any regulatory body unless based on substantial evidence, and in any case the evidence from the trial is reasonably strong, and so we adopt independent uniform priors on ϕ_1 and ϕ_2 (an alternative might be uniform on logit(ϕ_2) and on $\theta_e = \phi_2 - \phi_1$, but this has negligible impact).

For the log-cost distributions, we assume independent uniform priors for the $\lambda_{i0}, \lambda_{i1}$. Partly in view of the potential influence of individual observations, and because we might expect the variability in costs to be similar, O'Hagan and Stevens (2001) suggest assuming $\sigma_{10}, \sigma_{11}, \sigma_{20}, \sigma_{21}$ exchangeable in order to 'smooth' the four observed standard deviations towards a common value. We shall assume the $\log(\sigma)$s are normally distributed, such that

$$\log \sigma_{ij} \sim N[\mu_\sigma, \tau_\sigma^2]; \qquad i = 1, 2, \ j = 0, 1,$$

where μ_σ, τ_σ are given uniform priors.

Loss function or demands: No.

Computation/software: MCMC using WinBUGS.

Bayesian interpretation: Figure 9.5(a) plots the joint posterior distribution of θ_e and θ_c, showing they are reasonably independent: the posterior probability is 0.53 that Turbuhaler is cheaper, and 0.93 that it is more effective; the probability that it dominates pMDI is 0.51. Figure 9.5(b) shows the posterior distribution of the incremental net benefit assuming $K = £500$ per patient prevented from having exacerbations – a value at which there is approximate indifference as to the preferred treatment. The expected INB and 95% intervals are displayed in Figure 9.5(c), showing a steady preference for Turbuhaler as the willingness to pay for preventing exacerbations increases. The CEAC in Figure 9.5(d) suggests we can be 90% sure of the cost-effectiveness of Turbuhaler provided that K exceeds £5000. Estimates and intervals for relevant quantities are given in Table 9.3; comparison of the estimates of the σs with those shown in Table 9.3 reveals the shrinkage arising from the exchangeability assumption.

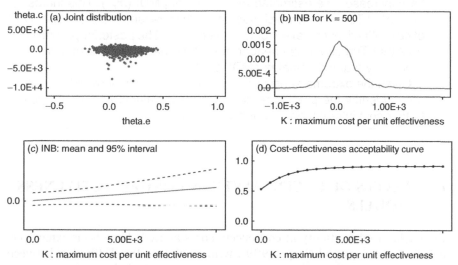

Figure 9.5 Plots of (a) joint distribution of incremental mean benefits θ_e and mean costs θ_c, (b) distribution of incremental net benefit assuming $K = £500$, (c) the expected INB and 95% interval, and (d) the CEAC for a range of K. These plots are direct output from WinBUGS.

Table 9.3 Prior-to-posterior cost-effectiveness analysis of Turbuhaler compared to pMDI: results are given assuming that the standard deviations of the log-costs are either exchangeable or independent.

Parameter		Posterior (exch.)		Posterior (indep.)	
		Median	95% interval	Median	95% interval
Effect of pMDI	ϕ_1	0.45	0.33 to 0.58	0.45	0.33 to 0.58
Effect of Turbuhaler	ϕ_2	0.58	0.45 to 0.70	0.58	0.45 to 0.70
Excess effect of Turbuhaler	$\theta_e = \phi_2 - \phi_1$	0.13	−0.04 to 0.30	0.13	−0.04 to 0.30
Mean cost of pMDI	m_{c1}	862	581 to 1620	983	625 to 2222
Mean cost of Turbuhaler	m_{c2}	835	626 to 1235	817	620 to 1225
Excess mean cost of Turbuhaler	$\theta_c = m_{c2} - m_{c1}$	−21	−801 to 455	−161	−1409 to 371
SD of log-costs, pMDI, exac.	σ_{10}	1.12	0.89 to 1.41	1.14	0.89 to 1.51
SD of log-costs, pMDI, no exac.	σ_{11}	1.37	1.08 to 1.84	1.52	1.17 to 2.08
SD of log-costs, Turbuhaler, exac.	σ_{20}	1.02	0.80 to 1.34	1.01	0.78 to 1.39
SD of log-costs, Turbuhaler, no exac.	σ_{21}	0.92	0.72 to 1.20	0.87	0.70 to 1.14
INB(500)		89	−394 to 851	438	−238 to 1652
INB(5000)		694	−350 to 1783	2834	−829 to 6455
INB(10 000)		1349	−528 to 3194	5423	−1685 to 12380
Q(500)		0.64		0.90	
Q(5000)		0.90		0.94	
Q(10 000)		0.92		0.93	

Sensitivity analysis: The assumption of exchangeable σs is the only form of informative prior that is currently being used. If we adopt independent uniform priors on the σs we obtain the results shown in the final two columns of Table 9.3. The independence assumption allows the two outlying costs to exert a strong influence on σ_{11}, which in turn substantially increases the estimated mean cost of pMDI (m_{c1}). This increases the INB of Turbuhaler, which substantially increases the probability $Q(K)$ of cost-effectiveness even for low values of K. The posterior probability is 0.72 that Turbuhaler is cheaper, and 0.93 that it is more effective: the probability that it dominates pMDI is 0.68.

   Given the extreme sensitivity to two outlying costs, it would be important to identify the precise reasons for these values, and ideally collect further cost information on additional patients.

9.7   LEVELS OF UNCERTAINTY IN COST-EFFECTIVENESS MODELS

Approaches to uncertainty in cost-effectiveness analysis have been extensively reviewed by Briggs and Gray (1999), who emphasise the distinction between conducting 'deterministic' sensitivity analysis in which inputs to a model are systematically varied within a reasonable range, and 'probabilistic' sensitivity analysis in which the relative plausibility of unknown parameters is taken into account.

We can relate these different approaches to analysis of sensitivity to different sources of uncertainty; similar taxonomies have been described by Briggs (2000) and the US Panel on Cost-Effectiveness (Manning *et al.*, 1996).

1. *Chance variability*. This is the unavoidable *within-individual* predictive uncertainty concerning specific outcomes, which will be empirically demonstrated by variability in outcomes between homogeneous individuals. We are usually not interested in this 'first-order' uncertainty (Briggs, 2000) since our focus is on the *expected* outcomes in homogeneous populations, but we shall illustrate its calculation in Section 9.8.

2. *Heterogeneity*. This source concerns *between-individual* variability in expected outcomes, due to either (a) identifiable subgroups of individuals with characteristics such as age, sex and other covariates, or (b) unmeasurable differences (latent variables). These are termed 'patient characteristics' by Briggs (2000). We shall generally want to use deterministic sensitivity analysis to see how expected outcomes vary between identifiable subgroups, possibly followed by probabilistic averaging over population subgroups according to their incidence.

3. *Parameter uncertainty*. This concerns *within-model* uncertainty as to the appropriate values for parameters. Parameters can be divided into two types:

   (a) *States-of-the-world*, which could, in theory, be measured precisely if sufficient evidence were available (e.g. risks, disease incidences): these have also been termed 'parameters that could be sampled' (Briggs, 2000). These can have distributions placed on them, corresponding to the 'second-order' uncertainty used in risk analysis (Burmaster and Wilson, 1996), and so be subject to probabilistic sensitivity analysis.

   (b) *Assumptions*, which are quantitative judgements placed in the model which can only be made precise through consensus agreement, for example discount rates for health benefits. These can be considered as one source of 'methodological uncertainty' (Briggs, 2000), and sensitivity to assumptions can only be carried out deterministically by rerunning analyses under different scenarios.

   The appropriate category for a quantity is not always clear. For example, whether values placed on quality-of-life scales are states-of-the-world or assumptions is a controversial point, and costs might also be placed in either category.

4. *'Ignorance'*. this *between-model* uncertainty describes our basic lack of knowledge concerning the appropriate qualitative structure of the model, for example, the dependence of hazard rates on background factors and history. This is also a component of 'methodological uncertainty' (Briggs, 2000). Deterministic sensitivity analysis takes the form of running through alternative models, although there is a Bayesian argument that model

structure can itself be considered as an unknown state-of-the-world and be subject to probabilistic sensitivity analysis (Draper, 1995).

In this chapter we shall primarily be concerned with probabilistic sensitivity analysis, although we will also illustrate deterministic sensitivity analysis with respect to parameter assumptions.

9.8   COMPLEX COST-EFFECTIVENESS MODELS

We have so far considered the situation in which the necessary estimates of effectiveness and costs are derived directly from clinical trial data. However, a clinical trial may neither address precisely the population of interest, nor last long enough for the rate of important long-term outcomes to be accurately assessed. In the former situation the trial results may need to be adjusted in order to generalise the cost-effectiveness analysis to other populations of interest (Rittenhouse, 1997), which may involve the type of adjustments used in cross-design synthesis (Section 8.4) and the explicit modelling of biases in observational studies (Section 7.3). In the latter case we will need a model for long-term outcomes, such as the Markov models that have been used extensively in cost-effectiveness analysis.

9.8.1   Discrete-time, discrete-state Markov models

These models are generally applied to the development of a disease process over time, and assume that in each 'cycle' an individual is in one of a finite set of states, and that there is a certain chance of transferring to a different state at the next cycle. The 'Markov' label refers to the assumption that the chance of entering a new state at the start of each cycle does not depend on the path the individual took to their current state (although the chance may depend on the cycle and other risk factors). There are obviously many extensions to this reasonably flexible framework (Briggs and Sculpher, 1997, 1998).

We shall first formally describe the generic structure of the model for a single homogeneous set of patients with common parameters. Assume a discrete-time model comprising N cycles labelled $t = 1, \ldots, N$, and that within each cycle t a patient remains in one of R states, and that all transitions occur at the start of each cycle. The probability distribution at the start of the first cycle $t = 1$ is represented by the row vector $\boldsymbol{\pi}_1$, and we assume a transition matrix $\boldsymbol{\Lambda}_t$ whose (i, j)th element $\Lambda_{t, ij}$ is the probability of moving from state i to state j between cycle $t - 1$ and t; thus the probability, for example, of being in state j during the second cycle is $\sum_i \pi_{1i} \Lambda_{2, ij}$. Hence, the marginal probability distribution $\boldsymbol{\pi}_t$ during cycle $t > 1$ obeys the recursive relationship

$$\boldsymbol{\pi}_t = \boldsymbol{\pi}_{t-1} \boldsymbol{\Lambda}_t. \tag{9.9}$$

Suppose the cost, at current prices, of spending a cycle in state r is C_r, $r = 1$, ..., R and there is a fixed entry cost C_0. It is standard practice in economic evaluations to discount costs that occur in future years, at rate δ_c (say) per cycle. Then the total cost acquired by each patient in the population is expected to be

$$ m_c = C_0 + \sum_{t=1}^{N} \frac{\pi_t \mathbf{C}'}{(1 + \delta_c)^{t-1}}. \tag{9.10} $$

Similarly, if the benefits associated with spending one cycle in each state are given by a row vector $\mathbf{b}$, discounted at rate δ_b per cycle, the total expected benefit for each patient is

$$ m_e = \sum_{t=1}^{N} \frac{\pi_t \mathbf{b}'}{(1 + \delta_b)^{t-1}}. \tag{9.11} $$

We note that different types of benefit may be reported, for example both life-years ($\mathbf{b} = 1$) and quality-adjusted life-years (QALYs), in which case $\mathbf{b}$ comprises a row vector of quality adjustments. A range of discount rates may also be explored: for example, guidance from NICE in the UK currently recommends that costs should be discounted at $\delta_c = 6\%$ per annum, while benefits are discounted at $\delta_b - 1.5\%$ (NICE, 2001). However, they add that sensitivity analyses should include assumptions of $\delta_b = 0\%$ and 6%.

Suppose there are S discrete subgroups labelled by s. The model described above can clearly be extended to allow, say, for different transition matrices within subgroups by extending the notation to Λ_{st}: this possibility is explored in detail in Spiegelhalter and Best (2003).

9.8.2 Micro-simulation in cost-effectiveness models

If we are using a more complex model in which it is not possible to write a formula for the expected outcomes, then it may be necessary to perform a much more complex simulation involving the trajectories of individual patients – this is known as *micro-simulation*. The sample mean of the simulations can be used as an estimate of the expected outcome in the population, and this approach does have the side-effect of giving the whole distribution of outcomes and, in particular, the variance among the population. This 'first-order simulation' approach is illustrated by Briggs (2000) and has been extensively exploited in the context of evaluating screening interventions (Cronin *et al.*, 1998).

For example, if we wished to explore this approach for the model described in Section 9.8.1, then we could simulate a starting state y_1 from the distribution π_1. We then simulate this individual's next state y_2 from the distribution comprising the y_1^{th} row of Λ_2, and so on. The discounted costs and benefits for the individual are then

$$C = C_0 + \sum_{t=1}^{N} \frac{C_{y_t}}{(1 + \delta_c)^{t-1}}, \tag{9.12}$$

$$B = \sum_{t=1}^{N} \frac{b_{y_t}}{(1 + \delta_b)^{t-1}}. \tag{9.13}$$

Averaging over many simulated patients (iterations) gives Monte Carlo estimates of the required expectations and also the variability of each outcome due to chance; Example 9.3 illustrates this process.

Note that if we simulate a patient under two treatments, then the incremental net benefit for that patient is estimated as

$$\text{INB} = K(B_2 - B_1) - (C_2 - C_1).$$

We could therefore estimate the proportion of the population for which the INB > 0 – this has been termed the 'probability of net benefit' (Willan, 2001). O'Hagan and Stevens (2002b) emphasise that this estimated population proportion must be carefully distinguished from the probability plotted in a CEAC, which reflects our uncertainty about the expectation over the whole population, and does not in any way take into account heterogeneity in benefit.

9.8.3   Micro-simulation and probabilistic sensitivity analysis

The previous section has described micro-simulation of individual patients, but this is all carried out for fixed parameters value ψ. Performing a probabilistic sensitivity analysis to allow for uncertainty in parameters is considerably more difficult in this context, and care must be taken. It would be tempting, but potentially misleading, to carry out a double simulation, in which a parameter value ψ^j is sampled from $p(\psi)$, followed by simulation of an outcome X^j conditional on ψ^j. The problem is that the variability in the subsequent X^js combines that due to parameter uncertainty and that due to chance variability; unfortunately the two cannot be easily disentangled.

We first note that the total variance of X can be written, using the identity (2.14) for conditional variances, as

$$V[X] = E_\psi[V(X|\psi)] + V_\psi[E(X|\psi)], \tag{9.14}$$

i.e. the expectation with respect to ψ of the conditional variance of X, plus the variance of the conditional expectations. For a probabilistic sensitivity analysis we are only really interested in the second term, since the first term is concerned with chance variability in the population of patients.

These two components may be separated using a time-consuming nested simulation procedure (Halpern *et al.*, 2000). We briefly discuss the necessary computations, when assuming a distribution $p(\psi)$ derived from either the two-stage or integrated approach. A value ψ^j for ψ is simulated from $p(\psi)$, followed by simulation of N (where N is large) values of the outcome $X_1^j, \ldots, X_N^j$ conditional on ψ^j. The sample mean $\overline{X}_N^j$ and variance V_N^j are stored. Monitoring $\overline{X}_N$ and V_N will allow estimation of the components of the overall variability shown in (9.14), since $V_\psi[\overline{X}_N]$ will estimate variability due to parameter uncertainty, while $E_\psi[V_N]$ gives that due to chance variability. This technique will be laborious, particularly when heterogeneity is present, although $E_\psi[V_N]$ may perhaps be reasonably estimated using only a limited set of ψ. See Cronin *et al.* (1998) for an application.

Example 9.3 *HIPS: Cost-effectiveness analysis using discrete-time Markov models*

References: Spiegelhalter and Best (2003) and Fitzpatrick *et al.* (1998).

Intervention: Prosthesis for total hip replacement (THR).

Aim of study: To model the costs and outcomes of THR in a specific subgroup, men aged 65–74, assuming a Charnley prosthesis as a baseline analysis.

Study design: Cost-effectiveness model.

Outcome measure: Effectiveness measured by life expectancy and QALYs, and costs of treatment measured in pounds sterling.

Statistical model: We assume a discrete-time, discrete-state Markov model with cycles of 1 year. Figure 9.6 illustrates the various states and possible transitions between states. Patients initially enter state 1 (primary THR) at time $t = 0$. The first cycle ($t = 1$) is assumed to start immediately following the primary operation; patients have either died at operation or post-operatively, in which case they enter state 5 (death), otherwise they remain in state 1. In each subsequent cycle, surviving patients remain in state 1 until they either die from other causes (progress to state 5) or their hip replacement fails and they require a revision THR operation. Since the need for revision and the operation are assumed simultaneous, patients undergoing a revision operation enter one of two states depending on whether they die at or post-operation (state 2) or survive (state 3). Surviving patients progress to state 4 (successful revision THR) in the following cycle, unless they die from other causes (progress to state 5). Patients in state 4 remain there until they either die from other causes (state 5) or require another revision THR operation, in which case they progress back to states 2 or 3 as before. We also assume a

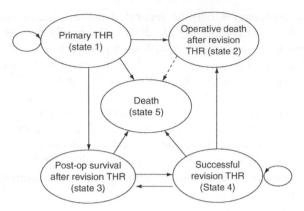

Figure 9.6 Markov model for outcomes following primary total hip replacement.

transition from state 2 to state 5 in the cycle following operative death after a revision THR. This is slightly artificial but is necessary to avoid multiple counting of revision costs if patients were to remain in state 2.

We assume λ_{op} is the operative mortality rate, γ_t is the chance of revision in year t, λ_t is the mortality rate t years after primary operation, and ρ is the re-revision rate which is assumed constant. The vector of state probabilities in cycle $t = 1$ is $\pi_1 = (1 - \lambda_{op}, 0, 0, 0, \lambda_{op})$. We shall only consider one stratum, men between 65 and 74, and take 25 cycles of the model assumed to run between ages 70 and 95. The transition matrix $\Lambda_{t, \, jk}$ is the probability of being in state j in year $t - 1$ and moving to state k at the start of year t; the transition probability matrix for $t = 2, \ldots, 25$ is given by

$$\begin{bmatrix} 1 - \gamma_t - \lambda_t & \lambda_{op}\gamma_t & (1 - \lambda_{op})\gamma_t & 0 & \lambda_t \\ 0 & 0 & 0 & 0 & 1 \\ 0 & 0 & 0 & 1 - \lambda_t & \lambda_t \\ 0 & \rho\lambda_{op} & \rho(1 - \lambda_{op}) & 1 - \rho - \lambda_t & \lambda_t \\ 0 & 0 & 0 & 0 & 1 \end{bmatrix}.$$

Baseline assumptions for the parameters of the model are given in Table 9.4; sources for these assumptions are provided in Fitzpatrick *et al.* (1998). Notable is the assumption that the revision risk increases linearly with time since operation, and constant re-revision risk. Health-related quality of life (HRQL) is measured in QALYs based on the degree of severity of pain patients would be likely to experience in different states of the model. Based on results from a Canadian study (Laupacis *et al.*, 1993), Fitzpatrick *etal.* (1998) assign values $v_1 = 1$, $v_2 = 0.69$, $v_3 = 0.38$ and $v_4 = 0.19$ for the HRQL of patients experiencing no, mild, moderate and severe pain, respectively. They then assume that after a successful THR operation, 80% of patients experience no pain and 20% experience

Table 9.4 Baseline parameters of total hip replacement model using a Charnley prosthesis: benefit weights b are 1 for life expectancy, $b = q_k$ for QALYs.

Parameter		Value
Operative mortality rate	λ_{op}	0.01
Revision rate	$\gamma_t = h(t-1)$	$0.0016(t-1)$
Re-revision rate	ρ	0.04
Mortality rate	λ_t	0.038 (65–74)
		0.091 (75–84)
		0.196 (84+)
Primary cost	C_0	£4052
Revision cost	C_2, C_3	£5290
Cost discount rate	δ_c	6%
Benefit discount rate	δ_b	1.5%
Quality weights	q_1	0.938
	q_2	−0.622
	q_3	−0.337
	q_4	0.938
	q_5	0

mild pain. For patients whose hip replacements fail, they assume that 15% experience severe pain and 85% experience moderate pain in the year preceding the year of the revision operation, with a 50–50 split between those experiencing moderate pain and severe pain in the year of operation. We therefore calculate quality weights for each state in our Markov model as follows:

$q_1 = 0.8v_1 + 0.2v_2 = 0.938,$

$q_2 = 0 + 1.06 \times (0.85v_3 + 0.15v_4 - 0.8v_1 - 0.2v_2) = -0.622,$

$q_3 = (v_3 + v_4)/2 + 1.06 \times (0.85v_3 + 0.15v_4 - 0.8v_1 - 0.2v_2) = -0.337,$

$q_4 = 0.8v_1 + 0.2v_2 = 0.938,$

$q_5 = 0.$

We note that the rather odd negative weights arise from the need to essentially 'subtract' quality from preceding years.

Prior distribution: One relevant state-of-the-world parameter in our model for prognosis following THR is the revision 'hazard' parameter h. It may be reasonable to assume uncertainty of ±50% about our assumed revision hazard which we now denote h_0. This gives an approximate 95% interval of $(h_0/1.5, h_0 \times 1.5)$ for h, which corresponds to a prior standard deviation on the log scale of around 0.2 (Table 5.2). We therefore specify the prior distribution for the log-hazard parameter as

$$\log(h) \sim N[\log(h_0), 0.2^2]. \tag{9.15}$$

Computation/software: MCMC methods implemented using WinBUGS.

Bayesian interpretation:

1. The closed-form calculation of expectations using (9.10) and (9.11) is shown in the 'closed-form' column of Table 9.5. Note that the expected life-years are around 10, and are not substantially reduced by quality adjustment.
2. The micro-simulation study showing variability among individuals is shown in 'population distribution' columns. The huge chance variability in the population is evident: however, as emphasised in Section 9.7, this between-individual variability is not of primary interest. The sampled means match the closed-form values up to Monte Carlo error − 100 000 iterations are used as the variability is so great, and even then the agreement for expected life-years is not good.
3. The final columns show the probabilistic sensitivity analysis by sampling from $p(\log(h))$ given in (9.15), and calculating the closed-form expectations at each iteration. This shows that the uncertainty about the revision hazard has a very limited effect on the expectations, particularly for life expectancy.

Table 9.5 Predicted outcomes from hip replacement in men aged 65–74 years. The baseline expectation is obtained in closed form assuming known parameters. The population distribution is obtained by micro-simulation of individuals. The probabilistic sensitivity analysis summarises the predictive distribution of the expectation, allowing for a subjective prior distribution on the hazard rate.

Parameter	Closed-form expectation	Population distribution		Prob. sens. analysis	
		Mean	SD	Median	95% interval
Life-years	9.939	9.954	5.426	9.939	9.936 to 9.941
QALYs	9.17	9.18	4.96	9.17	9.10 to 9.22
Costs	4458	4453	1220	4459	4334 to 4629

9.8.4   Comprehensive decision modelling

The primary advantage of a Bayesian approach is that it allows the synthesis of all available sources of evidence – whether from RCTs, databases, or expert judgement – into a single coherent and explicit model that can then be used to evaluate the cost-effectiveness of alternative policies. The approach has been termed 'comprehensive decision modelling', and can be thought of as extending the evidence synthesis methods described in Chapter 8 to allow for costs in particular and for utilities in general, and possibly incorporating a predictive model for the natural history of a disease. Alternatively, it can be thought of as extending standard economic modelling techniques such as decision or Markov models so that they are probabilistic.

Parmigiani (2002) discusses such models in detail, pointing out that models should be 'requisite', in the sense of only being as complex as necessary. Ideally such models should allow a variety of viewpoints to be considered and incorporate the 'best possible' evidence, while encouraging analysis of sensitivity to both deterministic inputs and uncertain parameters. From a computational perspective, comprehensive decision models might be implemented in spreadsheets if a two-stage Monte Carlo approach is being adopted, or using MCMC software if integrated evidence synthesis and predictions are desired.

A number of case studies have been reported. Parmigiani and Kamlet (1993) and Parmigiani (1999) apply the idea to screening for breast cancer, and many sources of evidence are brought together in a single model that predicts the consequences of alternative screening policies, while Cronin *et al.* (1998) use micro-simulation at the level of the individual patient to predict the consequences of different policy decisions on lowering expected mortality from prostate cancer. Samsa *et al.* (1999) consider ischaemic stroke and construct a model for natural history using data from major epidemiological studies, and a model for the effect of interventions based on databases, meta-analysis of trials, and Medicare claim records. They also use micro-simulation of the long-term consequences of different stroke-prevention policies in order to compare their cost-effectiveness. Matchar *et al.* (1997), Parmigiani *et al.* (1996, 1997), and Parmigiani (2002) consider further use of their Stroke Prevention Policy Model. Fully integrated applications using WinBUGS have also been reported by Cooper *et al.* (2002, 2003a, 2003b).

9.9 SIMULTANEOUS EVIDENCE SYNTHESIS AND COMPLEX COST-EFFECTIVENESS MODELLING

The previous section has illustrated the two-stage approach to incorporating uncertainty into a complex cost-effectiveness model, and we now consider the full integration with Bayesian prior-to-posterior analysis.

9.9.1 Generalised meta-analysis of evidence

Example 9.2 provided a simple case for the integrated framework using the evidence from a single study and without a complex cost-effectiveness model, but the common situation in which evidence is available from a variety of sources demands a more challenging statistical analysis of the kind discussed in detail in Chapter 8. If the evidence comprises a set of similar trials then a standard Bayesian random-effects meta-analysis may be sufficient. In more complex situations there may be multiple studies with relevance to the quantities in question but which may suffer from a range of potential inadequacies, such as being based on different populations, having non-randomised control groups, outcomes measured on different scales, and so on. As described in Section 8.4, it is natural to extend Bayesian random-effects modelling to allow variance components corresponding to different study designs (*i.e.* assuming study types are exchangeable), resulting in hierarchical models with a study type 'level'. There are clearly a number of issues in carrying out such potentially controversial modelling, such as when to judge studies or study types as 'exchangeable', how to put appropriate prior distributions on variance components, and how to carry out sensitivity analyses.

We shall consider as an illustration a somewhat simple formulation of such a model. Suppose we have a set of studies that are each intending to estimate a single parameter μ but, due to differences in populations studied and so on, any particular study (if carried out meticulously) would in fact be estimating a biased parameter θ_h. Here $\theta_h - \mu$ is the 'external bias', and a standard random-effects formulation might then assume $\theta_h \sim N[\mu, \tau^2]$ (note that the mean would not necessarily be μ if we suspected systematic bias in one direction). However, suppose that due to quality limitations there is additional 'internal bias' in the study, so that the true parameter being estimated is $\theta_h + \delta_h$. Then we might assume $\delta_h \sim N[0, \sigma_{\delta h}^2]$ if we did not suspect that the internal bias would favour one or other treatment. If we assume all the studies have the same potential for external bias, then we are left with a random-effects model in which, for study h, the data are estimating a parameter

$$\theta_h \sim N[\mu, \ \tau^2 + \sigma_{\delta h}^2]$$
$$\sim N[\mu, \ \tau_h^2 / q_h],$$

where $q_h = \tau^2 / (\tau^2 + \sigma_{\delta h}^2)$ can be considered the 'quality weight' for each study, being the proportion of between-study variability unrelated to internal biasing factors. Thus a high-quality randomised trial might have $q = 1$, while a non-randomised study may be downweighted by assigning $q = 0.1$. Note that if we assume all studies are of equal 'quality', then we have the standard random-effects meta-analysis.

Estimates or prior distributions of the between-study variance τ^2 and the quality weights q_h might be obtained from a possible combination of empirical random-effects analyses of RCTs of this intervention, historical 'similar' case studies, and judgement. Of course, sensitivity analysis of a range of assumptions about the quality weights can be carried out.

This technique is illustrated in Example 9.4.

Example 9.4 *HIPS (continued): Integrated generalised evidence synthesis and cost-effectiveness analysis*

Reference: Spiegelhalter and Best (2003).

Available evidence: In order to illustrate the trade-off between increased costs and benefits, we shall compare the cost-effectiveness of the Charnley prosthesis with a hypothetical alternative cemented prostheses costing an extra £350 but with some evidence for lower revision rates. We assume that all other costs (operating staff/theatre costs, length of hospital stay, X-rays etc.) are the same for both prosthesis types, and that the same method of QALY assessment is applicable for both types of prosthesis.

For illustration, we assume that the revision hazard for our hypothetical alternative is similar to that for the Stanmore prosthesis (a popular alternative to the Charnley in practice). Evidence on the relative revision hazards for the two prostheses is limited. The report by NICE on cost-effectiveness of different prostheses for THR (NICE Appraisal Group, 2000) cites three sources providing direct comparisons between Charnley and Stanmore revision rates:

1. The Swedish Hip Registry (Malchau and Herberts, 1998) provides non-randomised data submitted from all hospitals in Sweden from 1979, with record linkage to further procedures and death. Nine-year follow-up results are used for around 30 000 Charnley and 1000 Stanmore prostheses.
2. A British RCT (Marston *et al.*, 1996) randomised around 400 patients to each of Charnley or Stanmore and reported a mean follow-up of 6.5 years.
3. A case series (Britton *et al.*, 1996) of around 1200 patients in a single hospital with a mean follow-up of 8 years.

The available evidence from these three sources on revision hazards for Charnley and Stanmore prostheses is summarised in Table 9.6.

Statistical model: We assume the following model for pooling evidence on the revision hazard ratio for Stanmore versus Charnley prostheses. Let n_{ik} and r_{ik} denote the total number of patients receiving prosthesis i (1 = Charnley, 2 = Stanmore) in study k, and the number requiring a revision operation, respectively. We assume r_{ik} is binomially distributed with proportion p_{ik}, although a little care is required in relating these cumulative failure rates to a hazard ratio. From Section 2.4.2 we know that, assuming proportional hazards, the hazard ratio HR_k for Stanmore versus Charnley prostheses obeys

Table 9.6 Summary of evidence on revision hazards for Charnley and Stanmore prostheses: hazard ratios less than 1 are in favour of Stanmore.

Source	Charnley		Stanmore		Estimated hazard ratio	
	Number of patients	Revision rate	Number of patients	Revision rate	HR	(95% int.)
					Fixed-effects model	
Registry	28 525	5.9%	865	3.2%	0.55	(0.37 to 0.77)
RCT	200	3.5%	213	4.0%	1.34	(0.45 to 3.46)
Case Series	208	16.0%	982	7.0%	0.44	(0.28 to 0.66)
					Common-effect model	
					0.52	(0.39 to 0.67)
Quality weights [Registry, RCT, Case Series]					*Random-effects model*	
			[0.5, 1.0, 0.2]		0.61	(0.36 to 0.98)
			[1.0, 1.0, 1.0]		0.54	(0.37 to 0.78)
			[0.1, 1.0, 0.05]		0.82	(0.36 to 1.67)

$$HR_k = \frac{\log(1 - p_{2k})}{\log(1 - p_{1k})}$$

and hence

$$\log(HR_k) = \log(-\log(1 - p_{2k})) - \log(-\log(1 - p_{1k})).$$

Denoting the 'complementary log–log' parameter by $\log(-\log(1 - p_{1k})) = \psi_k$ leads to the following likelihood:

$$r_{ik} \sim \text{Bin}[p_{ik}, n_{ik}], \quad i = 1, 2,$$

$$\log(-\log(1 - p_{1k})) = \psi_k,$$
$$\log(-\log(1 - p_{2k})) = \psi_k + \log HR_k.$$

We consider three models: (a) fixed effects assuming independent intervention effects HR_k; (b) common effect in which $HR_k = HR$; and (c) random effects. The random-effects analysis with quality weights described in Section 8.4 leads to the model

$$\log(HR_k) \sim \text{N}\left[\log(\overline{HR}), \frac{\tau^2}{q_k}\right],$$

where $\overline{HR}$ is the overall estimate of the revision hazard ratio pooled across studies.

Prior distributions: For the fixed and common effects, independent uniform prior distributions are placed over the study effects ψ_k and $\log(HR_k)$ or $\log(HR)$. For the random-effects model, three studies do not provide sufficient evidence to accurately estimate the between-study standard deviation τ, and so substantial prior judgement is necessary. We would expect considerable heterogeneity in revision rates between studies, even if they are internally unbiased, and so assume τ has a normal distribution with mean 0.2 and standard deviation 0.05 (approximate 95% interval 0.1 to 0.3), corresponding to expecting $\pm 50\%$ variability in true hazard ratios between studies, with 95% uncertainty limits of 20% to 80% variability (e.g. at the upper end of the interval, $e^{1.96 \times 0.3} = 1.8$ or $\pm 80\%$ variability in hazard). Our knowledge of the potential biases of registries and case series suggests downweighting the non-randomised evidence. As a baseline assumption for the quality weights we take q_k equal to 0.5, 1.0 and 0.2 for the registry, RCT and case series studies, respectively. This corresponds to assuming that 'bias' in the registry and case series studies leads to a two- or fivefold increase in the revision rate variance, respectively, over and above the between-study variability expected for RCTs.

Computation/software: MCMC methods implemented using WinBUGS.

Bayesian interpretation: The results of the evidence synthesis are given in Table 9.6. The 'fixed-effects' estimates of the hazard ratio for each source are shown in the first three rows, revealing reasonable concordance between the non-randomised studies but with the randomised trial showing some evidence against the Stanmore. Forcing a common hazard ratio leads to the registry overwhelming the other sources (row 4 of Table 9.6). The results of a baseline random-effects analysis, with quality weights 0.5, 1, 0.2, are shown in row 5 of Table 9.6, with the hazard ratio estimated in favour of the Stanmore but with the 95% interval only just excluding 1.

Feeding these simulated parameter values into the cost-effectiveness model developed in Example 9.3 provides the estimated incremental changes in benefits and costs associated with a Stanmore rather than a Charnley prosthesis shown in Table 9.7. The estimated expected benefit is somewhat marginal, equivalent to 21 additional days (0.0579 × 365) of discounted quality-adjusted survival, but the CEAC suggests reasonable confidence of cost-effectiveness provided one is willing to pay more than around £10 000 per QALY.

Sensitivity analyses: As a sensitivity analysis, we consider two other choices of quality weights. First, we can further downweight all non-randomised evidence by taking q_k equal to 0.1, 1.0 and 0.05, respectively, which leads to an equivocal result with substantial uncertainty, as shown in Table 9.6. At the opposite extreme, setting all quality weights to 1 permits the domination of the registry data, leading to increased benefit.

The sensitivity of the final conclusions to the choice of quality weights is examined in Figure 9.7(a), which also illustrates the sensitivity to two different discount rates for health: 0% and 6%. It is clear that the choice of quality weights has a much stronger influence than the discount rates:

Table 9.7 Incremental changes in expected benefits and costs associated with using Stanmore rather than Charnley prostheses in men aged 65–74, assuming a synthesis of evidence using quality weights (0.5, 1.0, 0.2) for registry, RCT and case series data, respectively. INB(K) is the incremental net benefit per patient when the maximum acceptable cost per unit of effectiveness is K, and $Q(K) = P(\text{INB}(K) > 0)$ is the CEAC. Costs are discounted at 6% per annum, benefits at 1.5% per annum.

Parameter	Median	Prediction 95% interval
Incremental change in expected life-years	0.0026	0.0001 to 0.0049
Incremental change in expected QALYs	0.0579	0.0007 to 0.1078
Incremental change in expected costs	219	87 to 372
INB(5 000)	71	−362 to 452
INB(10 000)	360	−352 to 991
INB(15 000)	649	−344 to 1529
Q(5 000)	0.66	
Q(10 000)	0.87	
Q(15 000)	0.92	

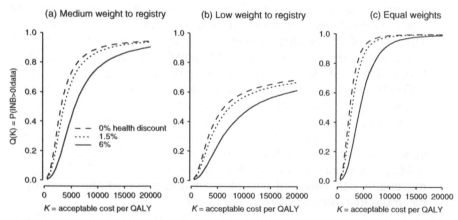

Figure 9.7 CEACs for a Stanmore compared to a Charnley prosthesis. (a) corresponds to the baseline analysis with quality weights (0,5, 1.0, 0.1) for registry, RCT and case series data, respectively, showing limited sensitivity to the annual discount rate for health benefits. (b) uses quality weights of (0.1, 1.0, 0.05); substantial downweighting the non-randomised evidence prevents a strong conclusion of cost-effectiveness. (c) weights all sources equally, and the increased role of the registry data leads to a high probability of cost-effectiveness.

if the non-randomised evidence is substantially downweighted (Figure 9.7(b)) the CEAC shows poor evidence for cost-effectiveness regardless of K, while equal weighting (Figure 9.7(c)) shows strong evidence for moderate K, even when discounting costs at 6%.

___

9.9.2 Comparison of integrated Bayesian and two-stage approach

To recap on Section 9.4, the integrated approach to evidence synthesis and cost-effectiveness analysis simultaneously derives the joint posterior distribution of all unknown parameters from a Bayesian probability model, and propagates the effects of the resulting uncertainty through the predictive model underlying the cost-effectiveness analysis. In contrast, the 'two-stage' approach would first carry out the evidence synthesis, summarising the joint posterior distribution parametrically, and then in a separate analysis use this as a prior distribution in a probabilistic sensitivity analysis in the cost-effectiveness model.

The advantages of the integrated approach include the following. First, there is no need to assume parametric distributional shapes for the posterior probability distributions, which may be important for inferences for smaller samples. Second, and perhaps more important, the appropriate probabilistic dependence between unknown quantities is propagated (Chessa *et al.*, 1999), rather than assuming either independence or being forced into, for example, multivariate normality. This can be particularly vital when propagating inferences which are likely to be strongly correlated, say when considering both baseline levels and treatment differences estimated from the same studies.

The disadvantages of the integrated approach are its additional complexity and the need for full MCMC software. The 'two-stage' approach, in contrast, might be implemented in a combination of standard statistical and spreadsheet programs. However, experience with such spreadsheets suggests that they might not be particularly transparent for complex problems, due to clumsy handling of arrays and opaque formula equations.

9.10 COST-EFFECTIVENESS OF CARRYING OUT RESEARCH: PAYBACK MODELS

9.10.1 Research planning in the public sector

Any organisation funding clinical trials must make decisions concerning the relative importance of alternative proposals, and hence there have been increased efforts to measure the potential 'payback' of expenditure on research. Buxton and Hanney (1998) review the issues and propose a staged semi-quantitative structure, while Eddy (1989) suggested a fully quantitative model based on assessing the future numbers to benefit and the expected benefit, with a subjective probability distribution over the potential benefits to be shown by the research. However, Eddy's limited approach was not adopted by its sponsors, the US Institute of Medicine, who preferred a more informal method that employed weights.

It is clearly possible to extend this broad approach to increasingly sophisticated models within a Bayesian framework, and Hornberger and Eghtesady (1998) state that 'by explicitly taking into consideration the costs and benefits of a trial, Bayesian statistical methods permit estimation of the value to a health care organisation of conducting a randomised trial instead of continuing to treat patients in the absence of more information'. Clearly this is a particular example of a decision-theoretic Bayesian approach, applied at the planning stage of a trial (Section 6.5) rather than at interim analyses (Section 6.6.4). Examples include Detsky (1985), Hornberger *et al.* (1995) and Hornberger and Eghtesady (1998) and others who explicitly calculate the expected utility of a trial in order to select sample sizes; such calculations can also, in theory, be used to rank studies that are competing for resources, and hence to decide whether the trial is worth doing in the first place.

The early analysis by Detsky (1985) assumed that a trial would need to achieve statistical significance in order to have an impact on future treatments, but Claxton (1999b) strongly argues that dependence on such inferential methods, whether classical or Bayesian, will lead to sub-optimal use of health resources. He recommends a full decision-theoretic approach to both fixed (Claxton and Posnett, 1996) and sequential (Claxton, 1999b) trials, basing his analysis on quantifying the expected benefit of further experimentation. This *value of information* approach is outlined briefly in Section 9.10.3.

9.10.2   Research planning in the pharmaceutical industry

Given the 'bottom line' of profitability in the pharmaceutical industry, it is natural to attempt to apply a decision-theoretic approach to individual trial design, designing a research programme for a specified intervention, and for selecting among competing research opportunities. Many of these ideas have already been discussed in the context of individual clinical trials, but here we are concerned with the 'corporate' context: a whole research programme in which there are multiple competing projects at different stages of drug development. Bergman and Gittins (1985) review quantitative approaches to planning a pharmaceutical research programme. Many of the proposed methods are sophisticated uses of bandit theory (Section 6.10) in order to allocate resources in a dynamically changing environment, but Senn (1996, 1997b) suggests a fairly straightforward scheme based on the Pearson index, which is the expected net present value divided by expected net present costs. He discusses the difficulties of eliciting suitable probabilities for the success of each stage of a drug development programme, conditional on the success of the previous stage, but suggests that formal Bayesian approaches involving subjective probability assessment and belief revision should be investigated in this context.

An integral part of this process is a realistic assessment of the chances of regulatory approval, and subsequent sales in the light of future competition and so on: although there must inevitable be a degree of speculation in these assessments, it still seems preferably to have explicit recognition of the relevant uncertainties when making decisions as to whether to pursue a particular development programme.

9.10.3 Value of information

Suppose we are deciding whether to adopt treatment 1 or treatment 2 as a policy, and wondering whether to fund further research to more accurately determine their relative advantages. The true costs and effectiveness are denoted by θ. Based on current information, the incremental net benefit $INB(\theta)$ is positive for θ in a region Θ_2, where treatment 2 would be preferred, and negative for θ in Θ_1, where treatment 1 would be preferred. We do not know θ, but suppose that we have a current posterior for which $E[INB(\theta)|data] > 0$ and so, on balance, treatment 2 is preferred. If, in fact, θ is in Θ_2 then we have made the right decision and there is no gain in knowing the exact value of θ, whereas if θ is truly in Θ_1 we have made the wrong decision and stand to lose $-INB(\theta)$. The *value of perfect information*, $VPI(\theta)$, is defined as the amount we would gain by knowing θ exactly: $VPI(\theta)$ is 0 when $INB(\theta) > 0$, and $-INB(\theta)$ when $INB(\theta) < 0$, which can be expressed as

$$VPI(\theta) = \max(-INB(\theta), 0).$$

Hence our expected value of perfect information, EVPI, is

$$EVPI_2 = E[\max(-INB(\theta), 0)|data], \tag{9.16}$$

where the subscript 2 indicates that treatment 2 is the currently preferred option. By symmetry, the EVPI when $E[INB(\theta)|data] < 0$, *i.e.* when treatment 1 is the preferred option, is

$$EVPI_1 = E[\max(INB(\theta), 0)|data].$$

This quantity is easy to calculate using MCMC by simulating values of θ, calculating $INB(\theta)$ and the VPI, and recording its Monte Carlo average over many iterations. However, we shall see in Example 9.5 that care must be taken with the Monte Carlo error.

We can obtain the EVPI in closed form if $INB(\theta)$ has a normal distribution, and this also sheds some light on the interpretation of this quantity. Suppose

$$INB(\theta) \sim N[\mu_I, \tau_I^2],$$

where the standardised statistic is denoted $z_I = \mu_I/\tau_I$; we assume $\mu_I > 0$ and hence treatment 2 is preferred. For simplicity of notation we shall temporarily drop the subscripts and denote INB by Y. Then $EVPI = E[\max(-Y, 0)]$, and therefore

$$
\begin{aligned}
EVPI &= \int_{-\infty}^{0} -y \frac{e^{-(y-\mu)^2/(2\tau^2)}}{\sqrt{2\pi}\tau} dy \\
&= \int_{-\infty}^{-\mu/\tau} (-t\tau - \mu) \frac{e^{-t^2/2}}{\sqrt{2\pi}} dt \qquad \text{(substituting } t = (y-\mu)/\tau) \\
&= -\tau \int_{-\infty}^{-z} t \frac{e^{-t^2/2}}{\sqrt{2\pi}} dt - \mu\Phi(-z) \\
&= \tau \left[\frac{e^{-z^2/2}}{\sqrt{2\pi}} - z\Phi(-z) \right].
\end{aligned}
\tag{9.17}
$$

The expression in square brackets is denoted $L(z)$ and is known as the 'unit normal loss function' (Claxton *et al.*, 2000). Figure 9.8 shows $L(z)$ plotted against the 'tail area' $\Phi(-z)$: the latter is $P(INB(\theta) < 0|data)$, the posterior probability that the wrong treatment is being preferred. The direct relationship in Figure 9.8 reveals that $L(z)$ is qualitatively equivalent to the tail area (being around 30–50% of its value in the region of interest), and hence EVPI in (9.17) is, approximately, proportional to the probability of making a wrong preference, weighted by τ, which reflects the potential importance of drawing a wrong conclusion. We also note that when $z_I = 0$, which occurs when K achieves its breakeven point, the EVPI reaches its maximum of $\tau/\sqrt{2\pi}$.

In terms of applying the EVPI to a population of current and future patients over the time horizon of a health-care intervention (T), the EVPI requires an adjustment to account for the incidence I_t of patients in each time period t and the discount rate δ_c, so that

$$EVPI_{POP} = EVPI \times \sum_{t=1}^{T} \frac{I_t}{(1 + \delta_c)^{t-1}}, \tag{9.18}$$

assuming no discounting in the first period.

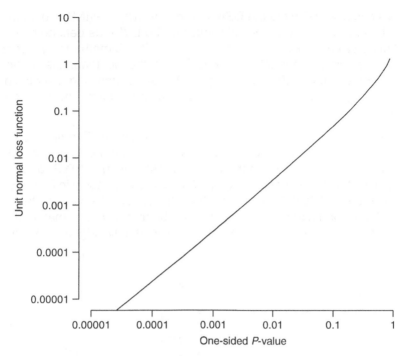

Figure 9.8 Plot of 'unit normal loss function' against P: the EVPI is the unit normal loss function multiplied by the standard deviation of the incremental net benefit.

Example 9.5 *HIV (continued): Calculating the expected value of perfect information*

Reference: Ades and Cliffe (2002) – see Example 8.7.

Costs and utilities: Ades and Cliffe (2002) specify the cost per test as $T = 3$, and the net benefit K per maternal diagnosis is judged to be around £50 000, with a range of £12 000 to £60 000. In this instance there is explicit net monetary benefit from maternal diagnosis and so it may be reasonable to take K as an unknown parameter, and Ades and Cliffe (2002) perform a *probabilistic* sensitivity analysis by giving K a somewhat complex prior distribution. In contrast, we prefer to continue to treat K as a willingness to pay for each unit of benefit, and therefore follow previous examples and conduct a *deterministic* sensitivity analysis in which K is varied up to £60 000.

The prenatal population in London is $N = 105\,000$, and hence the annual incremental net benefit is

$$\text{INB} = N(1 - a - b)(Ke(1 - h) - T(1 - eh)).$$

We can also calculate the CEAC, given by $Q(K) = P(\text{INB} > 0|\text{data})$.

Finally, we consider the calculation of the EVPI, as defined by (9.16). This is calculated in two ways: first, using MCMC methods; and second, by assuming a normal approximation to the posterior distribution of INB(K) and using (9.17). Taking a 10-year horizon and discounting at 6% per year gives a multiplier of 7.8 (not discounting the first year) in (9.18).

Bayesian interpretation: Following the findings in Example 8.7, the analysis is conducted without data source 4. Figure 9.9(a) shows the normal approximations to the posterior distributions of INB for different values of K. The expected INB and 95% limits are shown in Figure 9.9(b) for K up to £60 000, indicating that the policy of universal testing is preferred on balance provided that the benefit K from a maternal diagnosis is greater than around £10 000; K is certainly judged to exceed this

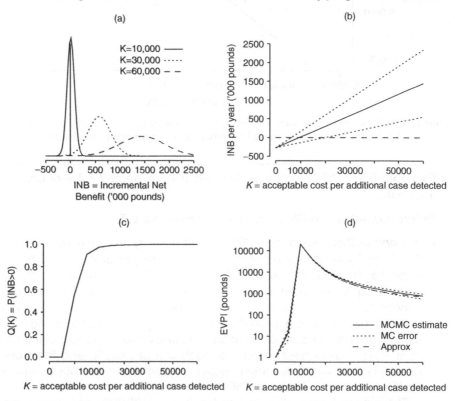

Figure 9.9   (a) and (b) show incremental net benefits, (c) cost-effectiveness acceptability curve, and (d) expected value of perfect information for universal versus targeted prenatal testing for HIV. Note that the EVPI is maximised at the threshold value of K at which the optimal decision changes.

value. The CEAC in Figure 9.9(c) points to a high probability of universal testing being cost-effective for reasonable values of K. Figure 9.9(d) shows the EVPI (± 2 Monte Carlo errors) calculated using 100 000 MCMC iterations and also using the normal approximation to the distribution of INB and (9.17). The Monte Carlo error is considerable even after 100 000 iterations and care must clearly be taken when using MCMC to calculate the EVPI. Nevertheless, (9.17) provides an adequate approximation. The EVPI is substantial for low values of K, but for values around £50 000 the EVPI is negligible. Hence, there appears to be little purpose in further research to determine the parameters more accurately.

The EVPI is intended for use in deciding whether to pursue a research programme, how to design it, and when to stop. First, the EVPI must be higher than the cost of research in order to pass the first 'hurdle' for a proposed programme to overcome, and this should continue to hold throughout the programme. Roughly, when the chance of making a wrong decision, weighted by its consequences, is sufficiently low then the programme can stop and a firm recommendation can be made. Another element of a value of information approach to research planning is that of partial expected value of perfect information (PEVPI), which considers each parameter in the cost-effectiveness analysis in turn, and thus informs the decision whether to conduct future research to yield more precise estimates of particular parameters. Claxton *et al.* (2001) provide a worked example.

In practice, no further research is going to lead to perfect information. Hence, the most relevant quantity may be the expected value of sample information (EVSI), which is essentially the EVPI allowing for the sampling error of a trial. This must exceed the sample costs to overcome the hurdle for a specific proposed trial, and the EVSI minus sample costs is known as the expected net benefit from sampling (ENBS). This model allows for unbalanced allocation of patients between arms, and the ability to revise design based on interim analyses (Claxton and Thompson, 2001; Claxton *et al.*, 2001), in order to optimise the ENBS. Felli and Hazen (1998, 1999) extend this utility perspective to sensitivity analysis, suggesting that an analysis should be considered sensitive to a particular uncertain input if the expected gain in utility from eliminating the uncertainty about that input exceeds a certain specified threshold.

9.11 DECISION THEORY IN COST-EFFECTIVENESS ANALYSIS, REGULATION AND POLICY

The debate about the formal role of decision theory in policy-making is continuing, and here we briefly run through some arguments for and against. Claims for its use include the following:

- Decision theory and economic argument clearly state that maximised expected utility is the sole criterion on which to choose between two options. Therefore measures of 'significance', posterior tail areas of incremental net benefit, and high probabilities on a CEAC are all irrelevant. (Claxton and Posnett, 1996). Claxton *et al.* (2000) point out that 'Once a price per effectiveness unit has been determined, costs can be incorporated, and the decision can then be based on (posterior) mean incremental net benefit measured in either monetary or effectiveness terms'.

- To maximise the health return from the limited resources available from a health budget, health-care purchasers should use rational resource allocation procedures. Otherwise the resulting decisions could be considered as irrational, inefficient and unethical.

- Uncertainty is taken into account through evaluating the benefit of further experimentation, as measured by a value of information analysis.

- This framework provides a formal basis for designing trials, assessing whether to approve an intervention for use, deciding whether an intervention is cost-effective, and commissioning further research.

- Specifying all necessary values may be difficult, but it is necessary for rational decision-making. Claxton (1999b) suggests the first step should be to establish a normative framework that best meets the needs of a system, and separately to conduct studies to see how to get the research into practice.

Among the arguments against are the following:

- The standard criticisms of decision-theoretic approaches to trials apply (Section 6.2): in particular, it is not realistic to specify a full model for the possible impact of research results (which may not even be 'significant') on clinical practice.

- The idea of a null hypothesis (the status quo), which lies behind the use of 'statistical significance' or posterior tail areas, is fundamentally different from that of an alternative hypothesis (a novel intervention). The consequences and costs of the former are generally established, whereas the impact of the latter must contain a substantial amount of judgement. Often, therefore, a choice between two treatments is not a choice between two equal contenders to be decided solely on the balance of net benefit – some convincing evidence is required before changing policy.

- A change in policy carries with it many hidden penalties: for example, it may be difficult to reverse if later found to be erroneous, and may hinder the development of other, better innovations. It would be difficult to explicitly model these phenomena with any plausibility.

- Value of information analysis is dependent on having the 'correct' model, which is never known and generally cannot be empirically checked. Sensitivity analysis can only compensate to some extent for this basic ignorance.

9.12   REGULATION AND HEALTH POLICY

9.12.1   The regulatory context

Regulatory bodies have a duty to protect the public from unsafe or ineffective therapies. Opinions on the relevance of Bayesian methods to drug or device regulation cover a broad spectrum: Whitehead (1997b, p. 204) and Koch (1991) see any use of priors as being controversial and inappropriate, while on the other hand Matthews (1998) claims that the use of sceptical priors 'should not be optional but mandatory'. Keiding (1994) criticises the 'ritual dances' currently prescribed for regulation, but wonders whether Bayesian methods will allow anything less ridiculous. O'Neill (1994), as a senior US Food and Drug Administration (FDA) statistician, acknowledges the appropriate conservatism arising out of the use of sceptical priors, and considers that Bayesian methods should be investigated in parallel with other techniques.

The full decision-theoretic approach (Section 9.11) takes an even more radical perspective. Claxton (1999a) and Claxton *et al.* (2000) suggest that agencies use decision theory for regulation, and evaluate the expected value of further investigation in order to assess whether sufficient evidence is available to permit approval. The crucial idea is that current demands for statistical significance (e.g. two independent studies with $P < 0.05$) is an inadequate criterion as it takes no account of the potential population at risk, the potential consequences of inappropriate approval, and the costs of obtaining more evidence.

9.12.2   Regulation of pharmaceuticals

The website of the FDA allows one to search for references to Bayesian methods among their published literature (Section A.2), although much of the discussion concerns medical devices (see Section 9.12.3). Guidelines for population pharmacokinetics are provided (US Food and Drug Administration, 1999a), which can be thought of as an empirical Bayes procedure (Section 6.12). There is also an interesting use of a Bayesian argument in the approval of the drug enoxaparin (Lovenox). The transcript of the Cardiovascular and Renal Drugs Advisory Committee meeting on 26 June 1997 (US Food and Drug Administration, 1986, pp. 212–218) shows the pharmaceutical company had been asked to make a statement about the effectiveness of enoxaparin plus aspirin as compared to placebo (aspirin alone), whereas their clinical trial had used an active control of heparin plus aspirin. They therefore used meta-analysis data comparing heparin plus aspirin with aspirin alone in order to produce a posterior distribution on the treatment comparison of interest: an example of indirect-comparison inference (Section 8.3). Analyses were repeated using the meta-analysis data directly, but also expressing scepticism about its relevance and reducing its influence, with results being expressed as posterior probabilities of treatment superiority over placebo. The committee welcomed this analysis and voted to approve the drug.

It is important to note that the latest international statistical guidelines for pharmaceutical submissions to regulatory agencies state that 'the use of Bayesian and other approaches may be considered when the reasons for their use are clear and when the resulting conclusions are sufficiently robust' (International Conference on Harmonisation E9 Expert Working Group, 1999). Unfortunately they do not go on to define what they mean by clear reasons and robust conclusions, and so it is still open as to what will constitute an appropriate Bayesian analysis for a pharmaceutical regulatory body.

9.12.3   Regulation of medical devices

The greatest enthusiasm for Bayesian methods appears to be in the FDA Center for Devices and Radiological Health (CDRH). They co-sponsored a workshop on Bayesian methods in November 1998, and have proposed a document *Statistical Guidance on Bayesian Methods in Medical Device Clinical Trials* (US Food and Drug Administration, 1998a).

Campbell (1999) described the potential for Bayesian methods in assessing medical devices, emphasising that devices differed from pharmaceuticals in having better-understood physical mechanisms, which meant that effectiveness was generally robust to small changes. Since devices tended to develop in incremental steps, a large body of relevant evidence existed and companies did not tend to follow established phases of drug development. The fact that an application for approval might include a variety of studies, including historical controls and registries, suggests that Bayesian methods for evidence synthesis might be appropriate. However, the standard conditions apply that the source and robustness of the prior information must be assessed, and that Bayesian analysis does not compensate for poor science and poor experimental design.

Campbell drew attention to the Transcan Breast Scanner, which was approved by the CDRH in April 1999 (US Food and Drug Administration, 1999b). A primary 'intended use' study on 72 women was supplemented by two additional studies of differing designs, using a hierarchical multinomial logistic regression model with study introduced as a random effect. MCMC simulation methods were used by means of the BUGS software. Searching the FDA website reveals a growing number of device submissions that exploit Bayesian reasoning.

9.13   CONCLUSIONS

In this chapter we have attempted to explore a range of concerns that arise in cost-effectiveness modelling, but acknowledge that there are a number of issues that we have passed over. In particular, we have not explored the sensitivity of the conclusions to 'ignorance' (Section 9.7) about the structure of the appropriate model: alternative models that could be used in this context include survival-type models with competing risks. It is vital to admit that even a reasonably complex model, such as that investigated in our example, cannot be assumed to be realistic and must be subject to careful criticism (Russell, 1999; Sculpher *et al.*, 2000).

As attempts are made towards evidence-based health policy in both clinical and public health contexts, models will inevitably become more complex and, while the methods described in this chapter may appear complicated, we feel that techniques such as these may well become commonplace in the future. If decisions made with the help of such analyses are to be truly accountable, it is important that the models and methods are transparent, easily updatable, and can be run by many parties in order to check sensitivity. Models implememented in spreadsheet programs have some of these characteristics, but we feel that user-friendly Bayesian simulation programs could contribute substantially to the field.

9.14 KEY POINTS

1. A Bayesian approach allows explicit recognition of multiple perspectives from the stakeholders involved.
2. Cost-effectiveness analyses fall naturally into a Bayesian framework, whether or not the evidence synthesis is carried out separately (the two-stage approach) or integrated in with the cost-effectiveness analysis.
3. Comprehensive decision modelling is likely to become increasingly important in making both healthcare and policy decisions.
4. Increased attention to pharmacoeconomics may lead decision-theoretic models for research planning to be explored, although this will not be straightforward.
5. There appears to be great potential for formal methods for planning in the pharmaceutical industry.
6. The regulation of devices is leading the way in establishing the role of evidence synthesis.
7. We expect this to be a significant area of research activity over the coming years.

EXERCISES

9.1. Consider the TACTIC study described in Example 9.2, and suppose we try to use the simple bivariate normal model of Section 9.5 to analyse this problem.

(a) Run the WinBUGS code for Example 9.2, and record the posterior correlation between θ_e and θ_c under the exchangeable model.
(b) Plot the joint posterior samples for θ_e θ_c and check whether bivariate normality might be a reasonable assumption.
(c) Making this assumption, use the methods of Section 9.5 to estimate the CEAC and INB, and hence check whether these analytical methods yield similar conclusions to those used in Example 9.2.

9.2. Gray *et al.* (2002) report the results of an economic analysis carried out alongside an RCT to evaluate the use of an intensive blood glocose control policy in patients with type 2 diabetes. Table 9.8 reports the results of the trial in terms of both costs and event-free years. They differentiate between the actual costs observed during the trial, and those adjusted for the fact that during the trial patients required additional clinical visits, and thus incurred additional costs above those seen in routine clinical practice. The latter estimate of costs is referred to as *non-trial*. Using the methods of Section 9.5, examine whether the policy of intensive glucose control is cost-effective for the different scenarios summarised in Table 9.8, *i.e.* whether to use trial costs or adjusted trial costs and/or whether to discount either costs or costs and life-years. Gray *et al.* (2002) did not report the correlation between costs and life-years, so consider assessing cost-effectiveness either (a) assuming specific values for the correlation ρ, or (b) placing a suitable prior distribution on ρ.

9.3. Consider the case of whether to use prophylactic antibiotics for women undergoing Caesarean sections described in Exercise 3.13. The problem may be formulated as a cost-effectiveness decision model and evaluated using WinBUGS, taking into account sources of uncertainty.

The odds ratio for infection (antibiotics vs. control) is estimated to be 0.40 (95% CI from 0.33 to 0.47) from a Cochrane systematic review, while the probability of wound infection without antibiotics is estimated to be

Table 9.8 Mean costs (£ at 1997 prices) and event-free life-years for intensive and conventional blood glucose control in patients with type 2 diabetes.

	Discount rate	Intervention ($n = 2729$)		Control ($n = 1138$)		Difference	
		Mean	SD	Mean	SD	Mean	95% CI
Costs (£)							
Total trial	0%	9608	8343	9869	120 222	−261	−1027 to +505
	6%	6958	5774	7170	8 689	−212	−761 to +338
Total non-trial	0%	8349	8153	7871	11 841	+478	−275 to +1232
	6%	6027	5674	5689	8 615	+338	−207 to +882
Event-free years							
Within trial	0%	14.89	6.93	14.29	7.06	+0.60	+0.12 to +1.10
	6%	9.17	3.20	8.88	3.44	+0.29	+0.06 to +0.53

Table 9.9 RCTs evaluating the effectiveness of using prophylactic antibiotics for women undergoing elective Caesarean sections in terms of infection rates. (Study quality: A=Good, B=OK, C=Poor.)

Study	Year	Antibotics Infections	Total	Control Infections	Total	Study quality
Dashow	1986	3	100	0	33	A
De Boer	1989	1	11	5	17	B
Duff	1982	0	42	0	40	B
Jakobi	1994	4	167	5	140	B
Lewis	1990	1	36	1	25	B
Mahomed	1988	12	115	15	117	A
Rothbard	1975	0	16	1	16	C

0.08, based on observing 60 infections in 750 women. The costs of administering antibiotics include a fixed cost of £10 plus between 4 and 7 minutes of consultant's time at £1 per minute. The hospital costs for Caesarean section without infection are £173 per day, and the average length of stay is 6.7 days (SE 0.33). If there is infection, the average length of stay rises to 8.8 days (SE 0.55) and the daily cost to £262. Utilities are assumed known at 0.95 QALYs without infection and 0.80 QALYs with infection.

(a) Obtain an algebraic expression for the incremental net benefit of using antibiotics for various choices of K, the acceptable cost per QALY.

(b) Use the information provided above to obtain the posterior distributions for the INB, and hence plot the cost-effectiveness acceptability curve.

9.4 Extend the model in Exercise 9.3 to take account of the actual meta-analysis of RCTs considering only elective Caesarean sections presented in Table 9.9 (Cooper *et al.*, 2002). Explore the sensitivity to downweighting studies according to their assessed quality.

9.5 In Example 9.5, Ades and Cliffe (2002) carried out a probabilistic sensitivity analysis for K, the net benefit of a maternal diagnosis. They adopted a distribution representing an estimate of £50 000, with a range from £12 000 to £60 000.

(a) What might be a suitable functional form for a prior distribution with these qualities?

(b) With such a prior distribution, carry out a probabilistic sensitivity analysis and estimate the incremental net benefit, the probability of cost-effectiveness and the EVPI.

9.6 In Example 9.4, what would be the effect of including a (hypothetical) additional randomised trial in which 28/400 (7%) of Charnley prostheses had needed revision, compared to 16/400 (4%) of Stanmore?

10

Conclusions and Implications for Future Research

10.1 INTRODUCTION

This book has described the general use of Bayesian methods in evaluation of health-care interventions, and has considered a number of specific areas of application. Whilst in many of these areas the advantages of adopting a Bayesian approach appear clear, a number of problems have also been identified. Section 10.2 summarises many of these advantages and disadvantages. Section 10.3 identifies areas requiring further research and makes a series of recommendations for the main participant groups in health-care evaluation. These conclusions are deliberately expressed in a 'list' style.

10.2 GENERAL ADVANTAGES AND PROBLEMS OF A BAYESIAN APPROACH

Potential advantages of Bayesian approaches in health-care evaulation

1. All evidence can potentially be taken into account.
2. Specification of a prior distribution requires sponsors, investigators and policy-makers to think carefully and be explicit about what external evidence and judgement they should include.
3. Hierarchical models, which also can be handled within a non-Bayesian framework, allow pooling of evidence and 'borrowing of strength' between multiple substudies.

Bayesian Approaches to Clinical Trials and Health-Care Evaluation   D. J. Spiegelhalter, K. R. Abrams and J. P. Myles
© 2004 John Wiley & Sons, Ltd ISBN: 0-471-49975-7

4. Potential biases can be explicitly modelled, allowing the synthesis of studies of varying designs.
5. The Bayesian approach focuses on the vital question: how should this piece of evidence change what we currently believe?
6. Probability statements can be made directly regarding quantities of interest, and predictive statements are easily derived.
7. Juxtaposition of current belief with clinical demands provide an intuitive and flexible mechanism for monitoring and reporting studies.
8. The inferential outputs from a Bayesian analysis feed naturally into a decision-theoretic and policy-making context.
9. Explicit recognition of the importance of context makes Bayesian methods particularly suitable for evaluation of health-care interventions, in which multiple parties may well interpret the same evidence in different ways.

Generic problems

1. Unfamiliarity with Bayesian techniques, perhaps along with their perceived mathematical complexity, and some conservatism on the part of potential users, has resulted in limited use of proper Bayesian methods to date.
2. The use of prior opinions acknowledges a subjective input into analyses, which may appear to contravene the scientific aim of objectivity.
3. Specification of priors, whether by elicitation or choice of defaults, is a contentious and difficult issue.
4. There are no established standards for design, analysis and reporting of Bayesian studies.
5. There is a danger that the additional complexity of Bayesian methods will lead to poor use.
6. A full decision-theoretic framework can lead to innovative but non-standard trial designs which may be very different from those currently in use.
7. Specification of expected utilities is difficult and may require extensive assumptions about future use of interventions.
8. Computational complexity of the methods has until recently been a major issue.
9. Software for implementation of the methods is still limited in availability and user-friendliness.

10.3   FUTURE RESEARCH AND DEVELOPMENT

We have claimed that Bayesian methods could be of great value when evaluating health-care interventions. For a realistic appraisal of the methodology, it is useful to distinguish the roles and requirements for six main participant groups: methodological researchers, sponsors, investigators, reviewers, policy-makers and consumers (see Sections 3.1 and 9.2). However, two common themes for all participants can immediately be identified. The first is the need for an extended set of case studies showing practical aspects of the Bayesian approach, in particular for prediction and handling multiple sub-studies, in which mathematical details are minimised but details of implementation are provided. We hope the examples in this book have contributed towards this goal. The second theme is the development of standards for the performance and reporting of Bayesian analyses, possibly derived from the checklist described in Section 3.21 and used throughout this book.

1. *Methodological researchers.* With regard to design, there is a need for transferable methods for sample-size calculation that are not based on Type I and Type II error, such as targeting precision, and realistic development of payback models, including modelling of dissemination. Simple and reliable elicitation methods for the priors of 'non-enthusiasts' require testing, as well as demonstrations of the use of empirical data as a basis for prior distributions. Reasonable default priors in non-standard situations need to be available. Methods for flexible model selection and robust MCMC analysis require development and dissemination, and there is a need for user-friendly software for clinical trials and evidence synthesis.

   It is essential to have appraisal criteria along the lines of the checklist used in this book, with possible reformulation as guidelines along the lines of 'How to read a Bayesian study' – it would also be useful to have the term 'Bayesian' in all relevant papers in order to aid literature searches. Finally, increased integration with a health-economic and policy perspective is highly desirable, together with flexible tools for implementation.

2. *Sponsors and investigators.* Both public sector and industry could extend their perspective beyond the classical Neyman–Pearson criteria, and in particular investigate quantitative payback models. The pharmaceutical industry might also investigate formal project prioritisation schemes. All sponsors could focus on the evidential basis for assumptions made concerning alternative hypotheses and the potential gains from technology, and use empirical reviews to establish reasonable prior opinions. There is also potential for 'open' studies in which interim results are reported to investigators.

   It would be valuable to gain experience in eliciting prior opinions from both enthusiasts and a general cross-section of the target community. There is great scope, when analysing data, to go beyond the usual limited list of models and consider a range of priors and structural assumptions. Finally, when reporting a study, it is vital that any Bayesian reporting allows future users to include the evidence in their synthesis or decision. The use of our checklist or a similar scheme for reporting should help in this.

3. *Reviewers/regulatory bodies.* Regulatory bodies could establish reasonable prior opinions based on past experience in order to provide default priors, and could take a more flexible approach to the use of data,

particularly in areas such as medical devices, and encourage efficient use of data by appropriate use of historical controls, evidence synthesis and so on. More experimental would be the explicit modelling of the consequences of decisions in order to decide evidential criteria.

4. *Policy-makers*. There is a need for careful case studies in which policy-makers explicitly go through the following stages in reaching a conclusion based on a full Bayesian analysis:

   • *Priors*. Specify prior opinions relevant at the time of decision-making.
   • *Modelling*. Pool all available evidence into a coherent model.
   • *Reporting*. Make predictive probability statements about the consequences of different policies.
   • *Decision-making*. Assign costs to potential consequences, and so assess (with sensitivity analysis) the expected value of different actions.

5. *Consumers*. Clinicians might be expected to exercise their subjective judgement concerning how their own prior beliefs are influenced by available evidence, while individual patients' utilities values can be elicited to see, for example, whether a population-based decision made by a health-care agency matches one based on their personal opinions.

Appendix A

Websites and Software

Here we give a selection of sites that currently provide useful material on Bayesian methods applicable to health-care evaluation and lists of links. This list is not exhaustive but should provide some entry into the huge range of material available on the internet. All sites were operational in June 2003. A good search engine is appropriate for specific topics.

A.1   THE SITE FOR THIS BOOK

http://www.mrc-bsu.cam.ac.uk/bayeseval/
This page contains downloads for all the examples that use WinBUGS. You can also download the BANDY (*Bayesian Analysis using Normal DYstributions*) program based on Excel, which allows simple analysis of odds-ratio and hazard-ratio data assuming normal priors and likelihoods. Many of the examples in the book are included with BANDY.

A.2   BAYESIAN METHODS IN HEALTH-CARE EVALUATION

http://www.fda.gov/cdrh/
This is the home page for the US Food and Drug Administration's Center for Devices and Radiological Health, which contains a number of items relating to Bayesian methods. To identify these use the Search facility with keyword 'Bayesian'.
http://www.shef.ac.uk/chebs/
The Centre for Bayesian Statistics in Health Economics (CHEBS) is a research centre in the University of Sheffield, UK, and its site provides recent research reports and news of events.

Bayesian Approaches to Clinical Trials and Health-Care Evaluation   D. J. Spiegelhalter, K. R. Abrams and J. P. Myles
© 2004 John Wiley & Sons,   Ltd ISBN: 0-471-49975-7

http://www.bayesian-initiative.com
The Bayesian Initiative in Health Economics and Outcome Research provides useful background material on Bayesian approaches to pharmacoeconomics, and a Bayesian 'primer' is provided.

http://lib.stat.cmu.edu/bayesworkshop/2001/BaSis.html
Provides a draft by the BaSiS group of *Standards for Reporting of Bayesian Analyses in the Scientific Literature.*

http://www.cochrane.org
The Cochrane Collaboration is not a Bayesian site, but is useful for its material on 'Preparing, maintaining and promoting the accessibility of systematic reviews of the effects of health care interventions'.

http://www.campbellcollaboration.org
The Campbell Collaboration is like the Cochrane Collaboration, but deals with evaluation of social policy.

A.3   BAYESIAN SOFTWARE

http://www.shef.ac.uk/~st1ao/1b.html
The First Bayes software is freely available and features good graphical presentation of conjugate analysis of basic data sets. It is suitable for teaching and is strong on predictive distributions.

http://www.mrc-bsu.cam.ac.uk/bugs/welcome.shtml
The BUGS software is designed for analysis of complex analysis using Markov chain Monte Carlo methods. The new WinBUGS version features an interface for specifying models as graphs. The software assumes familiarity with Bayesian methods and MCMC computation.

http://www.med.mcgill.ca/epidemiology/Joseph/software.html
Lawrence Joseph's Bayesian Software site provides downloadable code for a wide variety of sample-size calculations using prior opinion.

http://omie.med.jhmi.edu/bayes/
The Bayesian Communication page is hosted by Harold Lehmann, and features a prototype example in which a Bayesian analysis can be carried out on-line (Lehmann and Shachter, 1994; Lehmann and Nguyen, 1997).

http://www.research.att.com/~volinsky/bma.html
The Bayesian Model Averaging Home Page provides S-Plus and Fortran software for carrying out model averaging, as well as featuring reprints and links.

http://www.palisade.com/
The Palisade Corporation markets the @RISK software, which is an add-on to Excel that allows probability distributions to be placed over the inputs to spreadsheets. Predictive distributions over the outputs are then obtained by Monte Carlo simulation. Demonstration versions are available for downloading.

http://www.decisioneering.com/
  Decisioneering markets the Crystal Ball software, which is also an add-on to Excel and allows Monte Carlo inference using a range of prior distributions. Demonstration downloads are available.

http://www-math.bgsu.edu/~albert/mini_bayes/info.html
  This site is an adjunct to Jim Albert's (1996) book *Bayesian Computation Using Minitab* and features macros for carrying out a variety of analyses.

A.4   GENERAL BAYESIAN SITES

http://stat.rutgers.edu/~madigan/bayes_people.html
  The Bayesians Worldwide site has links to the home pages of many researchers in Bayesian methods. These provide a vast array of lecture notes, reprints and slide presentations.

http://www.bayesian.org/
  The International Society for Bayesian Analysis provides information on its activities and useful links.

http://www.amstat.org/sections/SBSS/
  The American Statistical Association Section on Bayesian Statistical Sciences (SBSS) has a preprint archive and links to other sites.

http://www.isds.duke.edu/sites/bayes.html
  This provides a list of Bayesian sites hosted from Duke University.

References

Abrams, K. (1998) Monitoring randomised controlled trials. Parkinson's disease trial illustrates the dangers of stopping early. *British Medical Journal*, **316** 1183–4.

Abrams, K. and Sansó, B. (1998) Approximate Bayesian inference for random effects meta-analysis. *Statistics in Medicine*, **17**, 201–18.

Abrams, K., Ashby, D. and Errington, D. (1994) Simple Bayesian analysis in clinical trials – a tutorial. *Controlled Clinical Trials*, **15**, 349–59.

Adar, R., Critchfield, G. C. and Eddy, D. M. (1989) A confidence profile analysis of the results of femoropopliteal percutaneous transluminal angioplasty in the treatment of lower-extremity ischemia. *Journal of Vascular Surgery*, **10**, 57–67.

Ades, A. E. and Cliffe, S. (2002) Markov chain Monte Carlo estimation of a multi-parameter decision model: Consistency of evidence and the accurate assessment of uncertainty. *Medical Decision Making*, **22**, 359–71.

Albert, J. (1996) *Bayesian Computation Using Minitab*. Wadsworth, Belmont, CA.

Albert, J. and Chib, S. (1996) Bayesian modelling of binary repeated measures data with application to crossover trials. In *Bayesian Biostatistics* (D. A. Berry and D. K. Stangl, eds), pp. 577–600. Marcel Dekker, New York.

Allen-Mersh, T. G., Earlam, S., Fordy, C., Abrams, K. and Houghton, J. (1994) Quality-of-life and survival with continuous hepatic-artery floxuridine infusion for colorectal liver metastases. *Lancet*, **344**, 1255–60.

Altman, D. G. (1994) Discussion of 'Bayesian approaches to randomised trials' by Spiegelhalter *et al*. *Journal of the Royal Statistical Society, Series A*, **157**, 387–416.

Altman, D. G. (2001). *Practical Statistics for Medical Research* (2nd edition) Chapman & Hall/CRC, Boca Raton, FL.

Altman, D. G., Babiker, A., Campbell, M. K., Clemens, F., Darbyshire, J., Elbourne, D., Grant, A. M., McLeer, S. K., Parmar, M., Pocock, S., Spiegelhalter, D. J., Walker, M. and Wallace, S. (2004) Issues in data monitoring and interim analysis of trials. *Health Technology Assessment*, **8**. To appear.

Anscombe, F. (1963) Sequential medical trials. *Journal of the American Statistical Association*, **58**, 365–83.

Arends, L., Hoes, A., Lubsen, J., Grobbee, D. and Stijnen, T. (2000) Baseline risk as predictor of treatment benefit: three clinical meta-re-analyses. *Statistics in Medicine*, **19**, 3497–518.

Armitage, P. (1985) The search for optimality in clinical trials. *International Statistical Review*, **53**, 15–24.

Armitage, P. (1989) Inference and decision in clinical trials. *Journal of Clinical Epidemiology*, **42**, 293–9.

Armitage, P. (1990) Discussion of 'Biostatistics and Bayes' by Breslow. *Statistical Science*, **5**(3).

Armitage, P. (1991a) Interim analysis in clinical trials. *Statistics in Medicine*, **10**, 925–37.

Armitage, P. (1991b) Letter to the editor. *Controlled Clinical Trials*, **12**, 345.

Armitage, P. (1993) A case for Bayesianism in clinical trials – discussion. *Statistics in Medicine*, **12**, 1395–404.

Armitage, P., McPherson, C. K. and Rowe, B. C. (1969) Repeated significance tests on accumulating data. *Journal of the Royal Statistical Society, Series A*, **132**, 235–44.

Ashby, D. and Hutton, J. (1996) Bayesian epidemiology. In *Bayesian Biostatistics* (D. A. Berry and D. K. Stangl, eds), pp. 109–38. Marcel Dekker, New York.

Ashby, D. and Smith, A. (2000) Evidence-based medicine as Bayesian decision-making. *Statistics in Medicine*, **19**, 3291–305.

Ashby, D., Hutton, J. and McGee, M. (1993) Simple Bayesian analyses for case–control studies in cancer epidemiology. *Statistician*, **42**, 385–97.

Ayanian, J., Landrum, M., Normand, S., Guadagnoli, E. and McNeil, B. (1998) Rating the appropriateness of coronary angiography – do practicing physicians agree with an expert panel and with each other? *New England Journal of Medicine*, **338**, 1896–1904.

Baraff, L. J., Oslund, S. and Prather, M. (1993) Effect of antibiotic therapy and etiologic microorganism on the risk of bacterial meningitis in children with occult bacteremia. *Pediatrics*, **92**, 140–3.

Barnett, V. (1982). *Comparative Statistical Inference* (2nd edition). John Wiley & Sons, Ltd, Chichester.

Bather, J. A. (1985) On the allocation of treatments in sequential medical trials. *International Statistical Review*, **53**, 1–13.

Baudoin, C. and O'Quigley, J. (1994) Symmetrical intervals and confidence intervals. *Biometrical Journal*, **36**, 927–34.

Baum, M., Houghton, J., Riley, D., MacIntyre, J., Berstock, D., McKinna, A., Jackson, I., Sainsbury, J. R. C., Wilson, A., Wheeler, T., Dobbs, J., Rees, G., Powles, T., Rubens, R., Haybrittle, J., McPherson, K. and Houghton, J. (1992) Results of the cancer-research campaign adjuvant trial for perioperative cyclophosphamide and long-term tamoxifen in early breast-cancer reported at the 10th year of follow-up. *Acta Oncologica*, **31**, 251–7.

Bayes, T. (1763). An essay towards solving a problem in the doctrine of chances. *Philosophical Transactions of the Royal Society*, **53**, 418.

Begg, C. B. (1989) Comments on 'Investigating therapies of potentially great benefit: ECMO' by J H Ware. *Statistical Science*, **4**, 320–2.

Begg, C. B. (1992). Book review of 'Cross Design Synthesis'. *Statistics in Medicine*, **12**, 1627–30.

Begg, C. B., DuMouchel, W., Harris, J., Dobson, A., Dear, K., Givens, G. H., Smith, D. D. and Tweedie, R. L. (1997) Publication bias in meta-analysis: a Bayesian data-augmentation approach to account for issues exemplified in the passive smoking debate – comments and rejoinders. *Statistical Science*, **12**, 241–50.

Belin, T. R., Elashoff, R. M., Leung, K., Nisembaum, R., Bastani, R., Nasseri, K. and Maxwell, A. (1995) Combining information from multiple sources in the analysis of non-equivalent control group design. In *Case Studies in Bayesian Statistics, Volume II* (C. Gatsonis, J. Hodges, R. Kass, and N. Singpurwalla, eds), pp. 241–60. Springer-Verlag, New York.

Ben-Shlomo, Y., Churchyard, A., Head, J., Hurwitz, B., Overstall, P., Ockelford, J. and Lees, A. J. (1998) Investigation by Parkinson's Disease Research Group of United Kingdom into excess mortality seen with combined levodopa and selegiline treatment in patients with early, mild Parkinson's disease: further results of randomised trial and confidential inquiry. *British Medical Journal*, **316**, 1191–6.

Bennett, C. C., Johnson, A., Field, D. J. and Elbourne, D. (2001) UK collaborative randomised trial of neonatal extracorporeal membrane oxygenation: follow-up to age 4 years. *Lancet*, **357**, 1094–6.

Benson, B. and Hartz, A. (2000) A comparison of observational studies and randomized controlled trials. *New England Journal of Medicine*, **342**, 1878–86.

Berger, J. (1985) *Statistical Decision Theory and Bayesian Inference*. Springer-Verlag, Berlin.

Berger, J. and Berry, D. A. (1988) Statistical analysis and the illusion of objectivity. *American Scientist*, **76**, 159–65.

Berger, J. and Wolpert, R. (1988) *The Likelihood Principle* (2nd edition) Institute of Mathematical Statistics, Hayward, CA.

Bergman, S. and Gittins, J. (eds) (1985) *Statistical Methods for Pharmaceutical Research Planning*. Marcel Dekker, New York.

Bernardinelli, L., Clayton, D., Pascutto, C., Montomoli, C., Ghislandi, M. and Songini, M. (1995) Bayesian analysis of space–time variation in disease risk. *Statistics in Medicine*, **14**, 2433–43.

Bernardo, J. M. and Smith, A. F. M. (1994) *Bayesian Theory*. John Wiley & Sons, Ltd, Chichester.

Berry, D. A. (1987) Interim analysis in clinical trials – the role of the likelihood principle. *American Statistician*, **41**, 117–22.

Berry, D. A. (1989a) Comments on 'Investigating therapies of potentially great benefit: ECMO' by J H Ware. *Statistical Science*, **4**, 306–10.

Berry, D. A. (1989b) Monitoring accumulating data in a clinical trial. *Biometrics*, **45**, 1197–1211.

Berry, D. A. (1991) Bayesian methods in phase III trials. *Drug Information Journal*, **25**, 345–68.

Berry, D. A. (1993) A case for Bayesianism in clinical trials. *Statistics in Medicine*, **12**, 1377–93.

Berry, D. A. (1994) Discussion of 'Bayesian approaches to randomised trials' by Spiegelhalter *et al*. *Journal of the Royal Statistical Society, Series A*, **157**, 387–416.

Berry, D. A. (1995) Decision analysis and Bayesian methods in clinical trials. *Cancer Treatment Research*, **75**, 125–54.

Berry, D. A. (1996a) *Statistics: A Bayesian Perspective*. Duxbury Press, Belmont, CA.

Berry, D. A. (1996b) When is a confirmatory randomised clinical trial needed? *Journal of the National Cancer Institute*, **88**, 1606–7.

Berry, D. A. (2001) Adaptive clinical trials and Bayesian statistics in drug development (with discussion). *Biopharmaceutical Report*, **9**(2), 1–11.

Berry, D. A. and Eick, S. G. (1995) Adaptive assignment versus balanced randomization in clinical trials – a decision analysis. *Statistics in Medicine*, **14**, 231–46.

Berry, D. A. and Pearson, L. M. (1985) Optimal designs for clinical trials with dichotomous responses. *Statistics in Medicine*, **4**, 597–608.

Berry, D. A. and Stangl, D. K. (1996a) Bayesian methods in health-related research. In *Bayesian Biostatistics* (D. A. Berry and D. K. Stangl, eds), pp. 3–66. Marcel Dekker, New York.

Berry, D. A. and Stangl, D. K. (eds) (1996b) *Bayesian Biostatistics*. Marcel Dekker, New York.

Berry, D. A., Wolff, M. C. and Sack, D. (1994) Decision making during a phase III randomised controlled trial. *Controlled Clinical Trials*, **15**, 360–78.

Berry, D. A., Mueller, P., Grieve, A., Smith, M., Parke, T., Blazck, R., Mitchard, N., and Krams, M. (2001a). Adaptive Bayesian designs for dose-ranging drug trials. In *Case Studies in Bayesian Statistics, Volume V*, (C. Gatsonis, B. Carlin and A. Carriquiry, eds), pp. 99–181. Springer-Verlag, New York.

Berry, G., Matthews, J. S. and Armitage, P. (2001b) *Statistics Methods in Medical Research* (4th edition). Blackwell Scientific Publications, Oxford.

Berry, S. (2000) Meta-analysis versus large trials: resolving the controversy. In *Meta-analysis in Medicine and Health Policy* (D. K. Stangl and D. A. Berry, eds), pp. 65–82. Marcel Dekker, New York.

Berry, S. and Kadane, J. B. (1997) Optimal Bayesian randomization. *Journal of the Royal Statistical Society, Series B*, **59**, 813–19.

Biggerstaff, B. J., Tweedie, R. L. and Mengersen, K. L. (1994) Passive smoking in the workplace: classical and Bayesian meta-analyses. *International Archives of Occupational and Environmental Health*, **66**, 269–77.

Black, N. (1996) Why we need observational studies to evaluate the effectiveness of health care. *British Medical Journal*, **312**, 1215–18.

Bland, J. M. and Altman, D. G. (1998) Statistics notes: Bayesians and frequentists. *British Medical Journal*, **317**(7166), 1151.

Box, G. E. P. (1980) Sampling and Bayes' inference in scientific modelling and robustness (with discussion). *Journal of the Royal Statistical Society, Series A*, **143**, 383–430.

Box, G. E. P. and Tiao, G. C. (1973) *Bayesian Inference in Statistical Analysis*. Addison-Wesley, Reading, MA.

Brant, L. J., Duncan, D. B. and Dixon, D. O. (1992) *K*-ratio *t*-tests for multiple comparisons involving several treatments and a control. *Statistics in Medicine*, **11**, 863–73.

Breslow, N. (1990) Biostatistics and Bayes. *Statistical Science*, **5**, 269–84.

Briggs, A. (2000). Handling uncertainty in cost-effectiveness models. *Pharmacoeconomics*, **17**, 479–500.

Briggs, A. and Gray, A. M. (1999) Handling uncertainty when performing economic evaluation of healthcare interventions. *Health Technology Assessment*, **3**(2), 1–134.

Briggs, A. and Sculpher, M. (1997) Markov models of medical prognosis – commentary. *British Medical Journal*, **314**, 354–5.

Briggs, A. and Sculpher, M. (1998) An introduction to Markov modelling for economic evaluation. *Pharmacoeconomics*, **13**, 397–409.

Bring, J. (1995) Stopping a clinical trial early because of toxicity – the Bayesian approach. *Controlled Clinical Trials*, **16**, 131–2.

Britton, A., Murray, D. W., Bulstrode, C. J., McPherson, K. and Denham, R. A. (1996) Long-term comparison of Charnley and Stanmore design total hip replacements. *Journal of Bone Joint Surgery*, **78B**, 802–8.

Britton, A., McKee, M., Black, N., McPherson, K., Sanderson, C. and Bain, C. (1998) Choosing between randomised and non-randomised studies: a systematic review. *Health Technology Assessment*, **2**(13), 1–124.

Brooks, S. P. (1998) Markov chain Monte Carlo method and its application. *The Statistician*, **47**, 69–100.

Brophy, J. M. and Joseph, L. (1995). Placing trials in context using Bayesian analysis – GUSTO revisited by Reverend Bayes. *Journal of the American Medical Association*, **273**, 871–5.

Brophy, J. M. and Joseph, L. (1997). Bayesian interim statistical analysis of randomised trials. *Lancet*, **349**, 1166–8.

Brophy, J. and Joseph, L. (2000) A Bayesian analysis of random mega-trials for the choice of thrombotic agents in acute myocardial infarction. In *Meta-analysis in Medicine and Health Policy* (D. K. Stangl and D. A. Berry, eds), pp. 83–104. Marcel Dekker, New York.

Brown, B. W., Herson, J., Atkinson, E. N. and Rozell, M. E. (1987) Projection from previous studies – a Bayesian and frequentist compromise. *Controlled Clinical Trials*, **8**, 29–44.

Brunier, H. C. and Whitehead, J. (1994) Sample sizes for phase II clinical trials derived from Bayesian decision theory. *Statistics in Medicine*, **13**, 2493–502.

Bryant, J. and Day, R. (2000) Clinical trials and sample size considerations: another perspective – comment. *Statistical Science*, **15**, 106–8.

Burmaster, D. E. and Wilson, A. M. (1996) An introduction to second-order random variables in human health risk assessments. *Human Ecological Risk Assessment*, **2**, 892–919.

Burton, P. R. (1994) Helping doctors to draw appropriate inferences from the analysis of medical studies. *Statistics in Medicine*, **13**, 1699–1713.

Burton, P. R., Gurrin, L. C. and Campbell, M. J. (1998) Clinical significance not statistical significance: a simple Bayesian alternative to *p* values. *Journal of Epidemiology and Community Health*, **52**, 318–23.

Buxton, M. and Hanney, S. (1998) Evaluating the NHS research and development programme: will the programme give value for money? *Journal of the Royal Society of Medicine*, **91**, 2–6.

Byar, D. P. (1980) Why data bases should not replace clinical trials. *Biometrics*, **36**, 337–42.

Byar, D. P., Simon, R. M., Friedewald, W. T., Schlesselman, J. J., DeMets, D. L., Ellenberg, J. H., Gail, M. H. and Ware, J. H. (1976) Randomized clinical trials. Perspectives on some recent ideas. *New England Journal of Medicine*, **295**, 74–80.

Byar, D. P., Schoenfeld, D. A., Green, S. B., Amato, D. A., Davis, R., De Gruttola, V., Finkelstein, D. M., Gatsonis, C., Gelber, R. D., Lagakos, S. *et al.* (1990) Design considerations for AIDS trials. *New England Journal of Medicine*, **323**, 1343–8.

Campbell, G. (1999) *A Regulatory Perspective for Bayesian Clinical Trials*. Food and Drug Administration, Rockville, MD.

Campbell, M., Fitzpatrick, R., Harnes, A., Kinmonth, A.L., Sandercock, P., Spiegelhalter, D. and Tyrer, P. (2000) Framework for design and evaluation of complex interventions to improve health. *British Medical Journal*, **34**, 694–6.

Carlin, B. P. and Louis, T. A. (2000) *Bayes and Empirical Bayes Methods for Data Analysis* (2nd edition) Chapman & Hall/CRC, Boca Raton, FL.

Carlin, B. P. and Sargent, D. J. (1996) Robust Bayesian approaches for clinical trial monitoring. *Statistics in Medicine*, **15**, 1093–1106.

Carlin, B. P., Chaloner, K., Church, T., Louis, T. A. and Matts, J. P. (1993) Bayesian approaches for monitoring clinical trials with an application to toxoplasmic encephalitis prophylaxis. *Statistician*, **42**, 355–67.

Carlin, B. P., Kadane, J. B. and Gelfand, A. E. (1998) Approaches for optimal sequential decision analysis in clinical trials. *Biometrics*, **54**, 964–75.

Carlin, J. B. (1992) Meta-analysis for 2×2 tables – a Bayesian approach. *Statistics in Medicine*, **11**, 141–58.

Carlin, J. B. (2000) Tutorial in biostatistics. Meta-analysis: formulating, evaluating, combining, and reporting – by SLT Normand, Statistics in Medicine, 18, 321–359 (1999). *Statistics in Medicine*, **19**, 753–9.

Casella, G. and George, E. (1992) Explaining the Gibbs sampler. *American Statistician*, **46**, 167–74.

Chalmers, I. (1997) What is the prior probability of a proposed new treatment being superior to established treatments? *British Medical Journal*, **314**, 74–5.

Chaloner, K. (1996) Elicitation of prior distributions. In *Bayesian Biostatistics* (D. A. Berry and D. K. Stangl, eds), pp. 141–56. Marcel Dekker, New York.

Chaloner, K. and Rhame, F. (2001) Quantifying and documenting prior beliefs in clinical trials. *Statistics in Medicine*, **20**, 581–600.

Chaloner, K. and Verdinelli, I. (1995) Bayesian experimental design – a review. *Statistical Science*, **10**, 273–304.

Chaloner, K., Church, T., Louis, T. A. and Matts, J. P. (1993) Graphical elicitation of a prior distribution for a clinical trial. *Statistician*, **42**, 341–53.

Chelimsky, E., Silberman, G. and Droitcour, J. (1993). Cross design synthesis. *Lancet*, **341**(8843), 498.

Chessa, A. G., Dekker, R., van Vliet, B., Steyerberg, E. W. and Habbema, J. D. F. (1999) Correlations in uncertainty analysis for medical decision making: an application to heart-valve replacement. *Medical Decision Making*, **19**, 276–86.

Choi, S. C. and Pepple, P. A. (1989). Monitoring clinical trials based on predictive probability of significance. *Biometrics*, **45**, 317–23.

Christiansen, C. and Morris, C. (1996). Fitting and checking a two-level Poisson model. In *Bayesian Biostatistics*, (D. A. Berry and D. K. Stangl, eds), pp. 467–502. Marcel Dekker, New York.

Christiansen, C. and Morris, C. (1997a) Improving the statistical approach to health care provider profiling. *Annals of Internal Medicine*, **127**, 764–8.

Christiansen, C. and Morris, C. (1997b) Hierarchical Poisson regression modeling. *Journal of the American Statistical Association*, **92**, 618–32.

Claxton, K. (1999a) Bayesian approaches to the value of information: implications for the regulation of new pharmaceuticals. *Health Economics*, **8**, 269–74.

Claxton, K. (1999b) The irrelevance of inference: a decision-making approach to the stochastic evaluation of health care technologies. *Journal of Health Economics*, **18**, 341–64.

Claxton, K. and Posnett, J. (1996) An economic approach to clinical trial design and research priority-setting. *Health Economics*, **5**, 513–24.

Claxton, K. and Thompson, K. M. (2001) A dynamic programming approach to the efficient design of clinical trials. *Journal of Health Economics*, **20**, 797–822.

Claxton, K., Lacey, L. F. and Walker, S. G. (2000) Selecting treatments: a decision theoretic approach. *Journal of the Royal Statistical Society, Series A*, **163**, 211–25.

Claxton, K., Neumann, P. J., Araki, S. and Weinstein, M. C. (2001) Bayesian value-of-information analysis – an application to a policy model of Alzheimer's disease. *International Journal of Technology Assessment in Health Care*, **17**, 38–55.

Clayton, D. G. and Hills, M. (1993). *Statistical Methods in Epidemiology*. Oxford Univerity Press, Oxford.

Cole, P. (1979). The evolving case–control study. *Journal of Chronic Diseases*, **32**, 15–27.

Collins, R., Peto, R., Flather, M. and ISIS-4 Collaborative Group (1995) ISIS-4 – a randomised factorial trial assessing early oral captopril, oral mononitrate, and intravenous magnesium sulphate in 58,050 patients with suspected acute myocardial infarction. *Lancet*, **345**, 669–85.

Concato, J., Shah, N. and Horwitz, R. I. (2000) Randomized, controlled trials, observational studies, and the hierarchy of research designs. *New England Journal of Medicine*, **342**, 1887–92.

Cooper, N. J., Sutton, A. J. and Abrams, K. R. (2002) Decision analytical economic modelling within a Bayesian framework: application to prophylactic antibiotics use for Caesarean section. *Statistical Methods in Medical Research*, **11**, 491–512.

Cooper, N., Abrams, K., Sutton, A., Turner, D. and Lambert, P. (2003a) Use of Bayesian methods for Markov modelling in cost-effectiveness analysis: an application to taxane use in advanced breast cancer. *Journal of the Royal Statistical Society, Series A*, **166**.

Cooper, N. J., Sutton, A. J., Abrams, K. R., Turner, D. and Wailloo, A. (2003b) Comprehensive decision analytical modelling in economic evaluation: A Bayesian approach. *Health Economics*. To appear

Cooper, N. J., Sutton, A. J., Mugford, M. and Abrams, K. R. (2003c). Use of Bayesian Markov chain Monte Carlo methods to model cost-of-illness data. *Medical Decision Making*, **23** 38–53.

Cornfield, J. (1966) Sequential trials, sequential analysis and the likelihood principle. *American Statistician*, **20**, 18–23.

Cornfield, J. (1969) The Bayesian outlook and its applications. *Biometrics*, **25**, 617–57.

Cornfield, J. (1976) Recent methodological contributions to clinical trials. *American Journal of Epidemiology*, **104**, 408–21.

Coronary Drug Project Research Group (1970) The Coronary Drug Project. Initial findings leading to a modification of its research protocol. *Journal of the American Medical Association*, **214**, 1301–13.

Coronary Drug Project Research Group (1975) Clofibrate and niacin in coronary heart disease. *Journal of the American Medical Association*, **231**, 360–81.

Cox, D. R. (1999) Discussion of 'Some statistical heresies' (Lindsey). *The Statistician*, **48**, 30.

Cox, D. R. and Farewell, V. T. (1997) Statistical basis of public policy – qualitative and quantitative aspects should not be confused. *British Medical Journal*, **314**, 73.

Craig, B. A., Fryback, D. G., Klein, R. and Klein, B. E. K. (1999) A Bayesian approach to modelling the natural history of a chronic condition from observations with intervention. *Statistics in Medicine*, **18**, 1355–72.

Cronin, K. A., Legler, J. M. and Etzioni, R. D. (1998) Assessing uncertainty in micro-simulation modelling with application to cancer screening interventions. *Statistics in Medicine*, **17**, 2509–23.

Cronin, K. A., Freedman, L. S., Lieberman, R., Weiss, H. L., Beenken, S. W. and Kelloff, G. J. (1999) Bayesian monitoring of phase II trials in cancer chemoprevention. *Journal of Clinical Epidemiology*, **52**, 705–11.

Daniels, M. J. (1999). A prior for the variance components in hierarchical models. *Canadian Journal of Statistics*, **27**, 569–80.

de Finetti, B. (1930). Funzione caratteristica di un fenomeno aleatorio. *Memorie dell'Accademia Nazionale dei Lincei*, **4**, 86–133.

Decisioneering (2000). Crystal Ball. http://www.decisioneering.com/crystal_ball.

DeGroot, M. H. (1970). *Optimal Statistical Decisions*. McGraw-Hill, New York.

DeMets, D. and Lan, K. K. G. (1994) Discussion of 'Bayesian approaches to randomised trials' by Spiegelhalter *et al*. *Journal of the Royal Statistical Society, Series A*, **157**, 387–416.

DeMets, D. L. (1984). Stopping guidelines vs stopping rules – a practitioner's point of view. *Communications in Statistics – Theory and Methods*, **13**, 2395–417.

Dempster, A. (1998). Bayesian methods. In *Encyclopedia of Biostatistics* (P. Armitage and T. Colton, eds), pp. 263–71. John Wiley & Sons, Ltd, Chichester.

Dempster, A., Selwyn, M. and Weeks, B. (1983). Combining historical and randomized controls for assessing trends in proportions. *Journal of the American Statistical Association*, **78**, 221–7.

DerSimonian, R. (1996). Meta-analysis in the design and monitoring of clinical trials. *Statistics in Medicine*, **15**, 1237–48.

DerSimonian, R. and Laird, N. (1986). Meta-analysis in clinical trials. *Controlled Clinical Trials*, **7**, 177–88.

Detsky, A. (1985) Using economic analysis to determine the resource consequences of choices made in planning clinical-trials. *Journal of Chronic Diseases*, **38**, 753–65.

Dignam, J. J., Bryant, J., Wieand, H. S., Fisher, B. and Wolmark, N. (1998) Early stopping of a clinical trial when there is evidence of no treatment benefit: Protocol B-14 of the National Surgical Adjuvant Breast and Bowel Project. *Controlled Clinical Trials*, **19**, 575–88.

Dixon, D. O. and Simon, R. (1991) Bayesian subset analysis. *Biometrics*, **47**, 871–81.

Dominici, F. (1998) Testing simultaneous hypotheses in pharmaceutical trials: a Bayesian approach. *Journal of Biopharmaceutical Statistics*, **8**, 283–97.

Dominici, F., Parmigiani, G., Wolpert, R. and Hasselblad, V. (1999) Meta-analysis of migraine headache treatments: Combining information from heterogeneous designs. *Journal of the American Statistical Association*, **94**, 16–28.

Donner, A. (1982). A Bayesian approach to the interpretation of subgroup results in clinical trials. *Journal of Chronic Diseases*, **35**, 429–35.

Dougherty, T. B., Porche, V. H. and Thall, P. F. (2000) Maximum tolerated dose of nalmefene in patients receiving epidural fentanyl and dilute bupivacaine for post-operative analgesia. *Anesthesiology*, **92**, 1010–16.

Draper, D. (1995) Assessment and propogation of model uncertainty (with discussion). *Journal of the Royal Statistical Society, Series B*, **57**, 45–97.

Droitcour, J., Silberman, G. and Chelimsky, E. (1993) Cross-design synthesis: a new form of meta-analysis for combining results from randomised clinical trials and medical-practice databases. *International Journal of Technology Assessment in Health Care*, **9**, 440–9.

DuMouchel, W. (1990). Bayesian meta-analysis. In *Statistical Methodology in the Pharmaceutical Sciences*, (D. Berry, ed.), pp. 509–29. Marcel Dekker, New York.

DuMouchel, W. and Harris, J. E. (1983) Bayes methods for combining the results of cancer studies in humans and other species (with comment). *Journal of the American Statistical Association*, **78**, 293–308.

DuMouchel, W. and Normand, S. (2000) Computer-modeling and graphical strategies for meta-analysis. In *Meta-analysis in Medicine and Health Policy* (D. K. Stangl and D. A. Berry, eds), pp. 127–78. Marcel Dekker, New York.

DuMouchel, W. and Waternaux, C. (1992). Discussion of 'Hierarchical models for combining information and for meta-analyses'. In *Bayesian Statistics 4*, (J. M. Bernardo, J. O. Berger, A. P. Dawid and A. F. M. Smith, eds), pp. 338–41. Clarendon Press, Oxford.

Dunn, D., Babiker, A., Hooker, M. and Darbyshire, J. (2002) The dangers of inferring treatment effects from observational data: a case study in HIV infection. *Controlled Clinical Trials*, **23**, 106–10.

Eddy, D. M. (1989). The confidence profile method – a Bayesian method for assessing health technologies. *Operations Research*, **37**, 210–28.

Eddy, D. M., Hasselblad, V., McGivney, W. and Hendee, W. (1988) The value of mammography screening in women under age 50 years. *Journal of the American Medical Association*, **259**, 1512–19.

Eddy, D. M., Hasselblad, V. and Shachter, R. (1990a) A Bayesian method for synthesizing evidence: The confidence profile method. *International Journal of Technology Assessment in Health Care*, **6**(1), 31–55.

Eddy, D. M., Hasselblad, V. and Shachter, R. (1990b). An introduction to a Bayesian method for meta-analysis – the confidence profile method. *Medical Decision Making*, **10**, 15–23.

Eddy, D. M., Hasselblad, V. and Shachter, R. (1992) *Meta-analysis by the Confidence Profile Method: The Statistical Synthesis of Evidence*. Academic Press, San Diego, CA.

Edwards, S., Lilford, R., Braunholtz, D. and Jackson, J. (1997) Why 'underpowered' trials are not necessarily unethical. *Lancet*, **350**, 804–7.

Edwards, S. J. L., Lilford, R. J., Jackson, J. C., Hewison, J. and Thornton, J. (1998) Ethical issues in the design and conduct of randomised controlled trials. *Health Technology Assessment*, **2**(15), 1–132.

Edwards, W., Lindman, H. and Savage, L. (1963) Bayesian statistical inference for psychological research. *Psychological Review*, **70**, 193–242.

Ellenberg, S., Fleming, T., and DeMets, D. L. (2002). *Data Monitoring Committees in Clinical Trials: a Practical Perspective*. John Wiley & Sons, Ltd, Chichester.

Errington, R. D., Ashby, D., Gore, S. M., Abrams, K. R., Myint, S., Bonnett, D. E., Blake, S. W. and Saxton, T. E. (1991) High-energy neutron treatment for pelvic cancers – study stopped because of increased mortality. *British Medical Journal*, **302**, 1045–51.

Etzioni, R. and Kadane, J. B. (1995) Bayesian statistical methods in public health and medicine. *Annual Review of Public Health*, **16**, 23–41.

Etzioni, R. and Pepe, M. S. (1994) Monitoring of a pilot toxicity study with 2 adverse outcomes. *Statistics in Medicine*, **13**, 2311–21.

Fayers, P. (1994) Discussion of 'Bayesian approaches to randomised trials' by Spiegelhalter *et al*. *Journal of the Royal Statistical Society, Series A*, **157**, 387–416.

Fayers, P. M., Ashby, D. and Parmar, M. K. B. (1997) Tutorial in biostatistics: Bayesian data monitoring in clinical trials. *Statistics in Medicine*, **16**, 1413–30.

Fayers, P. M., Cuschieri, A., Fielding, J., Craven, J., Uscinska, B. and Freedman, L. S. (2000) Sample size calculation for clinical trials: the impact of clinician beliefs. *British Journal of Cancer*, **82**, 213–9.

Feinstein, A. R. (1977) Clinical biostatistics XXXIX: The haze of Bayes, the aerial palaces of decision analysis, and the computerised Ouija board. *Clinical Pharmacology and Therapeutics*, **21**, 482–96.

Felli, J. C. and Hazen, G. B. (1998) Sensitivity analysis and the expected value of perfect information. *Medical Decision Making*, **18**, 95–109.

Felli, J. C. and Hazen, G. B. (1999). A Bayesian approach to sensitivity analysis. *Health Economics*, **8**, 263–8.

Field, D., Davis, C., Elbourne, D., Grant, A., Johnson, A., and Macrae, D. (1996). UK collaborative randomised trial of neonatal extracorporeal membrane oxygenation. *Lancet*, **348**, 75–82.

Fienberg, S. (1992). A brief history of statistics in three and one-half chapters: a review essay. *Statistical Science*, **7**, 208–25.

Fisher, L. D. (1996). Comments on Bayesian and frequentist analysis and interpretation of clinical trials – comment. *Controlled Clinical Trials*, **17**, 423–34.

Fitzpatrick, R., Shortall, E., Sculpher, M., Murray, D., Morris, R., Lodge, M., Dawson, J., Carr, A., Britton, A. and Briggs, A. (1998) Primary total hip replacement surgery: a systematic review of outcomes and modelling cost effectiveness associated with different prostheses. *Health Technology Assessment*, **2**(20), 1–64.

Fletcher, A., Spiegelhalter, D., Staessen, J., Thijs, L. and Bulpitt, C. (1993) Implications for trials in progress of publication of positive results. *Lancet*, **342**, 653–7.

Fluehler, H., Grieve, A. P., Mandallaz, D., Mau, J. and Moser, H. A. (1983) Bayesian approach to bioequivalence assessment – an example. *Journal of Pharmaceutical Sciences*, **72**, 1178–81.

Forster, J. J. (1994) A Bayesian approach to the analysis of binary crossover data. *Statistician*, **43**, 61–8.

Freedman, B. (1987) Equipoise and the ethics of clinical research. *New England Journal of Medicine*, **317**, 141–5.

Freedman, L. (1996) Bayesian statistical methods – a natural way to assess clinical evidence. *British Medical Journal*, **313**, 569–70.

Freedman, L. S. and Spiegelhalter, D. J. (1983) The assessment of subjective opinion and its use in relation to stopping rules for clinical trials. *The Statistician*, **32**, 153–60.

Freedman, L. S. and Spiegelhalter, D. J. (1989) Comparison of Bayesian with group sequential methods for monitoring clinical trials. *Controlled Clinical Trials*, **10**, 357–67.

Freedman, L. S. and Spiegelhalter, D. J. (1992) Application of Bayesian statistics to decision making during a clinical trial. *Statistics in Medicine*, **11**, 23–35.

Freedman, L. S., Lowe, D. and Macaskill, P. (1984) Stopping rules for clinical trials incorporating clinical opinion. *Biometrics*, **40**, 575–86.

Freedman, L. S., Spiegelhalter, D. J. and Parmar, M. K. B. (1994) The what, why and how of Bayesian clinical trials monitoring. *Statistics in Medicine*, **13**, 1371–83.

Freeman, P. (1993) The role of *p*-values in analyzing trial results. *Statistics in Medicine*, **12**, 1443–52.

Frei, A., Cottier, C., Wunderlich, P., and Ludin, E. (1987) Glycerol and dextran combined in the therapy of acute stroke. *Stroke*, **18**, 373–9.

Freidlin, B., Korn, E. L. and George, S. L. (1999) Data monitoring committees and interim monitoring guidelines. *Controlled Clinical Trials*, **20**, 395–407.

Fryback, D. G., Chinnis, J. O. and Ulvila, J. W. (2001a). Bayesian cost-effectiveness analysis – an example using the GUSTO trial. *International Journal of Technology Assessment in Health Care*, **17**, 83–97.

Fryback, D. G., Stout, N. K. and Rosenberg, M. A. (2001b). An elementary introduction to Bayesian computing using WinBUGS. *International Journal of Technology Assessment in Health Care*, **17**, 98–113.

Gardner, F. J. E., Konje, J. C., Abrams, K. R., Brown, L. J. R., Khanna, S., Al-Azzawi, F., Bell, S. C. and Taylor, D. J. (2000) Endometrial protection from tamoxifen-stimulated changes by a levonorgestrel-releasing intrauterine system: a randomised controlled trial. *Lancet*, **356**, 1711–17.

Gasparini, M. and Eisele, J. (2000) A curve-free method for phase I clinical trials. *Biometrics*, **56**, 609–15.

Gatsonis, C. and Greenhouse, J. B. (1992) Bayesian methods for phase I clinical trials. *Statistics in Medicine*, **11**, 1377–89.

Geddes, J., Freemantle, N., Harrison, P. and Bebbington, P. (2000) Atypical antipsychotics in the treatment of schizophrenia: systematic overview and meta-regression analysis. *British Medical Journal*, **321**, 1371–6.

Gelman, A. and Rubin, D. B. (1996) Markov chain Monte Carlo methods in biostatistics. *Statistical Methods in Medical Research*, **5**, 339–55.

Gelman, A., Carlin, J., Stern, H. and Rubin, D. B. (1995) *Bayesian Data Analysis*. Chapman & Hall, London.

General Accounting Office (1992) *Cross Design Synthesis: A New Strategy for Medical Effectiveness Research*. General Accounting Office, Washington, DC.

Genest, C. and Zidek, J. (1986) Combining probability distributions: a critique and an annotated bibliography (with discussion). *Statistical Science*, **1**, 114–48.

George, S. L., Li, C. C., Berry, D. A. and Green, M. R. (1994) Stopping a clinical trial early – frequentist and Bayesian approaches applied to a CALGB trial in non-small-cell lung cancer. *Statistics in Medicine*, **13**, 1313–27.

Gilbert, J. P., McPeek, B. and Mosteller, F. (1977) Statistics and ethics in surgery and anesthesia. *Science*, **198**, 684–9.

Gilks, W. R., Richardson, S. and Spiegelhalter, D. J. (1996) *Markov Chain Monte Carlo in Practice*. Chapman & Hall, London.

Gittins, J. and Pezeshk, H. (2000) How large should a clinical trial be? *The Statistician*, **49**, 177–87.

Givens, G. H., Smith, D. D. and Tweedie, R. L. (1997) Publication bias in meta-analysis: a Bayesian data-augmentation approach to account for issues exemplified in the passive smoking debate. *Statistical Science*, **12**, 221–40.

Goldstein, H. and Spiegelhalter, D. J. (1996) League tables and their limitations: statistical issues in comparisons of institutional performance (with discussion). *Journal of the Royal Statistical Society, Series A*, **159**, 385–443.

Goodman, S. N. (1999a) Towards evidence-based medical statistics: 1. The P value fallacy. *Annals of Internal Medicine*, **130**, 995–1004.

Goodman, S. N. (1999b) Towards evidence-based medical statistics: 2. The Bayes factor. *Annals of Internal Medicine*, **130**, 1005–13.

Goodman, S. N., Zahurak, M. L. and Piantadosi, S. (1995) Some practical improvements in the continual reassessment method for phase I studies. *Statistics in Medicine,* **14,** 1149–61.

Gore, S. M. (1987) Biostatistics and the Medical Research Council. *Medical Research Council News.*

Gould, A. L. (1991) Using prior findings to augment active-controlled trials and trials with small placebo groups. *Drug Information Journal,* **25,** 369–80.

Gould, A. L. (1998) Multi-centre trial analysis revisited. *Statistics in Medicine,* **17,** 1779–97.

Gray, A., Clarke, P., Farmer, A. and Holman, R. (2002) Implementing intensive control of blood glucose concentration and blood pressure in type 2 diabetes in England: cost analysis (UKPDS 63). *British Medical Journal,* **325,** 860–3.

Gray, R. J. (1994) A Bayesian analysis of institutional effects in a multicenter cancer clinical trial. *Biometrics,* **50,** 244–53.

GREAT Group (1992) Feasibility, safety and efficacy of domiciliary thrombolysis by general practitioners: Grampian region early anisteplase trial. *British Medical Journal,* **305,** 548–53.

Greenhouse, J. B. and Wasserman, L. (1995) Robust Bayesian methods for monitoring clinical trials. *Statistics in Medicine,* **14,** 1379–91.

Greenland, S. (2000) Principles of multilevel modelling. *International Journal of Epidemiology,* **29,** 158–67.

Greenland, S. and Robins, J. M. (1991) Empirical-Bayes adjustments for multiple comparisons are sometimes useful. *Epidemiology,* **2,** 244–51.

Grieve, A. P. (1985) A Bayesian analysis of the two-period crossover design for clinical trials. *Biometrics,* **41,** 979–90.

Grieve, A. P. (1988) Some uses of predictive distributions in pharmaceutical research. In *Biometry, Clinical Trials and Related Topics,* (T. Okuno, ed.), pp. 83–99. Elsevier Science, Amsterdam.

Grieve, A. P. (1991) Evaluation of bioequivalence studies. *European Journal of Clinical Pharmacology,* **40,** 201–2.

Grieve, A. P. (1994a) Bayesian analyses of two-treatment crossover studies. *Statistical Methods in Medical Research,* **3,** 407–29.

Grieve, A. P. (1994b) Discussion of 'Bayesian approaches to randomised trials'. *Journal of the Royal Statistical Society, Series A,* **157,** 387–416.

Grieve, A. P. (1995) Extending a Bayesian analysis of the two-period crossover to accommodate missing data. *Biometrika,* **82,** 277–86.

Grieve, A. P. (1998) Issues for statisticians in pharmaco-economic evaluations. *Statistics in Medicine,* **17,** 1715–23.

Grieve, A. P. and Senn, S. (1998) Estimating treatment effects in clinical crossover trials. *Journal of Biopharmaceutical Statistics,* **8,** 191–247.

Grossman, J., Parmar, M. K. B., Spiegelhalter, D. J. and Freedman, L. S. (1994) A unified method for monitoring and analysing controlled trials. *Statistics in Medicine,* **13,** 1815–26.

Gurrin, L. C., Kurinczuk, J. J. and Burton, P. R. (2000) Bayesian statistics in medical research: an intuitive alternative to conventional data analysis. *Journal of Evaluation of Clinical Practice,* **6,** 193–204.

Gustafson, P. (1989). Comments on 'Investigating therapies of potentially great benefit: ECMO' by J H Ware. *Statistical Science,* **4,** 310–17.

Halperin, M., Lan, K. K. G., Ware, J. H., Johnson, N. J. and DeMets, D. L. (1982) An aid to data monitoring in long-term clinical trials. *Controlled Clinical Trials,* **3,** 311–23.

Halpern, E. F., Weinstein, M. C., Hunink, M. G. M. and Gazelle, G. S. (2000) Representing both first- and second-order uncertainties by Monte Carlo simulation for groups of patients. *Medical Decision Making*, **20**, 314–22.

Hardy, R. and Thompson, S. (1998) Detecting and describing heterogeneity in meta-analysis. *Statistics in Medicine*, **17**, 841–56.

Harrell, F. E. and Shih, Y. C. T. (2001) Using full probability models to compute probabilities of actual interest to decision makers. *International Journal of Technology Assessment in Health Care*, **17**, 17–26.

Hasselblad, V. (1998) Meta-analysis of multi-treatment studies. *Medical Decision Making*, **18**, 37–43.

Healy, M. (1978) New methodology in clinical trials. *Biometrics*, **34**, 709–12.

Healy, M. J. R. (1994) Probability and decisions. *American Journal of Diseases in Children*, **71**, 90–4.

Hedges, L. V. (1998) Bayesian meta-analysis. In *Statistical Analysis of Medical Data: New Developments*, (B. S. Everitt and G. Dunn, eds), pp. 251–76. Arnold, London.

Heisterkamp, S. H., Doornbos, G. and Gankema, M. (1993) Disease mapping using empirical Bayes and Bayes methods on mortality statistics in The Netherlands. *Statistics in Medicine*, **12**, 1895–1913.

Heitjan, D. F. (1997). Bayesian interim analysis of phase II cancer clinical trials. *Statistics in Medicine*, **16**, 1791–1802.

Heitjan, D. F., Moskowitz, A. J. and Whang, W. (1999) Bayesian estimation of cost-effectiveness ratios from clinical trials. *Health Economics*, **8**, 191–201.

Henderson, W. G., Moritz, T., Goldman, S., Copeland, J. and Sethi, G. (1995) Use of cumulative meta-analysis in the design, monitoring, and final analysis of a clinical trial: a case study. *Controlled Clinical Trials*, **16**, 331–41.

Herson, J. (1979). Predictive probability early termination for phase II clinical trials. *Biometrics*, **35**, 775–83.

Higgins, J. P. and Spiegelhalter, D. J. (2002) Being sceptical about meta-analysis: a Bayesian perspective on magnesium trials in myocardial infarction. *International Journal of Epidemiology*, **31**, 96–104.

Higgins, J. P. and Whitehead, A. (1996) Borrowing strength from external trials in a meta-analysis. *Statistics in Medicine*, **15**, 2733–49.

Hill, G., Forbes, W., Kozak, J. and MacNeill, I. (2000) Likelihood and clinical trials. *Journal of Clinical Epidemiology*, **53**, 223–7.

Hilsenbeck, S. G. (1988) Early termination of a phase II clinical trial. *Controlled Clinical Trials*, **9**(3), 177–88.

Hlatky, M. A. (1991) Using databases to evaluate therapy. *Statistics in Medicine*, **10**, 647–52.

Hoes, A. W., Grobbee, D. E., Lubsen, J., Tveld, A. J. M. I., Vanderdoes, E. and Hofman, A. (1995) Diuretics, beta-blockers, and the risk for sudden cardiac death in hypertensive patients. *Annals of Internal Medicine*, **123**, 481–7.

Holland, J. (1962) The Reverend Thomas Bayes, F. R. S. (1702–61). *Journal of the Royal Statistical Society, Series A*, **125**, 451–61.

Hornberger, J. (2001) Introduction to Bayesian reasoning. *International Journal of Technology Assessment in Health Care*, **17**, 9–16.

Hornberger, J. and Eghtesady, P. (1998) The cost-benefit of a randomised trial to a health care organisation. *Controlled Clinical Trials*, **19**, 198–211.

Hornberger, J. C., Brown, B. W. and Halpern, J. (1995) Designing a cost-effective clinical trial. *Statistics in Medicine*, **14**, 2249–59.

Hornbuckle, J., Vail, A., Abrams, K. R. and Thornton, J. G. (2000) Bayesian interpretation of trials: the example of intrapartum electronic fetal heart rate monitoring. *British Journal of Obstetrics and Gynaecology*, **107**, 3–10.

Howson, C. and Urbach, P. (1989) *Scientific Reasoning: The Bayesian Approach*. Open Court, La Salle, IL.

Hughes, M. D. (1991) Practical reporting of Bayesian analyses of clinical trials. *Drug Information Journal*, **25**, 381–93.

Hughes, M. D. (1993) Reporting Bayesian analyses of clinical trials. *Statistics in Medicine*, **12**, 1651–63.

Human Fertilisation and Embryology Authority (1996) *The Patients' Guide to DI and IVF Clinics* (2nd edition). Human Fertilisation and Embryology Authority, London.

Hutton, J. L. (1996) The ethics of randomised controlled trials: a matter of statistical belief? *Health Care Analysis*, **4**, 95–102.

Ibrahim, J. G. and Chen, M. H. (2000) Power prior distributions for regression models. *Statistical Science*, **15**, 46–60.

International Conference on Harmonisation E9 Expert Working Group (1999) Statistical principles for clinical trials: ICH harmonised tripartite guideline. *Statistics in Medicine*, **18**, 1905–42. See also http://www.ich.org/ich5e.html.

Ioannidis, J. P. A., Haidich, A. B., Pappa, M., Pantazis, N., Kokori, S. I., Tektonidou, M. G., Contopoulos-Ioannidis, D. G. and Lau, J. (2001) Comparison of evidence of treatment effects in randomized and nonrandomized studies. *Journal of the American Medical Association*, **286**, 821–30.

Jeffreys, H. (1961) *Theory of Probability* (3rd edition). Oxford University Press, Oxford.

Jennison, C. (1990) Discussion of 'Biostatistics and Bayes' by Breslow. *Statistical Science*, **5**(3).

Jones, B., Teather, D., Wang, J. and Lewis, J. A. (1998) A comparison of various estimators of a treatment difference for a multi-centre clinical trial. *Statistics in Medicine*, **17**, 1767–77.

Jones, D. R. (1995) Meta-analysis: weighing the evidence. *Statistics in Medicine*, **14**, 137–49.

Joseph, L., Duberger, R. and Belisle, P. (1997) Bayesian and mixed Bayesian/likelihood criteria for sample size determination. *Statistics in Medicine*, **16**, 769–81.

Kadane, J. (1996) *Bayesian Methods and Ethics in a Clinical Trial Design*. John Wiley & Sons, Inc. New York.

Kadane, J. and Wolfson, L. (1996). Priors for the design and analysis of clinical trials. *Bayesian Biostatistics* (D. A. Berry and D. K. Stangl, eds), pp. 157–84. Marcel Dekker, New York.

Kadane, J. B. (1995) Prime time for Bayes. *Controlled Clinical Trials*, **16**, 313–18.

Kadane, J. B., Vlachos, P. and Wieand, S. (1998) Decision analysis for a data monitoring committee of a clinical trial. In *Applied Decision Analysis* (F. J. Girón and M. L. Martínez, eds), pp. 115–21. Kluwer, Boston.

Kadane, J. B. and Wolfson, L. J. (1997) Experiences in elicitation. *The Statistician*, **46**, 1–17.

Kass, R. E. and Greenhouse, J. B. (1989) Comments on 'Investigating therapies of potentially great benefit: ECMO' by J H Ware. *Statistical Science*, **4**, 310–17.

Kass, R. E. and Raftery, A. (1995) Bayes factors and model uncertainty. *Journal of the American Statistical Association*, **90**, 773–95.

Kass, R. E. and Wasserman, L. (1995) A reference Bayesian test for nested hypotheses and its relationship to the Schwarz criterion. *Journal of the American Statistical Association*, **90**, 928–34.

Kass, R. E. and Wasserman, L. (1996) The selection of prior distributions by formal rules. *Journal of the American Statistical Association*, **91**, 1343–70.

Keiding, N. (1994) Discussion of 'Bayesian approaches to randomised trials'. *Journal of the Royal Statistical Society, Series A*, **157**, 387–416.

Koch, G. G. (1991) Summary and discussion for 'Statistical issues in the pharmaceutical industry: analysis and reporting of phase III clinical trials including kinetic/dynamic analysis and Bayesian analysis'. *Drug Information Journal*, **25**, 433–7.

Korn, E. L. (1990) Projection from previous studies: a caution [letter; comment]. *Controlled Clinical Trials*, **11**, 67–9.

Korn, E. L. and Simon, R. (1996) Data monitoring committees and problems of lower than expected accrual or event rates. *Controlled Clinical Trials*, **17**, 526–35.

Korn, E. L., Yu, K. F. and Miller, L. L. (1993) Stopping a clinical trial very early because of toxicity – summarizing the evidence. *Controlled Clinical Trials*, **14**, 286–95.

Kunz, R. and Oxman, A. D. (1998) The unpredictability paradox: review of empirical comparisons of randomised and non-randomised clinical trials. *British Medical Journal*, **317**, 1185–90.

Lachin, J. M. (1981) Sequential clinical trials for normal variates using interval composite hypotheses. *Biometrics*, **37**, 87–101.

Lan, K. K. G. and Wittes, J. (1988) The *B*-value – a tool for monitoring data. *Biometrics*, **44**, 579–85.

Lancet (1992) Cross design synthesis: A new strategy for studying medical outcomes? *Lancet*, **340**, 944–6.

Lanctot, K. L., and Naranjo, C. A. (1995) Comparison of the Bayesian approach and a simple algorithm for assessment of adverse drug events. *Clinical Pharmacology and Therapeutics*, **58**, 692–8.

Lang, T. and Secic, M. (1997) Considering 'prior probabilities': reporting Bayesian statistical analyses. In *How to Report Statistics in Medicine* (T. Lang and M. Secic), pp. 231–5. American College of Physicians, Philadelphia.

Lange, N. and Ryan, L. (1989) Assessing normality in random effects models. *Annals of Statistics*, **17**, 624–42.

Larose, D. T. and Dey, D. K. (1997) Grouped random effects models for Bayesian meta-analysis. *Statistics in Medicine*, **16**, 1817–29.

Larsen, K., Petersen, J. H., Budtz-Jørgensen, E. and Endahl, L. (2000). Interpreting parameters in the logistic regression model with random effects. *Biometrics*, **56**, 909–14.

Lau, J., Schmid, C. H. and Chalmers, T. C. (1995) Cumulative meta-analysis of clinical trials builds evidence for exemplary medical care. *Journal of Clinical Epidemiology*, **48**, 45–57.

Laupacis, A., Bourne, R., Rorabeck, C., Feeny, D., Wong, C., Tugwell, P., Leslie, K. and Bullas, R. (1993) The effect of elective total hip replacement on health-related quality of life. *Journal of Bone and Joint Surgery*, **75A**, 1619–26.

Lee, P. M. (1997) *Bayesian Statistics: An Introduction* (2nd edition). Edward Arnold, London.

Lee, S. J. and Zelen, M. (2000) Clinical trials and sample size considerations: Another perspective. *Statistical Science*, **15**, 95–103.

Legler, J. M. and Ryan, L. M. (1997) Latent variable models for teratogenesis using multiple binary outcomes. *Journal of the American Statistical Association*, **92**, 13–20.

Lehmann, H. P. and Goodman, S. N. (2000) Bayesian communication: A clinically significant paradigm for electronic publication. *Journal of the American Medical Informatics Association* **7**, 254–66.

Lehmann, H. P. and Nguyen, B. (1997) Bayesian communication of research results over the World Wide Web. *M D Computing*, **14**, 353–9.

Lehmann, H. P. and Shachter, R. D. (1994) A physician-based architecture for the construction and use of statistical models. *Methods of Information in Medicine*, **33**, 423–32.

Lewis, J. (1994) Discussion of 'Bayesian approaches to randomized trials'. *Journal of the Royal Statistical Society, Series A*, **157**, 387–416.

Lewis, R. J. and Wears, R. L. (1993) An introduction to the Bayesian analysis of clinical trials. *Annals of Emergency Medicine*, **22**, 1328–36.

Li, Z. and Begg, C. B. (1994) Random effects models for combining results from controlled and uncontrolled studies in a meta-analysis. *Journal of the American Statistical Association*, **89**, 1523–7.

Liao, J. G. (1999) A hierarchical Bayesian model for combining multiple 2 × 2 tables using conditional likelihoods. *Biometrics*, **55**, 268–72.

Lilford, R. for the Fetal Compromise Group (1994) Formal measurement of clinical uncertainty: prelude to a trial in perinatal medicine. *British Medical Journal*, **308**, 111–12.

Lilford, R. J. and Braunholtz, D. (1996). The statistical basis of public policy: a paradigm shift is overdue. *British Medical Journal*, **313**, 603–7.

Lilford, R. J. and Jackson, J. (1995) Equipoise and the ethics of randomization. *Journal of the Royal Society of Medicine*, **88**, 552–9.

Lilford, R. J., Thornton, J. G. and Braunholtz, D. (1995) Clinical trials and rare diseases – a way out of a conundrum. *British Medical Journal*, **311**, 1621–5.

Lindley, D. V. (1957) A statistical paradox. *Biometrika*, **44**, 187–92.

Lindley, D. V. (1975) The effect of ethical design considerations on statistical analysis. *Applied Statistics*, **24**, 218–28.

Lindley, D. V. (1985) *Making Decisions* (2nd edition) John Wiley & Sons, Ltd, London.

Lindley, D. V. (1994) Discussion of 'Bayesian approaches to randomised trials'. *Journal of the Royal Statistical Society, Series A*, **157**, 387–416.

Lindley, D. V. (1997) The choice of sample size. *Statistician*, **46**, 129–38.

Lindley, D. V. (1998) Decision analysis and bioequivalence trials. *Statistical Science*, **13**, 136–41.

Lindley, D. V. (2000) The philosophy of statistics (with discussion). *The Statistician*, **49**, 293–337.

Lindley, D. V. and Scott, W. F. (1984) *New Cambridge Statistical Tables* (2nd edition). Cambridge University Press, Cambridge.

Louis, T. A. (1991). Using empirical Bayes methods in biopharmaceutical research. *Statistics in Medicine*, **10**, 811–29.

Luce, B. R. and Claxton, K. (1999) Redefining the analytical approach to pharmacoeconomics. *Health Economics*, **8**, 187–9.

Luce, B. R., Shih, Y. C. T. and Claxton, K. (2001) Bayesian approaches to technology assessment and decision making. *International Journal of Technology Assessment in Health Care*, **17**, 1–5.

Malchau, H. and Herberts, P. (1998). Prognosis of total hip replacement. Revision and re-revision rate in THR: a revision-risk study of 148,359 primary operations. Scientific exhibition presented to the 65th Annual Meeting of the American Academy of Orthopaedic Surgeons, New Orleans.

Manning, W. G., Fryback, F. G. and Weinstein, M. C. (1996) Reflecting uncertainty in cost-effectiveness analysis. In *Cost Effectiveness in Health and Medicine*, (M. R. Gold, J. R. Siegel, M. C. Weinstein and L. B. Russell, eds), pp. 247–75. Oxford University Press, New York.

Marshall, E. C. and Spiegelhalter, D. J. (1998) League tables of in-vitro fertilisation clinics: how confident can we be about the rankings? *British Medical Journal*, **317**, 1701–4.

Marshall, R. J. (1988) Bayesian analysis of case–control studies. *Statistics in Medicine*, **7**, 1223–30.

Marston, R. A., Cobb, A. G. and Bentley, G. (1996) Stanmore compared with Charnley total hip replacement – a prospective study of 413 arthroplasties. *Journal of Bone and Joint Surgery*, **78B**, 178–84.

Matchar, D. B., Samsa, G. P., Matthews, J. R., Ancukiewicz, M., Parmigiani, G., Hasselblad, V., Wolf, P. A., D'Agostino, R. B. and Lipscomb, J. (1997) The stroke prevention policy model: linking evidence and clinical decisions. *Annals of Internal Medicine*, **127**, 704–11.

Matsuyama, Y., Sakamoto, J. and Ohashi, Y. (1998) A Bayesian hierarchical survival model for the institutional effects in a multi-centre cancer clinical trial. *Statistics in Medicine*, **17**, 1893–1908.

Matthews, J. N. S. (1995) Small clinical trials – are they all bad? *Statistics in Medicine*, **14**, 115–26.

Matthews, R. (1998). Fact *versus* factions: the use and abuse of subjectivity in scientific research. Technical report, European Science and Environment Forum, Cambridge.

Matthews, R. A. J. (2001) Methods for assessing the credibility of clinical trial outcomes. *Drug Information Journal*, **35**, 1469–78.

McIntosh, M. W. (1996) The population risk as an explanatory variable in research synthesis of clinical trials. *Statistics in Medicine*, **15**, 1713–28.

McPherson, K. (1982) On choosing the number of interim analyses in clinical trials. *Statistics in Medicine*, **1**, 25–36.

Mehta, C. R. and Cain, K. C. (1984) Charts for the early stopping of pilot studies. *Journal of Clinical Oncology*, **2**, 676–82.

Meier, P. (1975) Statistics and medical experimentation. *Biometrics*, **31**, 511–29.

Miller, M. A. and Seaman, J. W. (1998) A Bayesian approach to assessing the superiority of a dose combination. *Biometrical Journal*, **40**, 43–55.

Moher, D., Schulz, K. F. and Altman, D. G. (2001) The CONSORT statement: revised recommendations for improving the quality of reports of parallel-group randomised trials. *Lancet*, **357**, 1191–4.

Morris, C. N. and Normand, S. L. (1992) Hierarchical models for combining infromation and for meta-analysis. In *Bayesian Statistics 4*, (J. Bernardo, J. O. Berger, D. V. Lindley, and A. F. M. Smith, eds), pp. 321–44. Oxford University Press.

Morrison, L. J., Verbeck, P. R., McDonald, A. C., Sawadsky, B. V. and Cook, D. J. (2000) Mortality and prehospital thrombolysis for acute myocardial infarction – a meta-analysis. *Journal of the American Medical Association*, **283**, 2686–92.

Murphy, A. and Winkler, R. (1977) Reliability of subjective probability forecasts of precipitation and temperature. *Applied Statistics*, **26**, 41–7.

Natarajan, R. and Kass, R. E. (2000) Reference Bayesian methods for generalized linear mixed models. *Journal of the American Statistical Association*, **95**, 227–37.

New York State Department of Health (1998). *Coronary Artery Bypass Surgery in New York State, 1994–1996*. New York State Department of Health, Albany, NY. http://www.health.state.ny.us/nysdoh/consumer/heart/homehear.htm.

Neyman, J. (1934) On the two different aspects of the representative method: the method of stratified sampling and the method of purposive selection (with discussion). *Journal of the Royal Statistical Society*, **97**, 558–625.

Neyman, J. (1935) Statistical problems in agricultural experimentation (with discussion). *Supplement to the Journal of the Royal Statistical Society*, **2**, 107–80.

NICE (2001) *Health Technology Assessment: Guidance for Manufacturers and Sponsors*. National Institute for Clinical Excellence.

NICE Appraisal Group (2000) The effectiveness and cost effectiveness of different prostheses for primary total hip replacement. Technical report, http://www.nice.org.uk.

Normand, S.-L. T. (1999) Meta-analysis: formulating, evaluating, combining and reporting. *Statistics in Medicine*, **18**, 321–59.

Normand, S.-L., Glickman, M. E. and Gatsonis, C. A. (1997) Statistical methods for profiling providers of medical care: issues and applications. *Journal of the American Statistical Association*, **92**, 803–14.

Nurminen, M. and Mutanen, P. (1989) Bayesian analysis of case–control studies. *Statistics in Medicine*, **8**, 1023–4.

O'Brien, P. C. (1998) Data and safety monitoring. In *Encyclopedia of Biostatistics* (P. Armitage and T. Colton, eds), pp. 1058–66. John Wiley & Sons, Ltd, Chichester.

O'Hagan, A. (1994) *Kendall's Advanced Theory of Statistics Vol 2B: Bayesian Inference*. Arnold, London.

O'Hagan, A. and Luce, B. R. (2003) *A Primer on Bayesian Statistics in Health Economics and Outcomes Research*. Centre for Bayesian Statistics in Health Economics, Sheffield. http://www.shef.ac.uk/~st1ao/pdf/primer.pdf.

O'Hagan, A. and Stevens, J. W. (2001) A framework for cost-effectiveness analysis from clinical trial data. *Health Economics*, **10**, 303–15.

O'Hagan, A. and Stevens, J. W. (2002a) Bayesian methods for the design and analysis of cost-effectiveness trials in the evaluation of health care technologies. *Statistical Methods in Medical Research*, **11**, 469–90.

O'Hagan, A. and Stevens, J. W. (2002b) The probability of cost-effectiveness. *BMC Medical Research Methodology*, **2**(5). http://biomedcentral.com/1471-2288/2/5.

O'Hagan, A., Stevens, J. and Montmartin, J. (2000) Inference for the cost-effectiveness acceptability curve and cost-effectiveness ratio. *Pharmacoeconomics*, **17**, 339–49.

O'Hagan, A., Stevens, J. W. and Montmartin, J. (2001) Bayesian cost-effectiveness analysis from clinical trial data. *Statistics in Medicine*, **20**, 733–53.

O'Neill, R. T. (1994) Conclusions: 2. *Statistics in Medicine*, **13**, 1493–9.

O'Quigley, J. (1992) Estimating the probability of toxicity at the recommended dose following a phase I clinical trial in cancer. *Biometrics*, **48**, 853–62.

O'Quigley, J., Pepe, M. and Fisher, L. (1990) Continual reassessment method – a practical design for phase I clinical trials in cancer. *Biometrics*, **46**, 33–48.

O'Rourke, K. (1996) Two cheers for Bayes. *Controlled Clinical Trials*, **17**, 350–2.

Palisade Europe (2001) @RISK 4.0. Technical report, http://www.palisade-europe.com.

Palmer, C. R. (1993) Ethics and statistical methodology in clinical trials. *Journal of Medical Ethics*, **19**, 219–22.

Palmer, C. R. (2002) Ethics, data-dependent designs, and the strategy of clinical trials: time to start learning as we go? *Statistical Methods in Medical Research*, **11**, 381–402.

Palmer, C. R. and Rosenberger, W. F. (1999) Ethics and practice: alternative designs for phase III randomized clinical trials. *Controlled Clinical Trials*, **20**, 172–86.

Papineau, D. (1994). The virtues of randomisation. *British Journal for the Philosophy of Science*, **45**, 437–50.

Parmar, M. K. B., Spiegelhalter, D. J. and Freedman, L. S. (1994) The CHART trials: Bayesian design and monitoring in practice. *Statistics in Medicine*, **13**, 1297–312.

Parmar, M. K. B., Ungerleider, R. S. and Simon, R. (1996) Assessing whether to perform a confirmatory randomised clinical trial. *Journal of the National Cancer Institute*, **88**, 1645–51.

Parmar, M. K. B., Griffiths, G. O., Spiegelhalter, D. J., Souhami, R. L., Altman, D. G. and van der Scheuren, E. (2001) Monitoring large randomised clinical trials – a new approach using Bayesian methods. *Lancet*, **358**, 375–81.

Parmigiani, G. (1999). Decision models in screening for breast cancer. In *Bayesian Statistics 6*, (J. Bernardo, J. Berger, A. Dawid, and A. Smith, eds), pp. 525–46. Oxford University Press, Oxford.

Parmigiani, G. (2002). *Modeling in Medical Decision Making: a Bayesian Approach*. John Wiley & Sons, Ltd, Chichester.

Parmigiani, G. and Kamlet, M. (1993). A cost-utility analysis of alternative strategies in screening for breast cancer. In *Case Studies in Bayesian Statistics*, (C. Gatsonis, J. Hodges, R. Kass, and N. Singpurwalla, eds), pp. 390–402. Springer-Verlag, Berlin.

Parmigiani, G., Anckiewicz, M. and Matchar, D. (1996). Decision models in clinical recommendations development: the stroke prevention policy model. In *Bayesian Biostatistics* (D. A. Berry and D. K. Stangl, eds), pp. 207–33. Marcel Dekker, New York.

Parmigiani, G., Samsa, G. P., Ancukiewicz, M., Lipscomb, J., Hasselblad, V. and Matchar, D. B. (1997). Assessing uncertainty in cost-effectiveness analyses: application to a complex decision model. *Medical Decision Making*, **17**, 390–401.

Pauler, D. K. and Wakefield, J. (2000) Modeling and implementation in Bayesian meta-analysis. In *Meta-analysis in Medicine and Health Policy* (D. K. Stangl and D. A. Berry, eds), pp. 205–30. Marcel Dekker, New York.

Pepple, P. A. and Choi, S. C. (1997) Bayesian approach to two-stage phase II trial. *Journal of Biopharmaceutical Statistics*, **7**, 271–86.

Peto, R. (1985) Discussion of 'On the allocation of treatments in sequential medical trials' by J. Bather. *International Statistical Review*, **53**, 1–13.

Peto, R. and Baigent, C. (1998) Trials: the next 50 years. *British Medical Journal*, **317**, 1170–1.

Peto, R., Pike, M. C., Armitage, P., Breslow, N. E., Cox, D. R., Howard, S. V., Mantel, N., McPherson, K., Peto, J. and Smith, P. G. (1976) Design and analysis of randomised clinical trials requiring prolonged observation of each patient. I. Introduction and design. *British Journal of Cancer*, **34**, 585–612.

Pocock, S. (1976) The combination of randomized and historical controls in clinical trials. *Journal of Chronic Diseases*, **29**, 175–88.

Pocock, S. (1992) When to stop a clinical trial. *British Medical Journal*, **305**, 235–40.

Pocock, S. (1994) Discussion of 'Bayesian approaches to randomized trials'. *Journal of the Royal Statistical Society, Series A*, **157**, 387–416.

Pocock, S. and Hughes, M. D. (1989) Practical problems in interim analyses, with particular regard to estimation. *Controlled Clinical Trials*, **10**, S209–21.

Pocock, S. and Hughes, M. D. (1990) Estimation issues in clinical trials and overviews. *Statistics in Medicine*, **9**, 657–71.

Pocock, S. and Spiegelhalter, D. (1992) Domiciliary thrombolysis by general practicioners. *British Medical Journal*, **305**, 1015.

Pocock, S. and White, I. (1999) Trials stopped early: too good to be true? *Lancet*, **353**, 943–4.

Prevost, T. C., Abrams, K. R. and Jones, D. R. (2000) Hierarchical models in generalised synthesis of evidence: an example based on studies of breast cancer screening. *Statistics in Medicine*, **19**, 3359–76.

Qian, J., Stangl, D. and George, S. (1996). A Weibull model for survival data: using prediction to decide when to stop a clinical trial. In *Bayesian Biostatistics* (D. A. Berry and D. K. Stangl, eds), pp. 187–205. Marcel Dekker, New York.

Racine, A., Grieve, A. P., Fluhler, H. and Smith, A. F. M. (1986) Bayesian methods in practice – experiences in the pharmaceutical industry. *Applied Statistics*, **35**, 93–150.

Racine-Poon, A. and Wakefield, J. (1996). Bayesian analysis of population pharmacokinetic and instantaneous pharmacodynamic relationships. In *Bayesian Biostatistics* (D. A. Berry and D. K. Stangl, eds), pp. 321–54. Marcel Dekker, New York.

Raghunathan, T. E. and Siscovick, D. S. (1996) A multiple-imputation analysis of a case-control study of the risk of primary cardiac-arrest among pharmacologically treated hypertensives. *Applied Statistics*, **45**, 335–52.

Raudenbush, S. W. and Bryk, A. S. (1985) Empirical Bayes meta-analysis. *Journal of Educational Statistics*, **10**, 75–98.

Reeves, B., MacLehose, R., Harvey, I., Sheldon, T., Russell, I. and Black, A. (2001) A review of observational, quasi-experimental and randomised study designs for the evaluation of the effectiveness of healthcare interventions. In *The Advanced Handbook of Methods in Evidence Based Healthcare* (A. Stevens, K. Abrams, J. Brazier, R. Fitzpatrick and R. Lilford, eds), pp. 116–35. Sage, London.

Richards, B., Blandy, J., Bloom, H. G. J. and MRC Bladder Cancer Working Party (1994) The effect of intravesical thiotepa on tumor recurrence after endoscopic treatment of newly-diagnosed superficial bladder cancer – a further report with long-term follow-up of a Medical Research Council randomised trial. *British Journal of Urology*, **73**, 632–8.

Richardson, S. and Gilks, W. R. (1993) A Bayesian approach to measurement error problems in epidemiology using conditional independence models. *American Journal of Epidemiology*, **138**, 430–42.

Richardson, S., Monfort, C., Green, M., Draper, G. and Muirhead, C. (1995) Spatial variation of natural radiation and childhood leukaemia incidence in Great Britain. *Statistics in Medicine*, **14**, 2487–501.

Rittenhouse, B. E. (1997) Exorcising protocol-induced spirits: making the clinical trial relevant for economics. *Medical Decision Making*, **17**, 331–9.

Rogatko, A. (1992) Bayesian approach for meta-analysis of controlled clinical-trials. *Communications in Statistics – Theory and Methods*, **21**, 1441–62.

Rosenbaum, P. R. and Rubin, D. B. (1984) Sensitivity of Bayes inference with data-dependent stopping rules. *American Statistician*, **38**, 106–9.

Rosner, G. L. and Berry, D. A. (1995) A Bayesian group sequential design for a multiple arm randomized clinical trial. *Statistics in Medicine*, **14**, 381–94.

Royall, R. (1986) The effect of sample size on the meaning of significance tests. *American Statistician*, **40**, 313–15.

Royall, R. and Berry, D. A. (1989) Comments on 'Investigating therapies of potentially great benefit: ECMO' by J H Ware. *Statistical Science*, **4**, 313–19.

Rubin, D. (1978) Bayesian inference for casual effects: the role of randomization. *Annals of Statistics*, **7**, 34–58.

Rubin, D. B. (1992) A new perspective. In *The Future of Meta-analysis*, (D. Rubin, K. Wachter and M. Straf, eds), pp. 155–65. Russell Sage Foundation, New York.

Russell, L. B. (1999) Modelling for cost-effectiveness analysis. *Statistics in Medicine*, **18**, 3235–44.

Ryan, L. (1993) Using historical controls in the analysis of developmental toxicity data. *Biometrics*, **49**, 1126–35.

Samsa, G. P., Reutter, R. A., Parmigiani, G., Ancukiewicz, M., Abrahamse, P., Lipscomb, J. and Matchar, D. B. (1999) Performing cost-effectiveness analysis by integrating randomized trial data with a comprehensive decision model: application to treatment of acute ischemic stroke. *Journal of Clinical Epidemiology*, **52**, 259–71.

Sanderson, C., McKee, M., Britton, A., Black, N., McPherson, K. and Bain, C. (2001). Randomised and non-randomised studies: threats to internal and external validity. In *The Advanced Handbook of Methods in Evidence Based Healthcare* (A. Stevens, K. Abrams, J. Brazier, R. Fitzpatrick and R. Lilford, eds), pp. 95–115. Sage, London.

Sargent, D. and Carlin, B. (1996) Robust Bayesian design and analysis of clinical trials via prior partitioning (with discussion). In *Bayesian Robustness* (J. O. Berger, ed.), IMS Lecture Notes – Monograph Series, 29, pp. 175–93. Institute of Mathematical Statistics, Hayward, CA.

Sasahara, A., Cole, T., Ederer, F., Murray, J., Wenger, N., Sherry, S. and Stengle, J. (1973) Urokinase Pulmonary Embolism Trial, a national cooperative study. *Circulation*, **47**(Suppl. 2), 1–108.

Savage, L. (1971) Elicitation of personal probabilities and expectations. *Journal of the American Statistical Association*, **66**, 783–801.

Schwartz, D., Flamant, R. and Lellouch, J. (1980) *Clinical Trials*. Academic Press, London.

Sculpher, M., Fenwick, E. and Claxton, K. (2000) Assessing quality in decision-analytic cost-effectiveness models. *Pharmacoeconomics*, **17**, 461–77.

Selwyn, M. R., Dempster, A. P. and Hall, N. R. (1981) A Bayesian approach to bioequivalence for the 2 × 2 changeover design. *Biometrics*, **37**, 11–21.

Selwyn, M. R. and Hall, N. R. (1984) On Bayesian methods for bioequivalence. *Biometrics*, **40**, 1103–8.

Senn, S. (1996) Some statistical issues in project prioritization in the pharmaceutical industry. *Statistics in Medicine*, **15**, 2689–702.

Senn, S. (1997a) Statistical basis of public policy – present remembrance of priors past is not the same as a true prior. *British Medical Journal*, **314**, 73.

Senn, S. (1997b) *Statistical Issues in Drug Development*. John Wiley & Sons, Ltd, Chichester.

Senn, S. (2002) Ethical considerations concerning treatment allocation in drug development trials. *Statistical Methods in Medical Research*, **11**, 403–11.

Shachter, R., Eddy, D. M. and Hasselblad, V. (1990) An influence diagram approach to medical technology assessment. In *Influence Diagrams, Belief Nets and Decision Analysis*, (R. M. Oliver and J. Q. Smith, eds), pp. 321–50. John Wiley & Sons, Ltd, Chichester.

Shakespeare, T. P., Gebski, V. J., Veness, M. J. and Simes, J. (2001). Improving interpretation of clinical studies by use of confidence levels, clinical significance curves, and risk-benefit contours. *Lancet*, **357**, 1349–53.

Sharp, S. J. and Thompson, S. G. (2000) Analysing the relationship between treatment effect and underlying risk in meta-analysis: comparison and development of approaches. *Statistics in Medicine*, **19**, 3251–74.

Sheiner, L. (1991) The intellectual health of clinical drug evaluation. *Clinical Pharmacology and Therapeutics*, **50**, 4–9.

Sheiner, L. and Wakefield, J. (1999) Population modelling in drug development. *Statistical Methods in Medical Research*, **8**, 183–93.

Shepherd, J., Blauw, G. J., Murphy, M. B., Bollen, E. L. E. M., Buckley, B. M., Cobbe, S. M., Ford, I., Gaw, A., Hyland, M., Jukema, J. W., Kamper, A. M., Macfarlane, P. W., Meinders, A. E., Norrie, J., Packard, C. J., Perry, I. J., Stott, D. J., Sweeney, B. J., Twomey, C. and Westendorp, R. G. J. (2002) Pravastatin in elderly individuals at risk of vascular disease (PROSPER): a randomised controlled trial. *Lancet*, **360**, 1623–30.

Silliman, N. P. (1997) Hierarchical selection models with applications in meta-analysis. *Journal of the American Statistical Association*, **92**, 926–36.

Simes, R. J. (1986) Application of statistical decision theory to treatment choices – implications for the design and analysis of clinical trials. *Statistics in Medicine*, **5**, 411–20.

Simon, R. (1977) Adaptive treatment assignment methods and clinical trials. *Biometrics*, **33**, 743–9.

Simon, R. (1994a) Problems of multiplicity in clinical trials. *Journal of Statistical Planning and Inference*, **42**, 209–21.

Simon, R. (1994b). Some practical aspects of the interim monitoring of clinical trials. *Statistics in Medicine*, **13**, 1401–9.

Simon, R. (2000). Meta-analysis of clinical trials: opportunities and limitations. *Statistical Science*, **15**, 305–20.

Simon, R. and Freedman, L. S. (1997) Bayesian design and analysis of two × two factorial clinical trials. *Biometrics*, **53**, 456–64.

Simon, R., Dixon, D. O., and Friedlin, B. (1996). Bayesian subset analysis of a clinical trial for the treatment of HIV infections. In *Bayesian Biostatistics* (D. A. Berry and D. K. Stangl, eds), pp. 555–76. Marcel Dekker, New York.

Smith, D., Givens, G. H. and Tweedie, R. L. (2000). Adjustment for publication bias and quality bias in Bayesian meta-analysis. In *Meta-analysis in Medicine and Health Policy* (D. K. Stangl and D. A. Berry, eds), pp. 277–304. Marcel Dekker, New York.

Smith, T., Spiegelhalter, D. J. and Parmar, M. K. B. (1996). Bayesian meta-analysis of randomized trials using graphical models and BUGS. In *Bayesian Biostatistics* (D. A. Berry and D. K. Stangl, eds), pp. 411–27. Marcel Dekker, New York.

Smith, T. C., Spiegelhalter, D. J. and Thomas, A. (1995) Bayesian approaches to random-effects meta-analysis: a comparative study. *Statistics in Medicine*, **14**, 2685–99.

Song, F., Altman, D., Glenne, A.-M. and Deeks, J. J. (2003) Validity of indirect comparison for estimating efficacy of competing interventions: empirical evidence from published meta-analyses. *British Medical Journal*, **326** (7387), 472–5.

Souhami, R. L., Craft, A. W., van der Eijken, J., Nooij, M., Spooner, D., Bramwell, V. H. C., Wierzbicki, R., Malcolm, A. J., Kirkpatrick, A., Uscinska, B. M., Van Glabbeke, M. and Machin, D. (1997). Randomised trial of two regimens of chemotherapy in operable osteosarcoma: a study of the European Osteosarcoma Intergroup. *Lancet*, **350** (9082), 911–17.

Spiegelhalter, D. J. (1998) Bayesian graphical modelling: a case-study in monitoring health outcomes. *Applied Statistics*, **47**, 115–33.

Spiegelhalter, D. (2001) Bayesian methods for cluster randomized trials with continuous responses. *Statistics in Medicine*, **20**, 435–52.

Spiegelhalter, D. J. and Best, N. G. (2003) Bayesian methods for evidence synthesis and complex cost-effectiveness models: an example in hip prostheses. *Statistics in Medicine*, In press.

Spiegelhalter, D. J. and Freedman, L. S. (1986) A predictive approach to selecting the size of a clinical trial, based on subjective clinical opinion. *Statistics in Medicine*, **5**, 1–13.

Spiegelhalter, D. J., Freedman, L. S. and Blackburn, P. R. (1986) Monitoring clinical trials: conditional or predictive power? *Controlled Clinical Trials*, **7**, 8–17.

Spiegelhalter, D. J., Freedman, L. S. and Parmar, M. K. B. (1993) Applying Bayesian ideas in drug development and clinical trials. *Statistics in Medicine*, **12**, 1501–17.

Spiegelhalter, D. J., Freedman, L. S. and Parmar, M. K. B. (1994) Bayesian approaches to randomized trials (with discussion). *Journal of the Royal Statistical Society, Series A*, **157**, 357–416.

Spiegelhalter, D., Myles, J., Jones, D. and Abrams, K. (2000) Bayesian methods in health technology assessment: a review. *Health Technology Assessment*, **4**(38), 1–130.

Stallard, N. (1998) Sample size determination for phase II clinical trials based on Bayesian decision theory. *Biometrics*, **54**, 279–94.

Stangl, D. K. (1996). Hierarchical analysis of continuous-time survival models. In *Bayesian Biostatistics* (D. A. Berry and D. K. Stangl, eds), pp. 429–50. Marcel Dekker, New York.

Stangl, D. K. and Berry, D. A. (1998) Bayesian statistics in medicine: where we are and where we should be going. *Sankhyā, Series B*, **60**, 176–95.

Stangl, D. K. and Berry, D. A. (eds) (2000). *Meta-analysis in Medicine and Health Policy*. Marcel Dekker, New York.

Stangl, D. K. and Greenhouse, J. B. (1998). Assessing placebo response using Bayesian hierarchical survival models. *Lifetime Data Analysis*, **4**, 5–28.

Stijnen, T. and Van Houwelingen, J. C. (1990). Empirical Bayes methods in clinical trials meta-analysis. *Biometrical Journal*, **32**, 335–46.

Strauss, N. and Simon, R. (1995). Investigating a sequence of randomized phase II trials to discover promising treatments. *Statistics in Medicine*, **14**, 1479–89.

Stroup, D. F., Berlin, J. A., Morton, S. C., Olkin, I., Williamson, G. D., Rennie, D., Moher, D., Becker, B. J., Sipe, T. A. and Thacker, S. B. (2000) Meta-analysis of observational studies in epidemiology: a proposal for reporting. *Journal of the American Medical Association*, **283**, 2008–12.

Su, X. Y. and Po, A. L. W. (1996) Combining event rates from clinical trials: comparison of Bayesian and classical methods. *Annals of Pharmacotherapy*, **30**, 460–5.

Sutton, A. and Abrams, K. R. (2001) Bayesian methods in meta-analysis and evidence synthesis. *Statistical Methods in Medical Research*, **10**, 277–303.

Sutton, A., Abrams, K. R., Jones, D. R., Sheldon, T. A. and Song, F. (2000) *Methods for Meta-analysis in Medical Research*. John Wiley & Sons, Ltd, Chichester.

Sutton, A., Abrams, K. R. and Jones, D. R. (2002) Generalized synthesis of evidence and the threat of dissemination bias: the example of electronic fetal heart rate monitoring (EFM). *Journal of Clinical Epidemiology*, **55**, 1013–24.

Tamura, R. N., Faries, D. E., Andersen, J. S. and Heiligenstein, J. H. (1994) A case study of an adaptive clinical trial in the treatment of out-patients with depressive disorder. *Journal of the American Statistical Association*, **89**, 768–76.

Tan, S., Machin, D., Tai, B. C., Foo, K. F. and Tan, E. H. (2002) A Bayesian re-assessment of two phase II trials of gemcitabine in metastatic nasopharyngeal cancer. *British Journal of Cancer*, **86**, 843–50.

Tan, S. B., Chung, Y. F. A., Tai, B. C., Cheung, Y. B. and Machin, D. (2003) Elicitation of prior distributions for a phase III randomized controlled trial of adjuvant therapy with surgery for hepatocellular carcinoma. *Controlled Clinical Trials*, **24**, 110–21.

Tarone, R. (1982) The use of historical control information in testing for a trend in proportions. *Biometrics*, **38**, 215–20.

Ten Centre Study Group (1987) Ten centre study of artificial surfactant (artificial lung expanding compound) in very premature babies. *British Medical Journal*, **294**, 991–6.

Teo, K. K., Yusuf, S., Collins, R., Held, P. H. and Peto, R. (1991) Effects of intravenous magnesium in suspected acute myocardial infarction: overview of randomised trials. *British Medical Journal*, **303**(6816), 1499–1503.

Thall, P. F. and Estey, E. H. (1993) A Bayesian strategy for screening cancer treatments prior to phase II clinical evaluation. *Statistics in Medicine*, **12**, 1197–1211.

Thall, P. F. and Russell, K. E. (1998) A strategy for dose-finding and safety monitoring based on efficacy and adverse outcomes in phase I/ II clinical trials. *Biometrics*, **54**, 251–64.

Thall, P. F. and Simon, R. (1990) Incorporating historical control data in planning phase II studies. *Statistics in Medicine*, **9**, 215–28.

Thall, P. F. and Sung, H. (1998) Some extensions and applications of a Bayesian strategy for monitoring multiple outcomes in clinical trials. *Statistics in Medicine*, **17**, 1563–80.

Thall, P. F., Simon, R. M. and Estey, E. H. (1996) New statistical strategy for monitoring safety and efficacy in single-arm clinical trials. *Journal of Clinical Oncology*, **14**, 296–303.

Thompson, S. G., Smith, T. C. and Sharp, S. J. (1997) Investigating underlying risk as a source of heterogeneity in meta-analysis. *Statistics in Medicine*, **16**, 2741–58.

Tsiatis, A. A. (1981) The asymptotic joint distribution of the efficient scores test for the proportional hazards model calculated over time. *Biometrika*, **68**, 311–15.

Tukey, J. (1977). Some thoughts on clinical trials, especially problems of multiplicity. *Science*, **198**, 679–84.

Tunis, S. R., Sheinhait, I. A., Schmid, C. H., Bishop, D. J. and Ross, S. D. (1997) Lansoprazole compared with histamine(2)-receptor antagonists in healing gastric ulcers: a meta-analysis. *Clinical Therapeutics*, **19**, 743–57.

Turner, R., Omar, R. and Thompson, S. (2001) Bayesian methods of analysis for cluster randomized trials with binary outcome data. *Statistics in Medicine*, **20**, 453–72.

Tversky, A. (1974). Assessing uncertainty. *Journal of the Royal Statistical Society, Series B*, **36**, 148–59.

Tweedie, R. L., Scott, D. J., Biggerstaff, B. J. and Mengersen, K. L. (1996) Bayesian meta-analysis, with application to studies of ETS and lung cancer. *Lung Cancer*, **14**, S171–94.

University Group Diabetes Program (1970) A study of the effects of hypoglycemic agents on vascular complications in patients with adult onset diabetes. *Diabetes*, **19** (Suppl. 2), 747–830.

Urbach, P. (1993) The value of randomization and control in clinical trials. *Statistics in Medicine*, **12**, 1421–31.

US Food and Drug Administration (1998a) Semiannual guidance agenda. *Federal Register*, **63** (212), 59 317–26.

US Food and Drug Administration (1998b) *Transcript of Cardiovascular and Renal Drugs Advisory Committee meeting, 26th June 1997.* http://www.fda.gov/ohrms/dockets/ac/97/transcpt/3320t1.pdf.

US Food and Drug Administration (1999a) *Guidance for Industry: Population Pharmacokinetics.* http://www.fda.gov.cder/guidance/index.htm.

US Food and Drug Administration (1999b) *Summary of Safety and Effectiveness Data for T-Scan Breast Scanner.* http://www.fda.gov/cdrh/pdf/p970033b.pdf.

Vail, A., Hornbuckle, J., Spiegelhalter, D. J. and Thornton, J. G. (2001) Prospective application of Bayesian monitoring and analysis in an 'open' randomized clinical trial. *Statistics in Medicine*, **20**, 3777–87.

van Hout, B. A., Al, M. J., Gordon, G. S. and Rutten, F. F. H. (1994) Costs, effects and C/E-ratios alongside a clinical trial. *Health Economics*, **3**, 309–19.

van Houwelingen, H. C. (1997) The future of biostatistics: expecting the unexpected. *Statistics in Medicine*, **16**, 2773–84.

van Houwelingen, H. and Senn, S. (1999) Investigating underlying risk as a source of heterogeneity in meta-analysis by S. G. Thompson, T. C. Smith and S. J. Sharp, Statistics in Medicine, 16, 2741–2758 (1997). *Statistics in Medicine*, **18**, 110–13.

van Houwelingen, H. C., Zwinderman, K., and Stijnen, T. (1993) A bivariate approach to meta-analysis. *Statistics in Medicine*, **12**, 2272–84.

Wakefield, J. and Bennett, J. (1996) The Bayesian modeling of covariates for population pharmacokinetic models. *Journal of the American Statistical Association*, **91**, 917–27.

Wakefield, J. and Walker, S. (1997) A population approach to initial dose selection. *Statistics in Medicine*, **16**, 1135–49.

Waller, S. E. and Duncan, D. (1969) A Bayes rule for the symmetric multiple comparison problem. *Journal of the American Statistical Association*, **64**, 1484–1503.

Ware, J. (1989) Investigating therapies of potentially great benefit: ECMO (with discussion). *Statistical Science*, **4**, 298–340.

Ware, J. H., Muller, J. E. and Braunwald, E. (1985) The futility index: an approach to the cost-effective termination of randomized clinical trials. *American Journal of Medicine*, **78**, 635–43.

Warn, D. E., Thompson, S. G. and Spiegelhalter, D. J. (2002). Bayesian random effects meta-analysis of trials with binary outcomes: methods for the absolute risk difference and relative risk scales. *Statistics in Medicine*, **21**, 1601–24.

Weiss, H. L., Urban, D. A., Grizzle, W. E., Cronin, K. A., Freedman, L. S., Kelloff, G. J. and Lieberman, R. (2001) Bayesian monitoring of a phase 2 chemoprevention trial in highrisk cohorts for prostate cancer. *Urology*, **57**, 220–3.

Wheatley, K. and Clayton, D. (2003) Be skeptical about unexpected large apparent treatment effects: the case of an MRC AML12 randomization. *Controlled Clinical Trials*, **24**, 66–70.

Whitehead, A. (2002) *Meta-analysis of Controlled Clinical Trials.* John Wiley & Sons, Ltd, Chichester.

Whitehead, J. (1986) Sample sizes for phase-II and phase-III clinical trials – an integrated approach. *Statistics in Medicine*, **5**, 459–64.

Whitehead, J. (1993) The case for frequentism in clinical trials. *Statistics in Medicine*, **12**, 1405–19.

Whitehead, J. (1997a) Bayesian decision procedures with application to dose-finding studies. *International Journal of Pharmaceutical Medicine*, **11**(4), 201–7.

Whitehead, J. (1997b) *The Design and Analysis of Sequential Clinical Trials* (2nd edition). John Wiley & Sons, Ltd, Chichester.

Whitehead, J. and Brunier, H. (1995) Bayesian decision procedures for dose determining experiments. *Statistics in Medicine*, **14**, 885–93.

Willan, A. R. (2001) On the probability of cost-effectiveness using data from randomised clinical trials. *BMC Medical Research Methodology*, **1**(8). http://biomedcentral. com/ 1471-2288/1/8.

Woods, K. L., Fletcher, S., Roffe, C. and Haider, Y. (1992) Intravenous magnesium sulphate in suspected acute myocardial infarction: results of the Second Leicester Intravenous Magnesium Intervention Trial (LIMIT-2). *Lancet*, **339** (8809), 1553–8.

Yao, T. J., Begg, C. B. and Livingston, P. O. (1996) Optimal sample size for a series of pilot trials of new agents. *Biometrics*, **52**, 992–1001.

Yusuf, S. (1997) Meta-analysis of randomised trials: Looking back and looking again. *Controlled Clinical Trials*, **18**, 594–601.

Yusuf, S., Teo, K. and Woods, K. (1993) Intravenous magnesium in acute myocardial infarction: an effective, safe, simple and inexpensive treatment. *Circulation*, **87**, 2043–6.

Zelen, M. (1969) Play the winner rule and the controlled clinical trial. *Journal of the American Statistical Association*, **64**, 131–46.

Zelen, M. (1990) Discussion of 'Biostatistics and Bayes' by Breslow. *Statistical Science*, **5** (3).

Zelen, M. and Parker, R. A. (1986) Case control studies and Bayesian inference. *Statistics in Medicine*, **5**, 261–9.

Zucker, D. R., Schmid, C. H., McIntosh, M. W., Agostino, R. B., Selker, H. P. and Lau, J. (1997) Combining single patient (*n*-of-1) trials to estimate population treatment effects and to evaluate individual patient responses to treatment. *Journal of Clinical Epidemiology*, **50**, 401–10.

Index

Page numbers in italics refer to examples in the text.

Statistics in Practice

Human and Biological Sciences

Brown and Prescott – Applied Mixed Models in Medicine
Ellenberg, Fleming and DeMets – Data Monitoring Committees in Clinical Trials:
A Practical Perspective
Lawson, Browne and Vidal Rodeiro – Disease Mapping with WinBUGS and
MLwiN
Lui–Statistical Estimation of Epidemiological Risk
Marubini and Valsecchi – Analysing Survival Data from Clinical Trials and
Observation Studies
Parmigiani – Modeling in Medical Decision Making: A Bayesian Approach
Senn – Cross-over Trials in Clinical Research, Second Edition
Senn – Statistical Issues in Drug Development
Spiegelhalter, Abrams and Myles – Bayesian Approaches to Clinical Trials and
Health-Care Evaluation
Whitehead – Design and Analysis of Sequential Clinical Trials, Revised Second
Edition
Whitehead – Meta-Analysis of Controlled Clinical Trials

Earth and Environmental Sciences

Buck, Cavanagh and Litton – Bayesian Approach to Interpreting Archaeological
Data
Glasbey and Horgan – Image Analysis in the Biological Sciences
Webster and Oliver – Geostatistics for Environmental Scientists

Industry, Commerce and Finance

Aitken – Statistics and the Evaluation of Evidence for Forensic Scientists
Lehtonen and Pahkinen – Practical Methods for Design and Analysis of Complex
Surveys, Second Edition
Ohser and Mücklich – Statistical Analysis of Microstructures in Materials Science